Toxicology of Reptiles

NEW PERSPECTIVES: TOXICOLOGY AND THE ENVIRONMENT

Series Editors

Donald E. Gardner, A. Wallace Hayes and John A. Thomas

Target Organ Toxicology in Marine and Freshwater Teleosts Volume 1 — Organs

Daniel Schlenk and William H. Benson (eds.), 2001

Target Organ Toxicology in Marine and Freshwater Teleosts Volume 2 — Systems

Daniel Schlenk and William H. Benson (eds.), 2001

Toxicology of Marine Mammals

Joseph G. Vos, Gregory D. Bossart, Michel Fournier and Tom O'Shea (eds.), 2003

Toxicology of Reptiles

Susan C. Gardner and Eva Oberdörster (eds.), 2006

New Perspectives: Toxicology and the Environment

Toxicology of Reptiles

Edited by

Susan C. Gardner
Centro de Investigaciones Biológicas del Noroeste
La Paz, Mexico

Eva Oberdörster
Southern Methodist University
Dallas, TX

CRC Press
Taylor & Francis Group
Boca Raton London New York

CRC Press is an imprint of the
Taylor & Francis Group, an informa business
A TAYLOR & FRANCIS BOOK

CRC Press
Taylor & Francis Group
6000 Broken Sound Parkway NW, Suite 300
Boca Raton, FL 33487-2742

ISBN-13: 978-0-8493-2715-5 (hbk)

Publisher's Note

The publisher has gone to great lengths to ensure the quality of this reprint but points out that some imperfections in the original copies may be apparent.

Library of Congress Cataloging-in-Publication Data

Toxicology of reptiles / edited by Susan C. Gardner and Eva Oberdörster.
p. cm. -- (New perspectives)
Includes bibliographical references and index.
ISBN 0-8493-2715-6
1. Reptiles--Effect of pollution on. 2. Reptiles--Ecophysiology. I. Gardner, Susan C. II. Oberdörster, Eva. III. New perspectives (Taylor & Francis)

QL669.8.T69 2005
571.9'5179--dc22 2005049107

Visit the Taylor & Francis Web site at
http://www.taylorandfrancis.com

and the CRC Press Web site at
http://www.crcpress.com

Contributors

A. Alonso Aguirre
Wildlife Trust
Palisades, New York

Joanna Burger
Rutgers University
Division of Life Sciences
Piscataway, New Jersey

Ian Callard
Biology Department
Boston University
Boston, Massachusetts

Malin Celander
Goteborg University
Department of Zoo Physiology
Goteborg, Sweden

Susan C. Gardner
Centro de Investigaciones Biológicas del Noroeste, S.C.
La Paz, Baja California Sur
México

William A. Hopkins
Department of Fisheries and Wildlife Sciences
Virginia Polytechnic and State University
Blacksburg, Virginia

Kelly Irwin
Arkansas Game & Fish Commission
Benton, Arkansas

Lisa Irwin
U.S. Fish and Wildlife Service
Arkansas Field Office
Conway, Arkansas

Jennifer M. Keller
National Institute of Standards and Technology
Hollings Marine Laboratory
Charleston, South Carolina

Noppadon Kitana
Department of Biology
Chulalongkorn University
Bangkok, Thailand

Emily Marquez
Biology Department
Boston University
Boston, Massachusetts

Patricia McClellan-Green
Department of Environmental and Molecular Toxicology
Center for Marine Sciences and Technology
North Carolina State University
Morehead City, North Carolina

Linda Meyers-Schöne
Blue World Environment
Albuquerque, New Mexico

Carys L. Mitchelmore
University of Maryland,
Center for Environmental Science,
Chesapeake Biological Laboratory
Solomons, Maryland

Apolonia Novillo
Biology Department
Boston University
Boston, Massachusetts

Eva Oberdörster
Department of Biology
Southern Methodist University
Dallas, Texas

Margie M. Peden-Adams
Medical University of
South Carolina
Department of Pediatrics
Marine Biomedicine &
Environmental Sciences Center
Charleston, South Carolina

Allen R. Place
Center of Marine Biology
University of Maryland
Baltimore, Maryland

Christopher L. Rowe
University of Maryland
Center for Environmental Science
Chesapeake Biological Laboratory
Solomons, Maryland

Kyle W. Selcer
Duquesne University
Pittsburgh, Pennsylvania

Emily Willingham
School of Medicine,
Department of Urology
University of California
San Francisco, California

Foreword

It is my honor to present the foreword of this book — a book that advances the fields related to reptilian biology, conservation, and ecological risk assessment. Historically, most reptilian toxicity information has been associated with tissue concentrations (e.g., Meyers-Schöne and Walton, 1994) with little cause-and-effect information available. In their book entitled *Ecotoxicology of Amphibians and Reptiles*, Sparling, Linder, and Bishop (2000) alerted us to the scarcity of relevant contaminant effects information on reptiles compared with other vertebrate taxa. The editors and authors of specific chapters stressed the need for research in areas such as the following:

- Physiology: to better understand the dynamics of chemical contaminant exposure, uptake, and elimination, with emphasis on reproductive physiology and endocrine modulators
- Pathology and disease: to better understand the influence of contaminants on humoral and cell-mediated responses
- Ecotoxicology: to better understand critical organ concentrations and regulatory capacities
- Population ecology: to better understand potential impacts of contaminants on the ability of reptiles to withstand perturbations that may affect the population

This book serves as a very useful text because it cohesively summarizes some of the cutting-edge research that is taking place in areas such as reptilian endocrinology, neurophysiology, immunology, and ecology. Conservation needs are also addressed as well as the issues related to complications associated with conducting population studies. This information is easily available for synthesis and use in the evaluation and understanding of potential risks of reptiles to environmental contaminants.

As we know, reptiles are often not included in the ecological risk assessment process. When present within an ecosystem, they frequently occupy key positions within the food web — one of the primary considerations when considering an ecological receptor. The dilemmas faced by risk assessors, however, are the absence of sufficient reptilian toxicity information and the lack of specific exposure information. Assumptions can be made with respect to the use of "default" values with uncertainty factors. One must determine what level of confidence such information would yield and what level would be acceptable. The selection of an appropriate receptor and utilization of technically defensible data to support the estimate of potential risk to that receptor are essential to both the scientific and regulatory management processes.

The question is likely to be asked whether reptiles are more toxicologically sensitive to chemical contaminants than birds or mammals. The few studies that have been conducted indicate metallothionein-metal metabolic systems in reptiles are similar to those in other vertebrate classes. However, the temperature-associated sex determination reproductive strategy in turtles and alligators may make them more susceptible to endocrine disruptors compared with other vertebrates. This comparison remains an area of fruitful study.

Significant progress has also been made in the establishment of at least one model test species, specifically, the Western fence lizard (*Sceloporus occidentalis*). Suitable wild populations have been located in the California Valley, the species is sexually dimorphic, females lay from 3 to 6 clutches of 8 to 15 eggs each, many assay endpoints have been identified for testing purposes, and the species is easily raised in the lab.

Historically, the reptilian ecotoxicology database has been extremely limited and has prevented us from sufficiently addressing the question "How similar are all reptiles?" From an evolutionary prospective, traditionally, crocodilians, turtles, snakes, lizards, amphisbaenids, and tuataras were grouped together under the class Reptilia. However, recent cladistic analysis proposes a classification system that places turtles and crocodilians in separate classes. Although this is not widely accepted among herpetologists, it does provide additional food for thought as we consider issues such as the most appropriate test model species and toxicological data that best represent "reptiles" as a whole.

This book, and the investigators who have supported the research discussed within it, prove that we are making great strides in the area of reptilian ecotoxicology. To the editors and authors — congratulations on a much needed and comprehensive effort. To the readers — give reptiles a try if you haven't already. We could use your help!

References

Meyers-Schöne, L. and Walton, B.T. 1994. Turtles as monitors of chemical contaminants in the environment. *Rev. Environ. Contam. Toxicol.*, 135, 93–153.

Sparling, D.W., Linder G., and Bishop C.A., Eds., *Ecotoxicology of Amphibians and Reptiles*, SETAC Press, Pensacola, FL, 2000.

Linda Meyers-Schöne, Ph.D.
Blue World Environment

Acknowledgments

We are indebted to Drs. Donald E. Gardner and Günter Oberdörster for the decades of service as scientific role models and for encouraging us to compile this book. We would also like to extend personal appreciation to Dr. Louis Guillette, Jr., for his contribution to the initial book outline and for his dedication to the study of comparative physiology, which helped initially propel the field of reptile toxicology to the current state of the science. We gratefully acknowledge the peer reviewers, including Drs. William S. Baldwin, Kym Campbell, David Hinton, Gerald LeBlanc, Michael Newman, Charles Rice, Jeffrey Seminoff, and Stacey Weiss, for their thoughtful suggestions for revision of the chapters. We would also like to thank the contributing authors for their extensive effort in the writing and revising of these chapters.

EO and SCG

Contents

chapter 1

Introduction to Reptilian Toxicology

Susan C. Gardner

Table of Contents

I. Book Overview

Compared with studies in other taxa, the field of reptilian toxicology is in its infancy. However, studies in other wildlife (fish and mammals in particular) have provided the foundation for reptilian studies, and many modern tools developed for toxicological assessment of other species have been successfully applied to reptiles. The purpose of this book is to summarize the information currently available on toxicology of reptiles within the context of what is known of nonreptilian species. The effects of contaminants on target organs are described, including examples of terrestrial, freshwater, and marine species. Major contaminant classes are covered within each chapter, with a focus on contaminants of greatest concern. Some overlap in the material covered among chapters is necessary and intentional so that each chapter is complete in its treatment of the material, whereas references to other chapters are included to direct the reader to more information on particular topics.

II. Reptilian Toxicology

The current population status of most reptilian species has not been evaluated. Of the species in which evaluations have been conducted, more than half are regarded as threatened. Chapter 2 provides an overview of the current phylogeny and status of reptile populations and the major threats to their stability, including examples of both terrestrial and aquatic species. Although habitat degradation is one of the most pronounced threats to reptile populations, Irwin and Irwin (Chapter 2) point out that population declines are generally the result of a combination of both anthropogenic and natural stressors, and rarely is one single factor operating in isolation. The response of a population to environmental stressors will differ according to each species' life history characteristics and the cumulative effects of the different stressors present. Therefore, the effects of natural stressors (e.g., drought, flood, disease) will be more pronounced when coupled with anthropogenic factors (e.g., contaminants, habitat destruction, harvest) and may eventually impede a species' ability to maintain stable populations.

In Chapter 2, Irwin and Irwin provide an overview of the life history characteristics that make many reptilian species difficult to study using short-term surveys, including long generation times, late maturation, and cryptic juvenile stages. Detecting declines in such species is often delayed and may not be possible until decreases in reproductive size classes are observed years or decades later. Therefore, the authors stress the importance of long-term herptofaunal surveys using standardized protocols to ensure the validity of comparisons of populations over time and space.

In Chapter 3, Hopkins illustrates the utility of reptiles as indicators of environmental contamination and emphasizes the potential of reptiles to serve as models in the study of environmental pollution. With high population densities and high efficiency of exploiting resources, reptiles may serve a major role as transporters of contaminants through food webs, the importance of which likely has been underestimated. To date, the majority of toxicological assessments in reptiles have been made up of tissue residue studies. Measurements of chemical residues in tissues are valuable because they provide an indication of the bioavailability of the contaminant in the environment and the ecologic and physiologic characteristics of the organism. Whereas the distribution of contaminants among tissues in reptiles is generally similar to those in other vertebrate wildlife, reptiles possess unique sets of traits that may influence accumulation patterns. High conversion efficiencies, a less developed enzymatic detoxification system, dietary patterns, and behavioral traits (as described in Chapter 7) are all factors that may influence uptake and accumulation of contaminants in reptiles. Likewise, Hopkins (Chapter 3) points out that differences exist in contaminant accumulation among individuals of the same species. For example, male–female differences in tissue residues observed in reptilian field surveys can be attributed to behavior, feeding ecology, reproductive state, or gender-based differences in physiology. Additionally, ontogenetic shifts in enzymatic metabolism and feeding strategies have also been observed in reptiles and could result in variation in the accumulation of contaminants throughout an individual's lifetime.

Despite a large body of literature reporting contaminant concentrations in reptilian tissues, interpretation of this information is difficult because of the lack of data on the corresponding concentrations of contaminants in the environment and their effects. In Chapter 3, Hopkins offers recommendations on where efforts in reptile tissue residue analyses should be focused to maximize the application of these results toward conservation initiatives. He proposes that field surveys be coupled with laboratory and field experiments and be designed to test well-conceived hypotheses to provide much needed information to improve understanding of contaminant effects on reptile populations.

Establishing relationships between contaminant exposure and biological effects in reptiles will likely involve an increased use of biomarkers. Biomarker research on fish has formed the basis for biomarker development in reptile species with the consequence that these tools have been applied in a similar fashion in both groups. Mitchelmore and coauthors (Chapter 4) describe the current state of reptilian biomarker research and the application of various tools ranging from the molecular level to the individual level. These tools are particularly valuable as compared with pure residue studies because they provide indications of synergistic effects among complex mixtures as well as the effects of contaminant metabolites not detected in routine analyses.

Measurements of xenobiotic induction or inhibition of enzymes are particularly useful as tools for biomonitoring. As in mammals, hepatic phase I enzymes in reptiles have been induced after contaminant exposure (e.g., polyaromatic hydrocarbon treatment in snakes and turtles), but much less work has been done on reptilian enzymes involved in phase II metabolism, which to date has consisted only of glutathione conjugation. Chapter 5 covers the basic anatomy of reptilian liver and kidney and their roles in metabolism and elimination of xenobiotics as compared with other vertebrates. Numerous cytochrome P450 isozymes known in mammals have been found in the reptile liver, whereas other mammalian isozymes may be completely absent in some reptiles (e.g., CYP4A in alligators). Additionally, unique mechanisms of defense, such as the production of melanins, provide a protective function against oxidative damage and may impart reptiles with resistance to environmental stressors despite the down-regulation of other enzyme systems.

In adult reptiles, kidney anatomy has been best studied in turtles, although in some aspects chelonian kidney function may be more primitive than in other reptilian species. Whereas some renal segments may be homologous to mammalian counterparts, other aspects, such as complex foldings of the proximal and distal tubules, are quite distinct from that of mammalian kidney. Male snakes and lizards have a sex segment of the kidney that displays seasonal variation in synthesis and secretion that is correlated with reproductive activity. In Chapter 5, McClellan-Green and coauthors demonstrate that because of the importance of the liver and kidneys in maintaining homeostasis and reproductive functions in reptiles, contaminant-induced effects on these organs may have serious consequences for the organism's fitness. As a result of the close proximity and interaction between the adrenal glands and the kidney, the effects of contaminants and handling stress on adrenal hormones (peptide and steroid) are also covered in Chapter 5.

Reptiles possess a broad diversity of reproductive characteristics (e.g., oviparity, viviparity, ovoviviparity), sex-determining mechanisms, methods of bearing offspring, and levels of parental care. Moreover, endocrine systems in reptiles can vary widely across species and result in varied responses to toxicants. In Chapter 6, Willingham describes how reptilian life history strategies can influence susceptibility to certain toxicants, especially those that disrupt endocrine pathways. Species that rely on temperature-dependent sex determination often experience greater susceptibility to contaminants than those in which sex is chromosomally determined. Additionally, despite the fact that both turtles and alligators use temperature-dependent sex determination via aromatase- and estrogen-mediated pathways, the pathways differ between the species. Estrogen receptors of different species also show different affinities to contaminants such as organochlorine compounds. Therefore, developmental effects of xenobiotics vary widely across reptilian species, making extrapolation of findings across species difficult, inappropriate, or both.

The developmental stage of the exposure to endocrine-disrupting chemicals is an important determinant of effects. Developmental processes that rely on endocrine signaling (including gonadogenesis, sex differentiation, neural development, and growth) are especially susceptible to endocrine-disrupting contaminants. Similarly, contaminants with endocrine activity also affect reptilian behavior (Chapter 7) and immune system function (Chapter 8). The magnitude of these effects varies with the developmental stage at which the exposure occurs.

Additional research is needed in the field of reptilian behavioral toxicology. Neurotoxicology involves relating differences in brain anatomy caused by toxicants to behavioral effects. In Chapter 7, Burger describes important considerations in designing studies of behavioral effects in reptiles. The choice of behavioral endpoints requires understanding the species ecology and life history, as well as their behavioral requirements for reproduction and survival. Likewise, the most useful behaviors for neurotoxicological assessments will be those that relate to survival and reproductive success in the wild (e.g., prey searching and capture, appropriate habitat use, mate recognition, and predator avoidance). For each of these behaviors there are measurement endpoints that include response time, behavior duration, and the accuracy of the behavior. Additionally, factors that affect an organism's morphology and physiology may have secondary effects on the behavior of the organism. Numerous neurotoxicologic effects have been reported in reptiles as a function of chemical exposure (e.g., neuroanatomic changes, suppressed mating behavior, and enhanced or suppressed locomotion), although there are some indications that reptiles may be less sensitive than birds and mammals to neurobehavioral impairment and show a narrow effects range for behavior endpoints. However, much more research is needed to arrive at conclusive comparisons across taxa. In Chapter 7 Burger emphasizes the need for better understanding of the mechanisms of neurobehavioral impairment, including corresponding brain contaminant concentrations; localization of xenobiotics within the brain; and their effects at the molecular, individual, and population levels.

Keller and coauthors (Chapter 8) describe the excellent potential for use of immunotoxicological tests in assessing contaminant effects in reptilian species. Although it is recognized that assays of immune function are often more sensitive than other toxicologic endpoints, examples of immunotoxicologic studies with reptiles are extremely limited and have been hindered by an incomplete understanding of cellular and molecular pathways of immune responses. The reptilian immune system is complex and highly developed, possessing all three categories of immune function (innate, cell-mediated, and humoral). Cell-mediated immunity has received the most attention in reptilian immunotoxicology studies, despite the fact that innate immunity also appears to be an important component of immune defenses in reptiles. To date, studies of immunotoxicologic responses of reptilian innate immunity have been limited to lysozyme activity. Hematology parameters (e.g., heterophil:lymphocyte ratio) are also useful indicators of stress to the immune system and have been associated with contaminant exposure in reptiles.

Keller and coauthors (Chapter 8) provide examples of studies demonstrating that exposure to environmentally relevant concentrations of contaminants can alter immune function in reptiles and possibly lead to decreased host resistance and increased vulnerability to disease. Suppression of immune functions (e.g., decreases in lysozyme activity and suppressed B-cell proliferation) by environmental exposures of Hg and organochlorines has been observed in reptiles and can be greatly exacerbated by xenobiotic mixtures. However, some contaminants can result in enhanced acquired immune functions, as observed with immunostimulatory effects of organochlorines on lymphocyte proliferation in sea turtles.

Because reptilian immune response is strongly affected by environmental factors, Keller and coauthors provide recommendations for standardization and validation of immunologic endpoints when assessing immunotoxicity in reptiles. Seasonal and reproductive effects in particular have been the topic of many studies and demonstrate that testosterone and other steroids dramatically alter reptilian immune functions. Developing a single standard reptilian immunotoxicologic model would be highly advantageous for comparative purposes, but because of species-specific differences in immunotoxic effects, any proposed model must consider the physiologic and life history differences between reptilian orders and species.

Environmental genotoxicology (the topic of Chapter 9) is a relatively new discipline, and its application to reptilian communities has been limited. Compounds are classified as genotoxic based on their ability to damage the DNA and induce heritable changes. Researchers in this field have benefited from the advancing technology of molecular biology applied to human health, but the number of studies of reptiles has been miniscule. Reptilian species appear to be effective monitors of genotoxic contaminants, although some reptile species (e.g., turtles) seem less sensitive to genotoxic effects than humans, based on endpoints related to the induction of chromosome interchange aberrations. Although the long-term effects of environmental genotoxic substances are largely unknown, it has been shown that longer-lived species will have a longer exposure time and more opportunities to accumulate harmful dosages, whereas less affected species are shorter lived, with earlier sexual maturity and high reproductive capacity (the importance of these traits is discussed in detail in Chapter 10). However, the ability to measure contaminant-induced DNA damage is directly dependent on accurate estimates of background levels of such alterations. These data are missing in reptiles, making it difficult to draw conclusions regarding the consequence of exposure to genotoxic agents. Nevertheless, Novillo and coauthors (Chapter 9) provide examples illustrating that sensitive and selective techniques are available that may be used to monitor genotoxic effects in reptiles despite the existence of information gaps and the need for much additional research.

Although there is a clear need for toxicological research to include more data on population effects, some examples of population-level studies in reptiles exist. Chapter 10 describes the difficulties of population-level research in ecotoxicology in general and for reptile species specifically. Selcer provides an overview of the major theoretical considerations involved in understanding the effects of contaminants on reptile populations and emphasizes how choosing reliable endpoints is critical in designing ecotoxicological studies. Separating the effects of contaminants from those of other proximate factors is challenging and usually requires historical knowledge of the demographics of the population or comparison with another closely related population in which the contaminant stress is absent. This is especially true for indirect effects of contaminants such as decreases in food availability or habitat quality. Selcer suggests that using life history parameters (e.g., growth rates, age structure, age-specific fecundity, and age-specific mortality) as the main focus of ecotoxicological studies may be more useful than traditional abundance estimates for assessing the status of a population. Selcer also highlights the need to identify indicator or sentinel species for reptiles and describes important attributes of various proposed reptile species.

III. Summary

During the past decades, reptilian toxicology has made up a disproportionately small percentage of toxicologic studies of vertebrates. Characteristics of some reptile species make them difficult to study, including long life span and generation time, low fecundity, and incompatibility with laboratory handling techniques. However, many of these same characteristics can also make these species more vulnerable to contaminant effects and therefore increase the need to broaden our understanding of reptilian toxicology. Reptile species are linked by a number of traits (e.g., ectothermia, pulmonary respiration, epidermal scales, internal fertility), yet possess a diverse array of life history characteristics and inter-specific difference (e.g., population distributions, migration patterns, diets, metabolic processes). These inter-specific differences impede extrapolation of information across reptilian species but simultaneously present excellent opportunities to use reptiles for comparative studies to advance our understanding of toxicologic mechanisms and ecological effects. Through this book, we endeavor to provide a comprehensive description of the current state of knowledge of reptilian toxicology from the perspective of target organ systems. Additionally, the authors highlight the most pressing information gaps and propose priority directions for the advancement of the field of toxicology of reptiles.

chapter 2

Global Threats Affecting the Status of Reptile Populations

Lisa Irwin and Kelly Irwin

Table of Contents

I. Introduction

The goals of this chapter are as follows: (1) to provide a global overview of the current threats affecting reptile populations, (2) to describe and provide examples of known threats, and (3) to provide examples of life history correlates that should be considered when studying the effects of contaminants on wild reptile populations. Wildlife populations are rarely exposed to only a single environmental contaminant or other stressor, so researchers must remain cognizant of the potential for other influencing factors when investigating toxic effects of contaminants on reptile populations. Although documented examples of population threats may not be available for each taxonomic group or region, we have attempted to illustrate these threats with specific examples, which can translate to other taxonomic groups or regions of the world. This chapter is by no means exhaustive in scope for every declining reptile species or every potential threat that reptile populations face today. However, it attempts to provide an overview of the most prevalent threats, with documented examples of the most significant impacts on reptile populations. It is hoped that this information will provide essential background information for researchers to consider when investigating contaminant-related issues and declining reptile populations.

Currently only 6% of the approximately 8000 known species of reptiles in the world have been evaluated for their conservation status by the World Conservation Union (IUCN)[1] (Table 2.1). The number of reptile species addressed by the IUCN is generally based on currently recognized full species and occasionally subspecies or geographic populations (e.g., Mediterranean green turtle). The IUCN produces a Red List, which lists all plants and animals currently known to be threatened. Categories used in the Red List are generally based on an evaluation of the best available evidence.[2] The categories include and are defined by the IUCN as follows: (1) extinct, "when there is no reasonable doubt that the last individual of a taxon has died"; (2) critically endangered, "when a taxon is facing an extremely high risk of extinction in the wild"; (3) endangered, "when the best available evidence indicates that a taxon is facing a very high risk of extinction in the wild"; (4) vulnerable, "when a taxon faces a high risk of extinction in the wild"; and (5) near threatened, "when a taxon has been evaluated against the criteria but does not yet qualify for

Table 2.1 Diversity and Conservation Status of Reptiles

	Total # known species	Total # and % of known species evaluated for conservation status	% threatened (of known species)	% threatened (of species evaluated)
Order Chelonia (turtles)	305	198 (65%)	42	65
Order Squamata (lizards, snakes, amphisbaenids)	7668	259 (3%)	2	62
Order Crocodylia (crocodylians)	23	14 (61%)	43	71
Order Rhynchocephalia (tuataras)	2	2 (100%)	50	50

Note: Numbers are based on 2003 IUCN Red List of Threatened Species.[1] "Threatened" as defined by the IUCN includes endangered, critically endangered, and vulnerable categories.

the higher categories but is close to or likely to qualify for a threatened category in the near future." Other categories IUCN uses include least concern, "when a taxon has been evaluated and does not qualify for any of the above categories." Least concern taxa are currently believed to be widespread and abundant. Another category, which is not considered a category of threat, is data deficient, "where there is inadequate information to make a direct, or indirect, assessment of its risk of extinction based on its distribution and/or population status." Although there may be abundant information on the biology of data-deficient taxa, there may not be any available data on the abundance or distribution; therefore, population status is unknown. The criteria for each of these categories are generally based on some measure of a reduction in population size over a given time period, extent of geographic range or level of fragmentation, and probability of extinction in the wild. The thresholds used to delineate each category will be specific to the taxon in question, such that abundance or population estimates are based on indices that are appropriate to that particular taxa. Actual or potential levels of exploitation and the effects of introduced taxa, hybridization, pathogens, pollutants, competitors, or parasites are all taken into consideration when estimating reduction in population size. Extent of fragmentation, area of occupancy, extent or quality of habitat, fluctuations in the number and locations of populations, and number of mature individuals are considered when evaluating geographic range.

Finally, there is another IUCN category for those taxa not yet evaluated against the criteria. Reptiles are one of the least known vertebrate taxa, with only approximately 6% of the known taxa evaluated to date.[1]

Birds; mammals; and, only recently, amphibians[3] have been completely evaluated on a global scale.[1] Of the reptiles evaluated to date (473 species), approximately 62% of them are currently believed to be imperiled (includes IUCN categories: critically endangered, endangered, and vulnerable).[1] Although this estimate is admittedly subject to some interpretation because of taxonomic changes, changes in threats, or lack of sufficient data for some species, it is likely an underestimate. It has been estimated that 22 species have gone extinct within the past 200 years.[1] Every order of reptile has experienced declines in some part of their geographic range. At present, one of the most imperiled groups of reptiles are the turtles, with more than 42% of known taxa classified as endangered, critically endangered, vulnerable, or near threatened.[1] Whereas crocodylians have 43% of known taxa classified as imperiled, the success of conservation programs, CITES (the Convention on International Trade in Endangered Species of Wild Fauna and Flora) protection, and captive breeding programs have significantly increased the stability of wild populations. Another 41 species of reptiles around the globe are considered at risk and classified by the IUCN as near threatened. All the causes for reptile population declines have not been identified; however, habitat loss and human exploitation have been identified as the primary causes.

Traditional classifications have historically placed crocodylians, turtles, snakes, lizards, amphisbaenids, and tuataras within the Class Reptilia. However, in recent years, some workers[4–7] have proposed a classification, based on cladistic analysis, that places turtles and crocodylians in separate classes. These proposed changes have not yet been widely accepted by the scientific community, so for the purposes of this chapter we will use the traditional classification, wherein turtles and crocodylians are referred to as members of the "Reptilia."

II. *Life History Characteristics and Ecologic Traits Related to Decline Susceptibility*

A. *Life Span and Sexual Maturity*

Delayed sexual maturity is one life history trait that is common to many long-lived vertebrates (e.g., many reptiles). This leads to more highly developed young and larger clutch sizes, but the trade-off is fewer reproductive events. Risks associated with delayed maturity are an increased chance of death before first reproductive event and longer generation times. In addition, substantially higher juvenile survivorship is required to maintain stable populations. The cryptic or secretive nature of juveniles, a key element in reaching reproductive age, may make it more difficult to accurately census populations with delayed sexual maturity. Thus, population estimates may greatly underestimate younger age classes, which can lead to the conclusion that a population appears to be in decline, when in actuality it may not be. Many late-maturing species (e.g., some turtles, crocodylians) do not reproduce every year; therefore, they usually have large clutch sizes in those years when they do reproduce. For example, it might appear that marine turtles, which may produce several large clutches in a single year, are quite prolific. However, marine turtles usually only lay eggs every 3 to 4 years and therefore have much lower lifetime fecundity than species that reproduce more frequently with smaller clutch sizes.[8] Sustained low hatchling or juvenile survivorship or increased adult mortality rates in long-lived, late-maturing species can lead to steady population decline. The life history strategies associated with long-lived species make it difficult for populations to recover from sustained harvest or other chronic population pressures. Any detectable population responses by these long-living species, whether negative (e.g., as a result of excessive harvest) or positive (e.g., as a result of head-start or other recovery programs) usually will be delayed.

Species with very low fecundity, such as tuataras (*Sphenodon* sp.), tortoises (*Gopherus* sp.), and some turtles (e.g., bog turtles, *Glyptemys muhlenbergii*) are particularly susceptible to indirect and direct anthropogenic stressors, such as increased number of predators (e.g., raccoons), commercial collection or human exploitation, and habitat destruction or alteration. The ability for such populations to recover from sustained external pressures is less likely than those with high fecundity. Tortoise populations, for example, are maintained primarily by high survivorship rates of adults. Adults are long lived and thus have long reproductive duration as a result of low adult predation rates. Even with fairly low egg production and low juvenile survivorship, populations can remain stable as long as external factors influencing adult survivorship do not exceed mortality rates. Habitat loss, disease, and commercial exploitation have all contributed to widespread population declines of tortoises, and their low fecundity cannot maintain sufficient juvenile recruitment to maintain stable populations. Condgon and colleagues[9] discuss the inherent difficulties associated with in-depth life history studies when dealing with long-lived species (e.g., turtles, crocodylians, tuatara, etc.). Very often, important life history information is lacking for one or more stages of their life history, which is essential for development of effective conservation strategies.

B. Detectability of Population Declines

Detecting population declines in reptile populations is inherently difficult as a result of cryptic or secretive nature, large home range size, low population densities, and lack or rarity of congregational behavior. Often, it is only the reproductive adults that are readily observed during egg laying (e.g., marine turtle nesting) or in search of mates during mating season (e.g., male timber rattlesnakes, *Crotalus horridus*, crossing roads). Observability of hatchlings and juveniles is often difficult because the young may use different niches than adults. Herpetofaunal survey methods may produce variable results depending on four degrees of effort: (1) survey technique, (2) duration, (3) timing of survey, and (4) environmental conditions. Thus, standardized survey methods are needed to be able to compare population estimates over space and time and subsequently be able to detect the true status of populations.

Gibbons and colleagues[14] advocated the use of standardized, long-term methods to monitor biodiversity (i.e., herpetofauna) to provide useful data on which to base land management or wildlife conservation decisions. Using short-term survey data can be problematic; not only do population indices change over time as a result of varying environmental conditions, but the degree of effort and sampling techniques used to conduct surveys may affect our ability to estimate population changes. Factors that can be misleading in making population estimates from short-term studies include the following: (1) behavioral differences, such as for species having a cryptic or fossorial nature; (2) temporal changes in activity levels; and (3) whether the extent of geographic distribution or habitat preferences are adequately known and monitored. A long-term survey requires commitment of time, resources, and properly trained personnel skilled in correctly identifying target individuals and recording data.

C. Geographic Distribution

Species with restricted geographic distributions are inherently susceptible to an increased risk of extinction or decline, whether continental endemics or insular. A number of factors may be involved, including lack of genetic diversity, increased susceptibility to disease, small spatial scale, and greater susceptibility to catastrophic events. Insular species are particularly vulnerable to human activity, and the link between human colonization of islands and subsequent declines in resident wildlife populations is well documented.[10,11]

Declines and extinctions are a result of habitat destruction, human exploitation, or the introduction of alien predator species. Insular species are generally naïve and have not evolved defense mechanisms against alien mammalian predators. Introduced mammalian predators have been implicated in the decline of a number of insular reptilian populations (see Case and Bolger[12]). Factors such as habitat suitability or availability of nonreptilian prey species for introduced predators may become critical to the survival of insular reptiles. Impacts to reptile populations by introduced predators are discussed in detail in Section VI of this chapter. Species restricted to islands or those with small geographic distributions are at highest risk for impacts from stochastic events as a result of a lack of nearby populations to recruit from or suitable habitat corridors to enable recolonization.

III. Threats Resulting from Alteration or Loss of Habitat

A. Conversion to Agriculture

Habitat loss and degradation are two of the primary causes for population declines in reptiles.[13–20] Lack of appropriate habitat can directly affect a species' ability to forage, hibernate, thermoregulate, or reproduce. Competition with feral or domesticated grazing livestock has been shown to affect the foraging habits of tortoises and lizards.[21–24] For example, conversion of large areas of forest habitat to agriculture in the Souss Valley of Morocco has negatively affected Mediterranean Spur-thighed tortoise (*Testudo graeca*) populations.[20] These tortoises are now restricted largely to irrigated, intensively farmed areas and are limited to nonnative *Opuntia* hedges that provide protection from predators and human collection. Intensive grazing by sheep, goats, and other livestock in other arid regions of the globe has affected the available forage for desert reptiles (e.g., tortoises, lizards[22–30]). Direct competition for or loss of forage has affected populations of desert tortoises (*Gopherus agassizi*) in the southwestern United States[16,24] and the giant tortoises (*Geochelone elephantopus*) of the Galapagos Islands.[27] Soil compaction as a result of the presence of livestock can affect plant community species composition in arid and semi arid ecosystems, leading to less desirable forage for desert fauna. Direct physical effects to tortoises, such as trampled young and burrows, are also believed to be significant stressors.[24] Other potential impacts of grazing to arid habitats include modified thermal regimes and reduced cover, which can increase the susceptibility of some species to predation.

B. *Deforestation*

Deforestation both directly and indirectly affects reptiles that require mature forest habitat. Some species migrate from wetlands to forested uplands to forage, nest, or find suitable hibernacula (e.g., chicken turtle, *Deirochelys reticularia*).[31] Essential microhabitats, such as moist areas associated with leaf litter or under fallen logs, are lost with large-scale deforestation. The loss of this microhabitat is the primary threat to species such as the Virgin Islands dwarf gecko (*Sphaerodactylus parthenopion*)[32] or the dwarf gecko (*S. ariasae*)[33] in the West Indies. Commonly cited as the world's smallest lizard, the endangered dwarf gecko quickly desiccates and dies when moist microhabitat is absent. Nearly 95% of Caribbean island forests have been cleared, and with increasing demands for farm lands and fuel wood, this and other forest-dependent species are likely to be extirpated.[12] Mainland species may withstand deforestation better than insular species, simply because there may be suitable habitat available in nearer proximity (i.e., easier access). However, extensive habitat fragmentation has largely reduced or obliterated suitable migration corridors for wildlife, making it difficult for them to relocate successfully, if at all.

C. *Habitat Fragmentation*

Lizard[34] and snake[35] communities have been shown to be vulnerable to habitat fragmentation. Habitat generalists can withstand drastic changes to landscapes better than habitat specialists. A species' ability to disperse can play an important role in dealing with loss or degradation of habitat. Some species may survive in fragmented habitats for several generations, whereas other cannot maintain viable populations without recruitment through immigration.

Although the creation of roads has probably stabilized in developed countries, the number of vehicles using these roads continues to increase, creating a hazard for a variety of species, even marine species (e.g., hatchlings often wander onto roads when disoriented by artificial lighting).[36–40] Roads create unnatural obstacles, adding to already fragmented habitats. The risks associated with surviving seasonal movements increase dramatically when habitat has been dissected by roads. Only rarely are efforts made to mitigate for potential risks from road crossings to wildlife, such as construction of wildlife crossing culverts[41,42] or road closings during seasonal migrations to and from hibernacula.[43]

D. *Aquatic Habitat Alteration*

Structural alteration of aquatic habitat has been shown to negatively affect freshwater aquatic turtles. Removal of snags in river systems reduces the availability of basking sites for aquatic turtles. Appropriate basking sites are necessary for thermoregulation, which is essential for digestion, egg development, and growth.[44–47] For example, the yellow blotched map (*Graptemys flavimaculata*)[48,49] and ringed map (*G. oculifera*) turtles are believed to have declined as a result of disturbance (e.g., channelization, gravel mining, and dam building) of the rivers throughout much of their limited range.[50] Sand mining for road maintenance and construction projects has become increasingly efficient even in remote areas of the world. The process of dredging channel and sand bank substrates removes suitable nesting habitat used by riverine turtles. Direct impacts from sand mining have been demonstrated in populations of river turtles in Bangladesh, India, Indonesia, Malaysia, and Thailand.[51] Not only are nesting females and their nests directly affected, but disturbed banks are subsequently destabilized, leading to further erosion and sedimentation of the rivers, contributing to poor water quality.

The creation of reservoirs can affect aquatic turtle populations by reducing or eliminating nest sites and basking habitat. Reservoir banks may not provide appropriate nesting substrate, and fluctuating water levels can flood nests. Increased water flows below the dams create steep cut banks, further reducing suitable basking and nesting habitats. To date, no long-term studies have been conducted to determine the effects of manmade reservoirs on aquatic turtle populations; however, it has been suggested[52] that some species will adapt adequately to modified reservoir habitat, whereas others may not.

E. *Coastal Development*

Coastal development has resulted in the alteration of nesting habitat of marine turtles.[53,54] Developed beach fronts are poor nest sites because of increased beach erosion, physical obstructions (because of beach armoring and sand compaction, resulting from beach nourishment[55]), and artificial lighting that disorients nesting females and dispersing hatchlings.[36–40] Hatchling disorientation extends time spent on the beach, thus increasing risk of mortality as a result of desiccation, predation, or interaction with vehicles. Vehicle traffic on developed beaches, for recreation or maintenance and cleaning, increases the risk of direct mortality of eggs or pre-emergent hatchlings.[56,57] Tire ruts can increase beach erosion and interfere with the ability of hatchlings to reach the ocean, and vehicle headlights can interrupt nesting females during the nesting season.[57]

F. *Loss of Wetlands*

Reptiles are among many wetland-dependent species that have been affected by extensive draining of wetlands for urban development[58] and conversion to agricultural lands.[20,59,60] For example, bog turtle populations have declined as a result of the extensive loss of bog habitat in the eastern United States,[61] and restoration efforts to create new bog wetlands have met with limited success. Other reptile species may not be directly affected by the loss of wetland habitat, but they may be influenced via other indirect effects. For example, loss of wetland habitat has been shown to reduce populations of amphibians[62,63] that make up a significant proportion of the diet of many reptiles.

G. *Other Habitat Issues*

Fire suppression reduces the amount of open under story required by open forest– savannah-dwelling reptilian species. For example, fire suppression in rocky cedar glade habitats of the Interior Highlands of Arkansas and Missouri allows for encroachment of cedar (*Juniperus*) and other woody species, resulting in overgrown glades, thereby reducing or eliminating habitat and preventing dispersal in some species, such as the collared lizard (*Crotaphytus collaris*).[64] Given their highly territorial nature, populations of collared lizards rely on the ability of young males to emigrate to new territories, away from older, established males, to maintain genetic diversity.[64] Appropriate fire management in glade habitats has been shown to increase collared lizard numbers.[65] Unchecked wildfires in arid regions can affect desert reptiles through direct mortality and loss of cover vegetation.[66,67]

IV. Threats Resulting from Nonanthropogenic Causes

A. *Disease*

The incidence of disease reported in wild reptile populations is largely limited to the Chelonians (see review[68]). For example, some North American gopher and desert tortoise (*Gopherus* sp.) populations have suffered population declines attributed in part to infections of the upper respiratory tract (i.e., commonly referred to as URTD); it is likely that this organism was introduced into wild populations through the release of infected captive animals.[69,70] Some examples of diseases with unknown cause that are affecting wild reptile populations include the shell disease (i.e., dyskeratosis) in desert tortoises, which have suffered high mortality rates,[70] and the fibropapilloma tumors, which are commonly found on green turtles (*Chelonia mydas*) and increasingly are being noted in loggerhead (*Caretta caretta*) and olive ridley turtles (*Lepidochelys olivacea*).[68,71–73] Fibropapillomatosis and other isolated cases of disease outbreak in some wild turtle populations[74–76] suggest the possibility of immunosuppression as a result of environmental contaminants (see Chapter 9).

B. Drought

Population responses of reptiles to short- and long-term drought are largely unknown. Given the general lack of long-term ecologic studies on reptile populations, distinguishing the causal effect on any short- or long-term fluctuation in a population is difficult. A few case studies have illustrated how differential responses to environmental cues, such as drought, exist among and within reptile species. Aquatic snakes from the same location, for example, respond differently to severe drought, with some species retreating from drying sites shortly after drought onset and others remaining near water bodies until indirect factors (e.g., lack of prey) begin to affect the population.[77] Seigel and colleagues[77] found that banded water snakes (*Nerodia fasciata*) retreated at initiation of drying of a site, whereas the black swamp snake (*Seminatrix pygea*) left only after their preferred prey items disappeared (*Gambusia* sp. and *Ambystoma talpoideum* larvae). Neonate response appeared to differ from that of adults, in that younger individuals of both species delayed retreat from drying sites compared with adults.[77]

Turtle and tortoise populations also demonstrate differential responses to drought. Some species respond by reducing reproductive energetic expenditures by decreasing the proportion of multiple clutches during drought years.[78] Some species increase emigration rates along with reduced reproductive output, whereas others respond by emigrating only after egg-laying is complete.[78] Differences in species response to drought also depend on the degree of terrestrial vs. aquatic orientation of the species. Of five aquatic and semiaquatic turtle species studied by Gibbons and colleagues,[78] the more terrestrial eastern mud turtle (*Kinosternon subrubrum*) appeared to be less affected by drought than did the more aquatic common musk turtle (*Sternotherus odoratus*). *Kinosternon* exhibited no apparent difference in reproductive output and movement patterns when compared with years with normal rainfall. Desert tortoises are able to survive periods of drought by reducing energetic needs and water loss by remaining dormant in burrows.[79] However, extended periods of drought, coupled with low biomass of annual succulents in the spring, appear to push them to their physiologic limits.[80] The presence of diseases, such as URTD that is now present in many *Gopherus* sp. populations, increases the risk of a population crash in those stressed by long-term drought.

Species may initially react differentially to environmental changes such as drought, but they also recover differentially. Some may rely directly on having sufficient moisture, whereas others are predicated on the recovery of prey species, which also recover differentially under the influence of predators. Availability of habitat corridors and nearby "source" populations for recolonization of a depleted site plays an important role in drought recovery. Again, this supports the caution needed for interpreting population declines as either a result of human activities vs. natural population fluctuations.

C. Stochastic Events

Insular populations in particular are inherently susceptible to effects from other stochastic events, such as hurricanes or tropical storms. (Also see earlier discussion in Section II on the susceptibility of insular populations.) Devastating events occurring in areas where introduced predators or habitat destruction have already affected resident populations could push them past the point of natural recovery.

V. Commercial Exploitation and Unsustainable Use

A. Consumptive Uses

Turtles in particular are sought after for a number of products, from consumable eggs and meat to their shells for use as ceremonial masks, jewelry, or medicinal products.[19,81–85] Alligator snapping turtles (*Macrochelys temminckii*) have been commercially harvested extensively in Louisiana, primarily for food, but the lack of historic baseline population data makes it difficult to determine the extent of population declines.[85]

Turtles have been harvested for human consumption or other uses for centuries. For example, Galapagos tortoises (*Geochelone* sp.) were traditionally harvested for food and oil by seafarers in the 1600s to 1800s (see historical review by Thorbjarnarson and colleagues[86]). When used for food or other uses by indigenous people, the level of harvest was presumably sustainable because of the small human population sizes. However, there has been a shift toward widespread commercial exploitation in the past four decades that has led to precipitous declines in a number of chelonians. The Asian turtle crisis provides a prime example of the dramatic increase in demand for turtles for food, pets, and medicinal trades. An upward shift in the Asian economy has facilitated this demand (see Van Dijk et al.[87]).

Other continents have experienced similar demands in the turtle trade and subsequent population declines. South American indigenous cultures have traditionally relied on riverine turtle eggs and adults for subsistence during the dry season.[88] Today, however, not only are the turtles still used for food by locals, but they are now increasingly used in bartering for other food and alcohol products as the widespread demand for turtle products has increased. This increased demand will likely exceed the natural ability for turtle populations to recover, particularly with the increased pressures being put on the adult life stage.[9] For example, in Venezuela, populations of the largest neotropical freshwater turtle (*Podocnemis expansa*) have dwindled to the point that in many areas the harvest for subsistence and for commercial trade has shifted to the smaller species of side-neck turtles (Family Pelomedusidae).[89] In Madagascar, a similar shift appears to have taken place with populations of the endemic pelomedusids.[90]

Although crocodylians inhabit and are sometimes harvested from the same habitats as turtles, they have maintained relatively stable populations. This is partly a result of their protection under CITES and Ecological Society of America (ESA) listings but also because of effective farming operations and regulated harvests. Although crocodylians tend to have higher market value than turtles, large breeding adults are taken from the wild much less frequently because of lower demand and greater risk in obtaining them. Turtles, however, are more easily exploited; they are a common by-catch species in fishing nets, are safer to handle, and are easily collected using a variety of means. Turtles incidentally collected in fishing nets in economically depressed areas, even those species protected by law, are likely to be consumed.[84] Although most of the data presented on population declines of turtles in remote areas come from anecdotal evidence gathered from local fishers or by the presence or absence of species in markets, downward population trends appear to be recurring around the globe.[87,91]

B. Commercial Trade

Many of the species being harvested for commercial trade are protected under CITES. However, the lack of enforcement in areas where animals are collected is a problem (i.e., lack of government-funded natural resource officers). The long-term survival of many species of turtle is questionable unless human population growth or the demand for turtle products is reduced or successful captive propagation programs are established in the very near future. Educating turtle trappers about the need for sustainable harvest and providing an impetus for them to protect their local natural resources will be imperative to the long-term survival of vulnerable species.

Crocodylians around the world are highly prized for their hides. Up until the 1960s and 1970s, before most crocodylians were protected, nearly all crocodylian species were at risk of extinction. With the inception of conservation and management plans, tightly regulated legal harvests, and the advent of commercial breeding farms, sustainable use of these long-lived, late-maturing reptiles has proved possible.[92,93]

Although the commercial trade in other reptile products, such as snake or lizard skins, is monitored through CITES, the impact of this trade on wild populations is largely unknown. This presents great uncertainty in the susceptibility of these species to declines attributable to trade of their hides. For example, the current population status of most sea snakes is virtually unknown throughout their ranges, but sea snakes are a significant by-catch in commercial shrimp trawlers and are subsequently harvested for their skins.[94]

Throughout its range, the greatest threat to diamondback terrapin (*Malaclemys terrapin*) populations is believed to be accidental drowning in commercial crab traps.[95,96] Although other threats such as habitat loss, predation, road kill, and pollution have been suggested as major population threats in some portions of its range, it appears that crab traps pose the greatest threat to the diamondback terrapin. Baseline population data are largely lacking for this species as with many turtle populations, often making it difficult to document significant declines.

C. Pet Trade

Reptiles have become increasingly popular in the pet trade, partly because of the relative ease of their care and husbandry, availability, and generally low prices. In the United States, herpetoculturalists are increasingly producing a variety of captive-bred lizards, snakes, and turtles, both native and exotic. The demand for turtles has grown widely and includes a variety of species ranging from the more common hatchling slider turtles (*Trachemys scripta*) to the relatively uncommon Blanding's (*Emydoidea blandingii*) or wood turtles (*Glyptemys insculpta*).[97] However, the demand still exists for rare species, even those designated as protected, as evidenced through repeated encounters of illegal shipments containing protected species such as the African endemic pancake tortoises (*Malacochersus tornieri*).[98,99]

Because it can take many years for some turtles to reach maturity, it is inefficient for farming operations to raise animals from hatchlings for their breeding stocks; thus, reproductive-age adults are inevitably taken from the wild to supply commercial turtle farming operations. The number of animals taken from the wild to initially supply and maintain captive breeding stocks is likely underestimated.[100] African pancake tortoises are a good example of a species vulnerable to extinction as a result of overcollection. With very narrow habitat requirements, low reproductive rates (one egg per clutch), and limited dispersal abilities, local populations have been easily exploited by commercial collectors to the point that populations in easily accessed areas have been extirpated.[98,101]

VI. Other Threats

A. Exotic and Invasive Species Introductions

As humans explored and colonized the world, particularly during the 17th to 20th centuries, the intentional or unintentional introduction of alien species of flora and fauna occurred, sometimes with devastating results. Many of these species are often able to exploit their new habitats by outcompeting or preying on native species. In general, the most successful introduced predators of reptiles have been mammals. For example, rats, cats, hogs, and dogs often accompanied humans as they colonized, either as pets, food, or inadvertent pests. Mongooses, for example, were historically introduced to islands as a means of controlling either native or introduced rats. However, more often than not, the diet of the introduced mongoose consisted primarily of native ground-dwelling birds and lizards (e.g., skinks) rather than the intended rodent pests.[103] Eradication programs of introduced mammals often relieves the pressure on affected native species; however. total eradication is usually not feasible or successful, leaving native populations vulnerable.[103]

Introduced mammals have contributed to declines of insular reptile species, and they have been implicated as the primary cause for extinctions. For example, the tuatara, endemic to New Zealand, consists of only two species (*Sphenodon punctatus* and *S. guntheri*) and are the only extant representatives of the Order Rhyncocephalia.[104] Tuataras have become extirpated on the two main islands of New Zealand within the past 200 years, partly as a result of habitat alteration and human exploitation but primarily as a result of the presence of nonnative mammalian predators. The remaining populations of tuatara, which are now restricted to the smaller islands off the northern coast of New Zealand, continue to be affected on all except the predator-free, uninhabited islands. It is clear that the effects on tuatara populations are related to the presence of the introduced Pacific rats (*Rattus exulans*).[105] However, it is still not clear whether these rats are directly affecting recruitment in the tuatara by predation of eggs and juveniles or whether the tuatara are being outcompeted for prey items (e.g., insects). Regardless of the mechanism, the effects of Pacific rats on tuatara populations appear to be much more dramatic on small islands, suggesting that habitat diversity, suitability, or both play a strong role in predator–prey population dynamics.

In addition to the tuatara, three species of lizards[106–108] and the three largest species of frogs[107] have also gone extinct in New Zealand, presumably as a result of the introduced Pacific rat. Ground foraging skinks have suffered extinctions in the Fiji Islands where cats and mongooses have been introduced,[109] yet these skinks continue to survive on other islands where only introduced rats are present. Other factors, including habitat suitability and alternative prey availability, are important determinants of the impact of an introduced predator on resident island fauna.

Introduced nonmammalian species have also taken their toll on native island reptile populations. Brown treesnakes (*Boiga irregularis*) are rear-fanged, venomous snakes native to New Guinea, the Solomon Islands, and northern Australia and were accidentally introduced via cargo shipments to the Pacific island of Guam after World War II.[110] These efficient terrestrial and arboreal predators were able to successfully invade a snake-free and otherwise predator-free island, where they attained incredibly dense populations by the mid-1980s. Juvenile brown tree snakes feed almost exclusively on lizards and have severely diminished the resident lizard fauna on Guam.[111] In addition, brown tree snakes have since nearly decimated many of the native bird species[112] and have negatively affected native fruit bat populations.[110] Efforts to eradicate this venomous invader are still in progress, not only for the benefit of the resident wildlife but for economic and human safety concerns as well.

Since its accidental introduction in Mobile, Alabama, a little more than 50 years ago, the fire ant (*Solenopsis invicta*) has spread throughout the southeastern United States in pastures and other disturbed sites. The introduction of this alien species has negatively affected a variety of native species,[113] including birds, reptiles, and amphibians, as well as native ant species.[18,114–122] Egg-laying reptiles appear to be the most susceptible to attacks from imported fire ants because the defenseless young provide an easy target during pipping or overwintering before emergence from the nests. American alligator (*Alligator mississippiensis*) eggs and hatchlings may be particularly vulnerable because the raised, disturbed substrates of the nest provide a suitable environment for fire ant colonization, accompanied by a readily accessible nutritional resource: the eggs and young. Studies in central Florida indicate that up to 20% of alligator nests contained fire ant colonies.[121] Given that the range of the fire ant now completely overlaps with that of the American alligator, this introduced species could become a serious threat to their survival. Although extensive studies have not yet been conducted, outright mortality of eggs, pipping young, recent hatchlings, or adults has been documented in American alligators,[121] gopher tortoises (*Gopherus polyphemus*),[123] eastern box turtles (*Terrapene carolina*),[120] loggerhead sea turtles (*Caretta caretta*),[122] racerunners (*Cnemidophorus sexlineatus*),[124] and Texas horned lizards (*Phrynosoma cornutum*).[125] Indirect effects, such as reduced survival rates, reduced body weights, and behavioral changes, resulting from fire ant envenomation are also evident in ground-nesting birds (e.g., northern bobwhite, or *Colinus virginianus*)[117] and American alligators.[121] Fire ants have been suggested as one of the primary threats and possible cause for the apparent declines in the semi-fossorial southern hognose snake (*Heterodon simus*).[18]

B. Indirect Impacts

The apparent global declines of amphibian populations suggest large-scale environmental degradation, which has the potential to affect virtually all living organisms. Specifically, amphibian declines may potentially lead to indirect effects on some reptilian populations, particularly specialists that feed primarily on amphibians. Hognose snakes (*Heterodon* sp.), for example, feed primarily on toads (*Bufo* spp.).[126] Southern hognose snake populations have apparently declined in the southeastern United States,[18] with the primary threats identified as habitat alteration and impacts from introduced fire ants. To our knowledge, toad populations in the southeastern United States have not been reported to be in decline, so it is likely that other factors are at work in the apparent decline of this snake species. However, the potential for indirect effects on reptiles that rely on amphibians as their primary prey items (e.g., garter snakes *Thamnophis* spp.,[127] *Xenodon severus*[128]) should not be discounted when studying reptile populations.

VII. Summary

The detection of population declines in reptiles requires an understanding of existing population status and consideration of natural population fluctuations over time. Most studies are short-term in duration and provide only a snapshot in time of a population. Population response to a natural event (e.g., drought, flood) may not be directly evident and may become apparent only after indirect effects such as decreased availability of prey. Population recovery rates may not be readily apparent or similar among species. Many long-lived, late-maturing species can have cryptic juvenile stages, which makes accurate population census difficult. Detection of declines in such populations may be delayed until the lack of reproductive size classes becomes evident. Most population declines are likely the result of a synergistic effect of both anthropogenic and natural stressors, and rarely is one single factor the primary cause of observed declines. Researchers must remain cognizant of cumulative effects on a population when investigating the effects of any single stressor on reptile populations.

Our current knowledge of environmental contaminants and their toxicologic effects on reptile populations clearly indicates that a number of chemical stressors are present in the environment. Whereas this chapter has attempted to illustrate the difficulties associated in determining reptile population declines, the following chapters will elaborate on the current scientific evidence of toxic effects of environmental contaminants on reptiles. Researchers cannot discount the role that environmental contaminants play in the global distress experienced by reptile populations. The presence of environmental contaminants may be the "straw that broke the camel's back" in an area where disease, alien predators, drought, human exploitation, and so on have affected a reptile population. Reptiles have evolved and adapted to survive a number of environmental challenges; however, the addition of human-induced contaminants within the past century may be seriously impairing their ability to continue to survive. It is apparent that many reptile populations around the world suffer from a variety of threats, and the cumulative effects eventually may supersede some species' ability to sustain viable populations.

Literature Cited

1. IUCN Red List of Threatened Species. Downloaded from http://www.redlist.org/ September 7, 2004. International Union for Conservation of Nature and Natural Resources, 2003.
2. IUCN Red List Categories and Criteria: Version 3.1. IUCN, Gland, Switzerland and Cambridge, UK, IUCN Species Survival Commission, 30 pp, 2001.
3. Young, B. E., Stuart, S. N., Chanson, J. S., Cox, N. A., and Boucher, T. M., *Disappearing Jewels: The Status of New World Amphibians*, NatureServe, Arlington, VA, 2004.
4. Ahlberg, P.E. and Milner, A.R., The origin and early diversification of tetrapods, *Nature*, 368, 507–514, 1994.
5. Hedges, S.B. and Poling, L., A molecular phylogeny of reptiles, *Science*, 283, 998–1001, 1999.
6. Laurin, M. and Reisz, R.R., A re-evaluation of early amniote phylogeny, *Zool. J. Linnaean Soc.*, 113, 105–225, 1995.
7. Zug, G.R., Vitt, L.J., and Caldwell, J.P., *Herpetology*, Academic Press, San Diego, 2001.
8. Ernst, C.H. and Barbour, R.W., *Turtles of the World*, Smithsonian Institution Press, Washington, D.C., 1989.
9. Congdon, J.D., Dunham, A.E., and Van Loben Sels, R.C., Delayed sexual maturity and demographics of Blanding's turtles (*Emydoidea blandingii*): Implications for conservation and management of long-lived organisms, *Cons. Biol.*, 7 (4), 826–833, 1993.
10. Pregill, G.K., Body size of insular lizards: a pattern of holocene dwarfism, *Evolution*, 40, 997–1008, 1986.
11. Case, T.J., Bolger, D.T., and Richman, A., *Reptilian Extinctions: The Last Ten Thousand Years*, Chapman & Hall, New York, 1991.
12. Case, T.J. and Bolger, D.T., The role of introduced species in shaping the distribution and abundance of island reptiles, *Evol. Ecol.*, 5, 272–290, 1991.
13. Buhlmann, K.A. and Gibbons, J.W., Imperiled aquatic reptiles of the southeastern United States: historical review and current conservation status, in *Aquatic Fauna in Peril: The Southeastern Perspective*, Benz, G.W. and Collins, D.E., Eds., Lenz Design & Communication., Decatur, GA, 1997, pp. 554.
14. Gibbons, J.W. et al., Perceptions of species abundance, distribution, and diversity: lessons from four decades of sampling on a government-managed reserve, *Environ. Manage.*, 21 (2), 259–268, 1997.
15. Gibbons, J.W. et al., The global decline of reptiles, deja vu amphibians, *BioScience*, 50 (8), 653–666, 2000.
16. Grover, M.C. and DeFalco, L.A., General Technical Report No. INT-GTR-316, 1995.
17. Henderson, R.W., Consequences of predator introductions and habitat destruction on amphibians and reptiles in the post-Columbus West Indies, *Caribb. J. Sci.*, 28 (1–2), 1–10, 1992.
18. Tuberville, T.D. et al., Apparent decline of the southern hog-nosed snake, *Heterodon simus*, *J. Elisha Mitchell Sci. Soc.*, 116 (1), 19–40, 2000.
19. Beshkov, V.A., On the distribution, relative abundance and protection of tortoises in Bulgaria, *Chel. Cons. Biol.*, 1 (1), 53–62, 1993.

20. Bayley, J.R. and Highfield, A.C., Observations on ecological changes threatening a population of *Testudo graeca graeca* in the Souss Valley, Southern Morocco, *Chel. Cons. Biol.*, 2 (1), 36–42, 1996.
21. Scott, N.J., Jr., *Evolution and management of the North American grassland herpetofauna*, D.M. Finch, ed. *Ecosystem disturbance and wildlife conservation in western grasslands: a symposium proceedings*, U.S. Forest Service, Rocky Mountain Forest and Range Experiment Station, Fort Collins, Colorado, 40–53, 1996.
22. Germano, D.J. and Hungerford, C.R., Reptile population changes with manipulation of Sonoron desert shrub, *Great Basin Nat.*, 41 (1), 129–138, 1981.
23. Ballinger, R.E. and Watts, K.S., Path to extinction: impact to vegetational change on lizard populations on Arapahoe Prairie in the Nebraska sandhills, *Am. Midl. Nat.*, 134, 413–417, 1995.
24. Avery, H.W. and Neibergs, A.G., Effects of cattle grazing on the desert tortoise, *Gopherus agassizii*: nutritional and behavioral interactions, in *Conservation, Restoration, and Management of Tortoises and Turtles — An International Conference*, Van Abbema, J., Ed., New York Turtle and Tortoise Society and the WCS Turtle Recovery Program, State University of New York, Purchase, NY, 1997, pp. 13–20.
25. Bury, R.B. and Busack, S.D., Some effects of off-road vehicles and sheep grazing on lizard populations in the Mohave desert, *Biol. Cons.*, 6, 179–183, 1974.
26. Bock, C.E., Smith, H.M., and Bock, J.H., The effect of livestock grazing upon abundance of the lizard, *Sceloporus scalaris*, in southeastern Arizona, *J. Herpetol.*, 24 (4), 445–446, 1990.
27. Hamann, O., On vegetation recovery, goats and giant tortoises on Pinta Island, Galapagos, Ecuador, *Biodiv. Cons.*, 2, 138–151, 1993.
28. Fleischner, T.L., Ecological costs of livestock grazing in Western North America, *Cons. Bio.*, 8 (3), 629–644, 1994.
29. Smith, G.T. et al., The effect of habitat fragmentation and livestock grazing on animal communities in remnants of gimlet *Eucalyptus salubris* woodland in the Western Australia wheatbelt. II. Lizards, *J. Appl. Ecol.*, 33, 1302–1310, 1996.
30. Read, J.L., Experimental trial of Australian arid zone reptiles as early warning indicators of overgrazing by cattle, *Aust. Ecol.*, 27, 55–66, 2002.
31. Buhlmann, K.A., Habitat use, terrestrial movements, and conservation of the turtle, *Deirochelys reticularia* in Virginia, *J. Herpetol.*, 29 (2), 1995.
32. MacLean, W.P., Water-loss rates of *Sphaerodactylus parthenopion* (Reptilia: Gekkonidae), the smallest amniote vertebrate, *Comp. Biochem. Physiol.*, 82A, 759–761, 1985.
33. Hedges, S.B. and Thomas, R., At the lower size limit in amniote vertebrates: a new diminutive lizard from the West Indies, *Caribb. J. Sci.*, 37 (3–4), 168–173, 2001.
34. Sarre, S., Wiegand, K., and Henle, K., The conservation biology of a specialist and a generalist gecko in the fragmented landscape of the western Australian wheatbelt, in *Species Survival in Fragmented Landscapes*, Settele, J., Margules, C., Poschlod, P., and Henle, K., Eds. Kluwer Academic Publishers, Dordrecht, 1996, pp. 39–51.
35. Kjoss, V.A. and Litvaitis, J.A., Community structure of snakes in a human-dominated landscape, *Biol. Cons.*, 98 (3), 285–292, 2001.
36. Witherington, B.E., Behavioral responses of nesting sea turtles to artificial lighting, *Herpetologica*, 48 (1), 31–39, 1992.

37. McFarlane, R.W., Disorientation of loggerhead hatchlings by artificial road lighting, *Copeia*, 1963, 153, 1963.
38. Philibosian, R., Disorientation of hawksbill turtle hatchlings, *Eretmochelys imbricata*, by stadium lights, *Copeia*, 1976, 824, 1976.
39. Peters, A. and Verhoeven, K.J.F., Impact of artificial lighting on the seaward orientation of hatchling loggerhead turtles, *J. Herpetol.*, 28, 112–114, 1994.
40. Salmon, M. and Witherington, B.E., Artificial lighting and seafinding by loggerhead hatchlings: evidence for lunar modulation, *Copeia*, 1995, 931–938, 1995.
41. Yanes, M., Velasco, J.M., and Suarez, F., Permeability of roads and railways to vertebrates: the importance of culverts, *Biol. Cons.*, 71, 217–222, 1995.
42. Dodd, C.K., Jr., Barichivich, W.J., and Smith, L.L., Effectiveness of a barrier wall and culverts in reducing wildlife mortality on a heavily traveled highway in Florida, *Biol. Cons.*, 118, 619–631, 2004.
43. Hupe, S., *Land and Resource Management Plan: Monitoring and Evaluation Report*, Fiscal Year 2001. Shawnee National Forest, U.S. Forest Service, Shawnee National Forest, Harrisburg, IL, 94 pp., 2001.
44. Boyer, D.R., Ecology of the basking habit in turtles, *Ecology*, 46, 99–118, 1965.
45. Crawford, K.M., Spotila, J.R., and Standora, E.A., Operative environmental temperatures and basking behavior of the turtle *Pseudemys scripta*, *Ecology*, 64, 989–999, 1983.
46. Lindeman, P.V., Aerial basking by hatchling freshwater turtles, *Herpetol. Rev.*, 24 (3), 84–87, 1993.
47. Lindeman, P.V., Surveys of basking map turtles *Graptemys* spp. in three river drainages and the importance of deadwood abundance, *Biol. Cons.*, 88 (1999), 33–42, 1999.
48. Jones, R.L., *Technical Draft: Yellow-Blotched Map Turtle* (Graptemys flavimaculata) *recovery plan*, U.S. Fish & Wildlife Service, Jackson, MS, 1992.
49. Seigel, R.A. and Brauman, R., *Reproduction and Nesting of the Yellow-Blotched Map Turtle*, U.S. Fish & Wildlife Service and the Mississippi Department of Wildlife, Fisheries and Parks, 1995.
50. Kofron, C.P., Aspects of ecology of the threatened ringed sawback turtle, *Graptemys oculifera*, *Amphibia-Reptilia*, 12, 161–168, 1991.
51. Moll, E.O., Effects of habitat alteration on river turtles of tropical Asia with emphasis on sand mining and dams, in *Conservation, Restoration, and Management of Tortoises and Turtles — An International Conference*, Van Abbema, J., Ed. New York Turtle and Tortoise Society and the WCS Turtle Recovery Program, State University of New York, Purchase, NY, 1997, pp. 37–41.
52. Dickerson, D.D., Reine, K.J., and Herrmann, K.L., Riverine turtle habitats potentially impacted by USACE reservoir operations, *EMRRP Technical Notes Collection (TN EMRRP-SI-05)*, U.S. Army Engineer Research and Development Center, Vicksburg, MS, 11, 1999.
53. Cheng, I., Tourism and the green turtle in conflict on Wan-an Island, Taiwan, *Mar. Turtle Newsl.*, 68, 5–6, 1995.
54. Drake, D.L., Marine turtle nesting, nest predation, hatch frequency, and nesting seasonality on the Osa Peninsula, Costa Rica, *Chel. Cons. Biol.*, 2 (1), 89–92, 1996.
55. Nelson, D.A. and Dickerson, D.D., Hardness of nourished and natural sea turtle nesting beaches on the east coast of Florida, Unpubl. Report, U.S. Army Waterways Experiment Station, Vicksburg, MS, 1988.

56. Mann, T.M., *Impact of developed coastline on nesting and hatchling sea turtles in southeastern Florida*, Florida Atlantic University, 1977.
57. Hosier, P.E., Kochhar, M., and Thayer, V., Off-road vehicle and pedestrian track effect on the sea-approach of hatchling loggerhead turtles, *Environ. Cons.*, 8, 158–161, 1981.
58. Hays, D.W. et al., *Washington State Recovery Plan for the Western Pond Turtle*, Washington Department of Fish and Wildlife, Olympia, WA, 1999.
59. Saumure, R.A. and Bider, J.R., Impact of agricultural development on a population of wood turtles (*Clemmys insculpta*) in southern Quebec, Canada, *Chel. Cons. Biol.*, 3 (1), 37–45, 1998.
60. Germano, D.J. and Bury, B.R., Western pond turtles (*Clemmys marmorata*) in the Central Valley of California: status and population structure, *Trans. West. Sec. Wildl. Soc.*, 37, 22–36, 2001.
61. Herman, D.W. and Tryon, B.W., Land use, development, and natural succession and their effects on bog turtle habitat in the southeastern United States, in *Conservation, Restoration, and Management of Tortoises and Turtles — An International Conference*, Van Abbema, J., Ed. New York Turtle and Tortoise Society and WCS Turtle Recovery Program, State University of New York, Purchase, NY, 1997, pp. 364–371.
62. Blaustein, A.B. and Wake, D.B., Declining amphibian populations: a global phenomenon? *Trends Ecol. Evol.*, 5 (7), 203–204, 1990.
63. Blaustein, A.B., Wake, D.B., and Sousa, W.P., Amphibian declines: judging stability, persistence, and susceptibility of populations to local and global extinctions, *Cons. Biol.*, 8 (1), 60–71, 1994.
64. Brisson, J.A., Strasburg, J.L., and Templeton, A.R., Impact of fire management on the ecology of collared lizard (*Crotophytus collaris*) populations living on the Ozark plateau, *Anim. Cons.*, 6, 247–254, 2003.
65. Templeton, A.R. et al., Disrupting evolutionary processes: the effect of habitat fragmentation on collared lizards of the Missouri Ozarks, *Proc. Natl. Acad. Sci.*, 998, 5426–5432, 2001.
66. Esque, T.C. et al., Effects of desert wildfires on desert tortoise (*Gopherus agassizii*) and other small vertebrates, *Southwest. Nat.*, 48 (1), 103–111, 2003.
67. Moore, J.E., Potential threats to tortoise populations in Parc National de "W", Niger, West Africa, in *Conservation, Restoration, and Management of Tortoises and Turtles — An International Conference*, Van Abbema, J., Ed., New York Turtle and Tortoise Society and the WCS Turtle Recovery Program, State University of New York, Purchase, NY, 1997, pp. 28–30.
68. Jacobson, E.R., Diseases in wild populations of turtles and tortoises: the chelonian charisma vs. coincidence conundrum, in *Conservation, Restoration, and Management of Tortoises and Turtles — An International Conference*, Van Abbema, J., Ed., New York Turtle and Tortoise Society and the WCS Turtle Recovery Program, State University of New York, Purchase, NY, 1997, pp. 87–90.
69. Seigel, R.A., Smith, R.B., and Seigel, N.A., Swine flu or 1918 pandemic? Upper respiratory tract disease and the sudden mortality of gopher tortoises (*Gopherus polyphemus*) on a protected habitat in Florida, *J. Herpetol.*, 37 (1), 137–144, 2003.
70. Berry, K.H., Demographic consequences of disease in two desert tortoise populations in California, USA, in *Conservation, Restoration, and Management of Tortoises and Turtles — An International Conference*, Van Abbema, J., Ed., New York Turtle and Tortoise Society and the WCS Turtle Recovery Program, State University of New York, Purchase, NY, 1997, pp. 91–99.

71. Herbst, L.H., Fibropapillomatosis of marine turtles, *Annu. Rev. Fish Dis.*, 4, 389–425, 1994.
72. Jacobson, E.R., Mansell, J.P., and Sundberg, J.P., Cutaneous fibropapillomas of green turtles (*Chelonia mydas*), *J. Comp. Pathol.*, 101, 39–52, 1989.
73. Lackovich, J.K. et al., Association of herpesvirus with fibropapillomatosis of the green turtle *Chelonia mydas* and the loggerhead turtle *Caretta caretta* in Florida, *Dis. Aquat. Org.*, 37, 89–97, 1999.
74. Dodd, C.K., Jr., Disease and population declines in the flattened musk turtle *Sternotherus depressus*, *Am. Midl. Nat.*, 119 (2), 394–401, 1988.
75. Fonnesbeck, C.J. and Dodd, C.K., Jr., Estimation of flattened musk turtle (*Sternotherus depressus*) survival, recapture, and recovery rate during and after a disease outbreak, *J. Herp.*, 37 (3), 602–607, 2003.
76. Lovich, J.E. et al., Prevalence and histopathology of shell disease in turtles from Lake Blackshear, GA, *J. Wildl. Dis.*, 32, 259–265, 1996.
77. Seigel, R.A., Gibbons, J.W., and Lynch, T.K., Temporal changes in reptile populations: effects of severe drought on aquatic snakes, *Herpetologica*, 51 (4), 424–434, 1995.
78. Gibbons, J.W., Greene, J.L., and Congdon, J.D., Drought-related responses of aquatic turtle populations, *J. Herpetol.*, 17 (3), 242–246, 1983.
79. Duda, J.J., Krzysik, A.J., and Freilich, J.E., Effects of drought on desert tortoise movement and activity, *J. Wildl. Manage.*, 63, 1181–1192, 1999.
80. Longshore, K.M., Jaeger, J.R., and Sappington, J.M., Desert tortoise (*Gopherus agassizii*) survival at two eastern Mojave desert sites: death by short-term drought?, *J. Herpetol.*, 37 (1), 169–177, 2003.
81. Rhodin, A.G.J., Mittermeier, R.A., and Hall, P.M., Distribution, osteology, and natural history of the Asian giant softshell turtle, *Pelochelys bibroni*, in Papua New Guinea, *Chel. Cons. Biol.*, 1 (1), 19–30, 1993.
82. Lawson, D.P., Local harvest of hingeback tortoises, *Kinixys erosa* and *K. homeana*, in southwestern Cameroon, *Chel. Cons. Biol.*, 3 (4), 722–729, 2000.
83. Mitchell, J.C. and Rhodin, A.G.J., Observations on the natural history and exploitation of the turtles of Nepal, with life history notes on *Melanochelys trijuga*, *Chel. Cons. Biol.*, 2 (1), 66–72, 1996.
84. Kuchling, G. and Mittermeier, R.A., Status and exploitation of the Madagascan big-headed turtle, *Erymnochelys madagascariensis*, *Chel. Cons. Biol.*, 1 (1), 13–18, 1993.
85. Sloan, K. and Lovich, J.E., Exploitation of the alligator snapping turtle, *Macroclemys temminckii*, in Louisiana: a case study, *Chel. Cons. Biol.*, 1 (3), 221–222, 1993.
86. Klemens, M.W., *Turtle Conservation*, Smithsonian Institution Press, 2000.
87. Van Dijk, P.P., Stuart, B.L., and Rhodin, A.G.J., Asian turtle trade: proceedings of a workshop on conservation and trade of freshwater turtles and tortoises in Asia, in *Conservation and Trade of Freshwater Turtles and Tortoises in Asia*, Chelonian Research Foundation, Phnom Penh, Cambodia, 2000, pp. 164.
88. Klemens, M.W. and Thorbjarnarson, J., Reptiles as a food source, *Biodiv. Cons.*, 4, 281–298, 1995.
89. Thorbjarnarson, J.B., Perez, N., and Escalona, T., Biology and conservation of aquatic turtles in the Cinaruco-Capanaparo National Park, Venezuela, in *Conservation, Restoration, and Management of Tortoises and Turtles — An International Conference*, Van Abbema, J., Ed., New York Turtle and Tortoise Society and the WCS Turtle Recovery Program, State University of New York, Purchase, NY, 1997, pp. 109–112.

90. Kuchling, G., Patterns of exploitation, decline, and extinction of *Erymnochelys madagascariensis*: implications for conservation, in *Conservation, Restoration, and Management of Tortoises and Turtles — An International Conference*, Van Abbema, J. Ed., New York Turtle and Tortoise Society and the WCS Turtle Recovery Program, State University of New York, Purchase, NY, 1997, pp. 113–117.
91. Van Abbema, J. et al., Proceedings: *Conservation, Restoration, and Management of Tortoises and Turtles — An International Conference*, Van Abbema, J. Ed., New York Turtle and Tortoise Society and the WCS Turtle Recovery Program, State University of New York, Purchase, NY, 1993.
92. Ross, J.P., *Crocodiles. Status Survey and Conservation Action Plan*, 2nd ed., IUCN-SSC Crocodile Specialist Group, IUCN, Gland, Switzerland and Cambridge, UK, 1997.
93. Thorbjarnarson, J., Crocodile tears and skins: international trade, economic constraints, and limits to the sustainable use of crocodilians, *Cons. Biol.*, 13 (3), 465–470, 1999.
94. Heatwole, H., *Sea Snakes and Humans*, Kreiger Publishing Company, Malabar, FL, 1999.
95. Seigel, R.A. and Gibbons, J.W., Workshop on the ecology, status, and management of the diamondback terrapin (*Malaclemys terrapin*), Savannah River Ecology Laboratory, 2 August 1994: final results and recommendations, *Chel. Cons. Biol.*, 1 (3), 240–243, 1995.
96. Hoyle, M.E. and Gibbons, J.W., Use of a marked population of diamondback terrapins (*Malaclemys terrapin*) to determine impacts of recreational crab pots, *Chel. Cons. Biol.*, 3 (4), 735–737, 2000.
97. Levell, J.P., Commercial exploitation of Blanding's turtle, *Emydoidea blandingii*, and the wood turtle, *Clemmys insculpta*, for the live animal trade, *Chel. Cons. Biol.*, 3 (4), 665–674, 2000.
98. Klemens, M.W. and Moll, D., An assessment of the effects of commercial exploitation on te pancake tortoise, *Malacochersus tornieri*, in Tanzania, *Chel. Cons. Biol.*, 1 (1), 197–206, 1993.
99. Thorbjarnarson, J. et al., Human use of turtles: a worldwide perspective, in *Turtle Conservation*, Klemens, M.W., Ed., Smithsonian Institution, 2000, pp. 33–84.
100. Reed, R.N. and Gibbons, J.W., *Conservation status of live U.S. nonmarine turtles in domestic and international trade*, Report to U.S. Department of the Interior, U.S. Fish & Wildlife Service, 2004.
101. Moll, D. and Klemens, M.W., Ecology and exploitation of the pancake tortoise in Tanzania, in *Conservation, Restoration, and Management of Tortoises and Turtles — An International Conference*, Van Abbema, J., Ed., New York Turtle and Tortoise Society and the WCS Turtle Recovery Program, State University of New York, Purchase, NY, 1997, pp. 135–148.
102. Pimentel, D., Biology of the Indian mongoose in Puerto Rico, *J. Mammal.*, 36, 62–68, 1955.
103. Cayot, L.J., Rassmann, K., and Trillmich, F., Are marine iguanas endangered on islands with introduced predators? *Notic. de Galapagos*, 53 (April), 13–15, 1994.
104. Fraser, N.C., The osteology and relationships of *Clevosaurus* (Reptilia: Sphenodontida), *Phil. Trans. Royal Soc. London, B*, 321, 125–178, 1988.
105. Cree, A., Daugherty, C.H., and Hay, J.M., Reproduction of a rare New Zealand reptile, the tuatara *Sphenodon punctatus*, on rat-free and rat-inhabited islands, *Cons. Biol.*, 9 (2), 373–383, 1995.

106. Bauer, A.M. and Russell, A.P., *Hoplodactylus delcourti* n. sp. (Reptilia: Gekkonidae), the largest known gecko, *New Zeal. J. Zoo.*, 13, 141–148, 1986.
107. Worthy, T.H., Paleoecological information concerning members of the frog genus *Leiopelma*: Leiopelmatidae in New Zealand, *J. Royal Soc. New Zeal.*, 17, 409–420, 1987b.
108. Hardy, G.S., The New Zealand Scincidae (Reptilia: Lacertilia); a taxonomic and zoogeographic study., *N. Z. J. Zool.*, 4, 221–325, 1977.
109. Zug, G.R., The lizards of Fiji: natural history and systematics, *Bishop Mus. Bull. Zool.*, (2), 1991.
110. Greene, H.W., The brown tree snake (*Boiga irregularis*) a disastrous vagabond, in *Snakes: The Evolution of Mystery in Nature*, University of California Press, Berkeley, CA, 1997, pp. 179.
111. Engbring, J. and Fritts, T.H., Demise of an insular avifauna: the brown tree snake of Guam, *Trans. West. Sec. Wildl. Soc.*, 24, 31–37, 1988.
112. Savidge, J.A., Extinction of an island avifauna by an introduced snake, Ecol., 68, 660–668, 1987.
113. Callcott, A.M.A. and Collins, H.L., Invasion and range expansion of imported fire ants (Hymenoptera: Formicidae) in North America from 1918–1995, *Fla. Entomol.*, 79, 240–251, 1996.
114. Mount, R.H., The red imported fire ant, *Solenopsis invicta*, (Hymenoptera: Formicidae), as a possible serious predator of some native southeastern vertebrates; direct observations and subjective impressions, *J. Ala. Acad. Sci.*, 52, 71–78, 1981.
115. Vinson, S.B. and Sorensen, A.A., *Imported Fire Ants: Life History and Impact*, Texas Dept. Agric., Austin, TX, 1–28, 1986.
116. Allen, C.R., Demarais, S., and Lutz, R.S., Red imported fire ant impact on wildlife: an overview, *Tex. J. Sci.*, 46, 51–59, 1994.
117. Allen, C.R., Lutz, R.S., and Demarais, S., Red imported fire ant impacts on northern bobwhite populations, *Ecol. Appl.*, 5, 632–638, 1995.
118. Freed, P.S. and Neitman, K., Notes on predation on the endangered Houston toad, *Bufo houstonensis*, *Tex. J. Sci.*, 40, 454–456, 1988.
119. Donaldson, W.A., Price, A.H., and Morse, J., The current status and future prospects of the Texas horned lizard (*Phrynosoma cornutum*) in Texas, *Tex. J. Sci.*, 46, 97–113, 1994.
120. Montgomery, W.B., Predation by the fire ant, *Solenopsis invicta*, on the three-toed box turtle, *Terrepene carolina triunguis*, *Bull. Chicago Herpetol. Soc.*, 31, 105–106, 1996.
121. Allen, C.R. et al., Effect of red imported fire ant envenomization on neonatal American alligators, *J. Herpetol.*, 31 (2), 318–321, 1997.
122. Moulis, R.A., Predation of the imported fire ant (*Solenopsis invicta*) on loggerhead sea turtle (*Caretta caretta*) nests on Wassaw National Wildlife Refuge, Georgia, *Chel. Cons. Biol.*, 2, 433–436, 1997.
123. Landers, J.L., Garner, J.A., and McRae, W.A., Reproduction of gopher tortoises (*Gopherus polyphemus*) in southwestern Georgia, *Herpetologica*, 36, 353–361, 1980.
124. Mount, R.H., Trauth, S.E., and Mason, W.H., Predation by the red imported fire ant, *Solenopsis invicta* (Hymenoptera; Formicidae), on eggs of the lizard, *Cnemidophorus sexlineatus* (Squamata: Teiidae), *J. Ala. Acad. Sci.*, 52, 66–70, 1981.
125. Price, A.H., *Phrynosoma cornutum*, *Catalog of American Amphibians & Reptiles*, 469, 1–7, 1990.

126. Carr, A.F., Jr., A contribution to the herpetology of Florida, *University of Florida Publications of the Biological Society Series*, 3 (1), 1–118, 1940.
127. Jennings, W.B., Bradford, DF., and Johnson, D.F., Dependence of the garter snake *Thamnophis elegans* on amphibians in the Sierra Nevada of California, *J. Herpetol.*, 26 (4), 503–505, 1992.
128. Duellman, W.E., *The Biology of an Equatorial Herpetofauna in Amazonian Ecuador*, University of Kansas Publications, Lawrence, KS, 1978.

chapter 3

Use of Tissue Residues in Reptile Ecotoxicology: A Call for Integration and Experimentalism

William A. Hopkins

Table of Contents

I. *Introduction*

> "The classical approach to establish the hazard of toxic chemicals to wildlife is to determine the amount of a chemical present and then compare that value with those known to do harm in experimental animals."
>
> **D. Peakall, 1992**

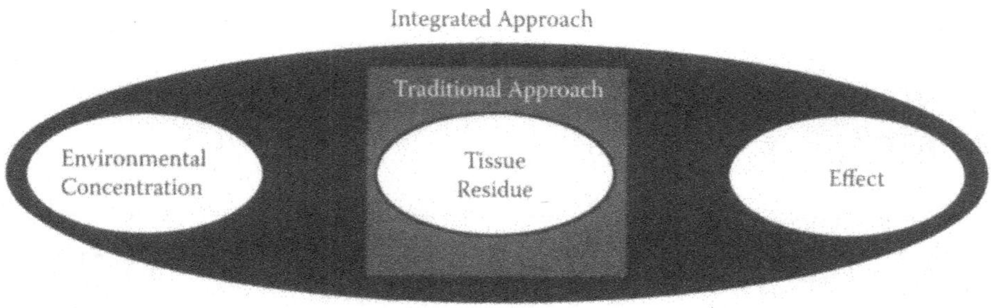

Figure 3.1 Simple depiction of the three key elements that should be included in integrative studies evaluating the importance of reptilian tissue residues. The shaded area represents the traditional approach that fails to provide important links among environmental conditions, accumulation, and effects.

Hazard and risk assessments in wildlife can be achieved by measuring contaminant levels in a variety of media, including water, sediment, soil, and biological tissues. Tissue residues (i.e., tissue concentrations) have been advocated by some scientists because residues often provide greater insight into the exposure conditions experienced by an organism compared with concentrations of contaminants in surrounding environmental matrices. The power of using tissue residues lies in their integrative nature; they account for bioavailability resulting from physico-chemical characteristics of the site and chemical speciation of the contaminant, as well as ecologic and physiologic characteristics of organisms that may influence uptake and accumulation. However, accumulation of contaminants is not necessarily hazardous to organisms. Expression of effects will depend on concentrations accumulated, characteristics of the compound, the duration of chemical exposure, the sensitivity of the organism under study, and a host of other ecologic and physical variables that can influence responsiveness. Thus, quantifying tissue residues alone has limited utility in conservation and risk-oriented initiatives. For tissue residues to be maximally useful within a conservation framework, one should understand the following: (1) the tissue residue concentration relative to exposure concentration and exposure duration, and (2) the effect (or lack thereof) of these tissue residues on the health of the organism or closely related species (Figure 3.1).

Tissue residue analyses currently represent the predominant segment of the reptile ecotoxicologic literature. Because of the wealth of information on this topic, numerous thorough reviews have RECENTLY emerged.[1–6] When used appropriately, quantification of tissue residues is, in fact, of central importance in reptile ecotoxicology and risk assessment. However, despite the hundreds of studies that document contaminant residues in reptiles, much of the existing literature falls short of providing the insight needed to protect the health of reptile populations in the face of anthropogenic change. In most cases, residue studies do not report environmental concentrations or effects (Figure 3.1), and only a handful of studies report both. Even in turtles, the most widely studied reptiles in ecotoxicology, the vast majority of tissue residue studies do not provide information regarding effects or concentrations of contaminants in environmental media such as food, water, soil, or sediment.[4] Additionally, risk assessors usually must rely on comparisons with other vertebrates because the relationships among tissue residues and responses have not been experimentally determined in reptiles. However, the uncertainties involved with extrapolating from other vertebrates (e.g., fish and birds) to reptiles are too large to produce risk assessments with a reasonable margin of error. Because of these knowledge gaps, interpreting tissue residues in reptiles remains difficult. Put simply, residue analyses by themselves are minimally informative toward reptile conservation efforts, and new approaches should be adopted to advance reptile ecotoxicology and risk assessment.

In this chapter I offer my perspective on where efforts in reptile tissue residue analyses should be focused to maximize our impact on conservation initiatives. Rather than summarize what has been thoroughly reviewed by others,[1–5] I concentrate on ways to redirect the current use of tissue residues based on a critical review of the literature. Many authors have advocated the use of tissue residues (e.g., concentrations in eggs) of field-collected, long-lived reptiles (e.g., turtles) as bioindicators of environmental contamination. Indeed, tissues from these organisms may be well suited for this purpose. However, reptiles deserve attention in their own right, not just as tools for monitoring habitat quality. Herein, I advocate redirecting toward a more integrated approach to reptile toxicology because we are now faced with a critical phase in the conservation of these organisms.[7] Clearly, effects-based work will be the most important step toward achieving this conservation agenda, and this is the focus of most other chapters in this text. However, as indicated in the quote by David Peakall at the beginning of the chapter, coupling effects-based work with measurements of tissue residues and assessment of exposure conditions will provide a much-needed framework to ensure future risk assessments have a higher degree of certainty. Because of logistical constraints (including funding and time) and limitations of individual expertise, it is obviously not always pragmatic for a researcher to evaluate effects of contaminants as well as contaminant concentrations in media and tissues. However, I draw attention to the limitations of the traditional approach to emphasize the importance of interdisciplinary efforts in the field. My intention is to generate discussion and to outline problems that would benefit from more integrative approaches.

II. A Comment on Biological Effects

Discussion of biological effects of contaminants is beyond the scope of this chapter and is covered thoroughly in other chapters in this volume. However, because I continually refer to residues as an intermediary between environmental conditions and effects, a brief comment is necessary on the types of effects that deserve priority attention.

Conservation initiatives and ecologic risk assessments primarily focus on the protection of complex systems (e.g., populations and communities), but in many cases ecotoxicologists focus their efforts on effects of contaminants on individual organisms (for exceptions, see references 8 and 9). Thus, ecotoxicologists must make important decisions regarding what responses to measure in individuals to maximize their contribution to conservation and risk initiatives. Although molecular and biochemical responses to contaminants are useful indices of exposure and provide insight into the mechanistic basis for toxicity, the expression of these endpoints at the organism level (let alone at the population or community levels) remains tenuous in most cases. Thus, from a conservation perspective, studies on reptiles that include integrated, organism-level responses will provide the greatest inference to population-level processes. If our ultimate objective is the conservation of healthy reptile populations, responses with demonstrated or strong theoretical implications for fitness (Figure 3.2) will provide information that is most useful in modeling and risk assessment. Although risk assessments generally focus on reproduction, growth, and survival,[10] other processes and traits such as development, performance (e.g., locomotion), behavior (e.g., predator–prey interactions), and key life history characteristics (e.g., age or size at first reproduction) can influence lifetime reproductive output and should also be prioritized in assessments of reptilian responses to contaminants. Chapter 10 provides more information on the effects of contaminants on populations.

III. A Call for Integration and Experimentalism

Reptile ecotoxicologists, like all scientists, must determine the most effective method of addressing hypotheses of interest. Whereas some

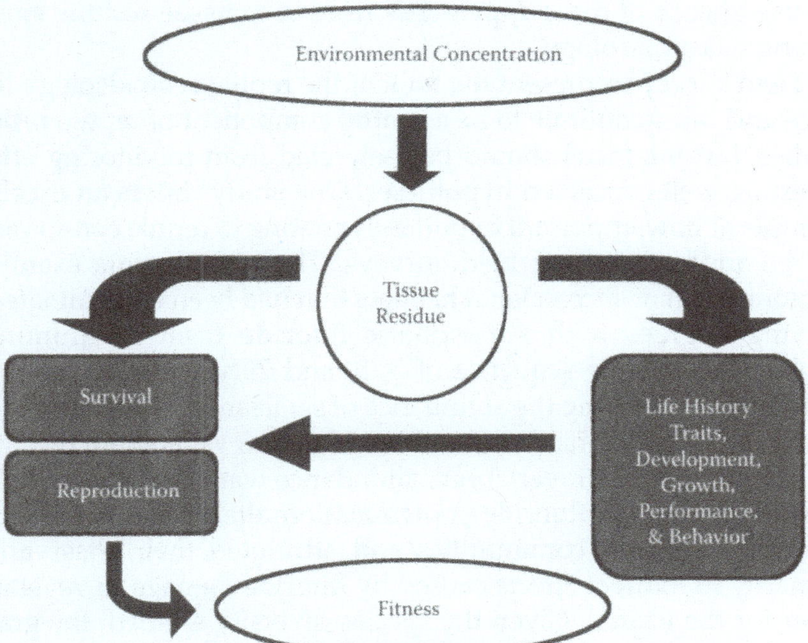

Figure 3.2 Conceptual diagram depicting how tissue residues can form the important links between contaminant concentrations in the environment and biological effects in organisms. Biological effects with demonstrated or strong theoretical implications for fitness are emphasized because they provide the clearest linkage between responses in individuals and effects at the population and community levels, the targets of most conservation and risk-based initiatives.

problems lend themselves to evaluation in the field, other hypotheses are clearly best tested under more controlled laboratory conditions. The greatest strength of field surveys is the ability to describe what is actually occurring at a given place and time, but ironically, this strength can also be a weakness. The complexity of natural systems often makes it difficult to distinguish between variables of interest (e.g., those relating to contamination) and other nontoxicologic variables.[11] Thus, field surveys often result in correlations rather than identification of causative factors. In contrast, laboratory studies can identify cause–effect relationships, but they sacrifice environmental realism in the process.[12] Because of these tradeoffs, some have advocated a pluralistic approach in which field surveys, laboratory experiments, and intermediary approaches, such as field and mesocosm experiments, are used in concert to gain a more holistic perspective on environmental issues.[11,13–15] The tradeoffs among these various approaches are well recognized by most scientists, yet too often the best approach is not adopted for the problem at hand. Here, I discuss the use of each of these approaches relative to tissue residue studies in reptile ecotoxicology.

Field surveys represent the bulk of the reptile ecotoxicology literature and must continue to be a central component of reptile residue studies, but the focus should be redirected from monitoring efforts to testing well-conceived hypotheses. One study[8] offers an excellent example of how important hypotheses relating to reptile conservation can be addressed using field surveys. The investigators examined recolonization of successional habitats that had been contaminated to varying degrees with atmospheric fluoride from an aluminum smelter over a time sequence of 3, 8, and 20 years of recovery. In addition to comparing the abundance of six lizard species at this suite of sites, a variety of habitat variables related to vegetation, leaf litter, soil conditions, and invertebrate abundance were characterized. The authors found that fluoride contamination altered the successional trajectory of lizard communities and attributed their observations primarily to indirect effects caused by fluoride damage to vegetative cover for the lizards. Given the species diversity studied, the gradation of exposure intensities, and the temporal component considered, the conclusions of this innovative study could have been strengthened if fluoride residues in lizards and their prey were examined. Such residues could allow a definitive distinction between direct effects of fluoride toxicity and the indirect effects described on habitat variables. Complementary laboratory feeding studies examining bioaccumulation and sensitivity of lizards to fluoride would also strengthen the conclusions of the authors' extensive fieldwork.

Field manipulations are exceedingly rare in reptile ecotoxicology but in many cases may provide greater insight than field surveys alone. Manipulations can be as straightforward as surveying a site before and after a preplanned application of a contaminant (e.g., pesticide overspray[16,17]) or as sophisticated as field transplantations of reptiles to contaminated sites. The power of field manipulations lies in their ability to reduce uncertainties associated with an animal's exposure history and to achieve replication in design. For example, in a classic set of studies by Turner and Lannom,[18] Turner et al.,[19,20] Medica et al.,[21] and Tuner and Medica,[22] lizard communities within large, replicated field enclosures were used to assess the effects of radiation on population parameters. Whereas some species underwent population declines most likely as a result of sterility, other species appeared to be less susceptible. The authors attributed such differential population-level responses to differences in life history characteristics among species; longer-lived species with delayed sexual maturity were more sensitive than shorter-lived species that mature earlier and have greater annual reproductive output.[21] Although these studies are more than 30 years old, they remain some of the most innovative approaches adopted in reptile ecotoxicology to date. The underutilization of such experimental field methods is unfortunate because they clearly provide insights into the relationships among environmental concentrations, tissue residues, and effects, and they also facilitate the linkage of individual-level responses to changes in population- and community-level parameters. In relation to contaminant accumulation, such methods represent a realistic approach to evaluating the complexities of contaminant bioavailability. Additionally, by coupling field manipulations with laboratory experiments, the direct effects of contaminants can be distinguished from indirect effects on environmental parameters such as changes in food resources or vegetative cover[11,17,23,24] that can ultimately influence the health of local reptile populations.

As technologic improvements continue to reduce the size of radiotransmitters, transplants of reptiles from uncontaminated to nearby contaminated sites (and vice versa to examine elimination and recovery) become increasingly plausible. Such approaches may be powerful but should also be adopted cautiously because of the potential ecologic effects of transplanting animals. Radiotelemetry enables one to follow individuals through time, examine their behavior and dietary patterns, and determine uptake and effects of contaminants at predetermined intervals. Additionally, transplanted animals can be compared with native conspecifics. Although some adult reptiles might be excessively disturbed by transplantation, juveniles of some species can thrive after transplant (juvenile tortoises[25]). In a rare example of using site transplantation with reptiles, Littrell[26] transplanted garter snakes (*Thamnophis elegans elegans*) in small cages at a field site immediately after treatment with the herbicide thiobencarb. Snakes were also examined in the laboratory after oral administration of the compound. By combining field and laboratory manipulations, Littrell suggested that thiobencarb probably posed little threat of acute toxicity to garter snakes. However, chronic exposure, sublethal effects, and tissue residues were not quantified. In a variation of site transplantation, Tinkle[27] irradiated lizards in the laboratory before releasing them in the field. As a result of reductions in fertility,[28] population declines occurred the year after irradiation. It is unfortunate that such innovative approaches have not been revisited by reptile ecotoxicologists in decades.

As more compelling hypotheses are tested in reptile ecotoxicology, laboratory experiments will become increasingly important

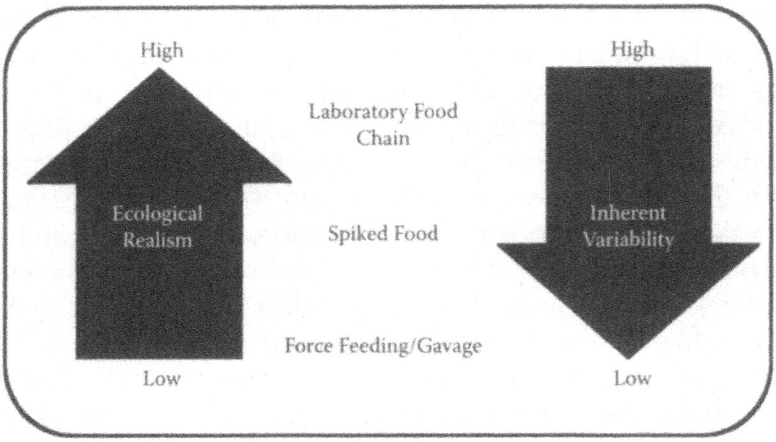

Figure 3.3 Conceptual diagram outlining tradeoffs among approaches to laboratory feeding studies. Note that approaches that reduce variability in dose administration tend to be less ecologically realistic, whereas more realistic approaches have greater inherent variability. Reproduced from reference 31.

complements to field investigations. For example, laboratory feeding studies are extremely valuable for understanding the process of trophic transfer in reptiles. Although field surveys can provide valuable information regarding contaminant burdens relative to trophic position, many uncertainties remain associated with this approach. In most field surveys the precise exposure route and history of the organism remain uncertain.[29] In contrast, laboratory studies allow isolation of trophic exposure from other exposure pathways; facilitate control of the dose, duration, and chemical form of contaminant exposure; and allow rigorous quantification of effects associated with exposure.[11,30,31] Such approaches will be particularly beneficial in comparative studies seeking to understand the toxicokinetics of contaminants in reptiles vs. other vertebrates, a very important consideration when attempting to understand the relative sensitivity of reptiles to pollutants. Controlling dosages in a laboratory setting will facilitate these taxonomic comparisons, something that is difficult in the field because of the unusual exposure scenarios experienced by some reptiles (e.g., infrequent ingestion of prey followed by long periods of digestive quiescence in snakes).

We have outlined the types of laboratory approaches that can be used to evaluate uptake and accumulation of contaminants in reptiles.[31] We identified the advantages and disadvantages of different methods of oral administration and reproduce a summary in Figure 3.3. Forced ingestion will clearly be the method of choice in circumstances where precise and repeatable doses, control of chemical form, and rapid administration are favored. However, such approaches are unrealistic because they remove voluntary ingestion as a variable, an important consideration given the fact that food aversion can be a natural defense mechanism that reduces exposure to contaminants. Snakes offer novel experimental opportunities because they swallow organisms whole and do not chew their prey. Thus, precise doses and chemical forms of compounds can be injected into previously killed prey (e.g., rodents) and voluntarily ingested by snakes under controlled conditions.[11] Laboratory food chains are the most realistic alternative to force feeding but are more labor intensive and tend to have greater uncertainty associated with dose administration.[31] Approaches to laboratory food chains range from provision of prey collected from the field to rearing multiple trophic levels in the laboratory.[31]

Each of the previously mentioned field and laboratory approaches has advantages and disadvantages, and in some cases a combination of these approaches is most informative. For example, in the late 1990s we began studying a population of banded water snakes (*Nerodia fasciata*) inhabiting a series of wetlands receiving wastes from a coal-burning power plant. Initial field surveys indicated that snakes collected from the contaminated area had extremely high tissue concentrations of As and Se (and to a lesser degree Cd). Water concentrations of contaminants were low at the site, making diet the most likely route of exposure. Collection of potential prey items at the site demonstrated that fish and amphibians had elevated tissue concentrations of As, Se, and Cd. In addition, we found that snakes with elevated tissue residues had increased metabolic rates, suggesting that they might expend greater-than-normal energy to maintain basal metabolism.[29] However, a series of laboratory studies[30,32] later suggested that the effects observed were largely dependent on which prey items were ingested at the site. When snakes were fed fish collected from the wetland, trophic transfer resulted in histopathologic abnormalities,[33] but accumulation of contaminants was only a fraction of what we documented in the field and there were no clear effects on growth, metabolism, or survival in laboratory-exposed snakes.[30,32] Additional fieldwork now suggests that amphibians may be more important dietary components for snakes than fish at this site. Some amphibians have contaminant concentrations three to five times higher than fish at the site,[34,35] which could explain the graded accumulation and responses observed among studies. Differences in chemical speciation among different prey types[36] could also influence bioavailability and further confound comparisons of uptake and toxicity in the field and laboratory. This example emphasizes the importance of pluralism in study design. Although our field surveys and laboratory experiments each addressed environmental exposure, tissue residues, and effects, either of these approaches used singularly would result in very different conclusions. Better characterization of trophic structure at field sites, perhaps by using stable isotopes,[37] might allow more effective interpretation of field surveys.

As the importance of laboratory manipulations increases, so will the utility of adopting a few good reptilian models for ecotoxicologic studies. The use of models has been adopted in most other branches of ecotoxicology (as well as in most scientific disciplines) and would prove equally important for reptiles as long as their limitations are also recognized. One of the largest obstacles to developing models in reptile ecotoxicology is the problem of reptilian diversity; reptiles represent a paraphyletic group that pools widely divergent evolutionary lineages into a single "artificial" taxon. The incredible diversity of reptiles undoubtedly results in immense diversity in ecologic exposure, physiologic processes that regulate uptake and elimination rates, and resulting responses. Clearly, responses of snakes should not be considered indicative of responses expected of crocodilians. Although it is impossible to develop a model species to capture such immense diversity, a few good models offer a strong starting point for experimental work. Development of a few tractable models would promote interdisciplinary collaboration, something that is greatly needed in reptile ecotoxicology. In Chapter 10, various reptile species with promise as model taxa for toxicologic study are proposed. At least one laboratory (L. Talent and colleagues, Oklahoma State University) has taken important steps toward development of a model (fence lizards, *Sceloporus*),[38,39] and other scientists from academia and government have become engaged in this promising effort. Introduced reptiles such as the brown tree snake and brown anole, which are ecologic pests in Guam[40–42] and Florida,[43–46] respectively, may make good models to address hypotheses that require sacrificing large numbers of animals.[47] Several studies on brown tree snakes offer a glimpse at the potential utility of introduced species for toxicity testing,[48–51] but only two studies to date have considered tissue residues of toxicants in brown tree snakes.[52,53] In perhaps the most valuable of these studies to date, Brooks and colleagues[48] controlled dosages and compared the dermal and oral toxicity of 18 chemicals. Because the study focused on pest management, it emphasized acute toxicity. However, future studies could use brown tree snakes for chronic studies in which tissue residues and sublethal endpoints related to fitness are considered. Because of their presence on military installations on Guam, it seems logical that studies on contaminants of concern to the U.S. Department of Defense (e.g., Pb or perchlorates) could be a top priority for testing on brown tree snakes. Although these exotic species may be a good starting point for lethal testing (e.g., LD50 experiments) of compounds, caution should still be exercised when extrapolating these results to other reptile species that may be more or less sensitive.

IV. Applying the Integrative Approach

To this point, I have focused on providing a framework for redefining current approaches for collecting and interpreting tissue residues in reptile ecotoxicology. Although there is a tremendous amount of literature on tissue residues in reptiles, most of it is phenomenologic, leaving major knowledge gaps in even the most fundamental processes of uptake and accumulation. Here, I outline some of the major research questions pertaining to contaminant uptake and accumulation in reptiles. In all cases, consideration should be given to how tissue residues relate to environmental conditions and effects (Figure 3.1).

A. Mechanisms of Contaminant Uptake in Juveniles and Adults

Because of a lack of experimental studies, we know surprisingly little about the mechanisms of contaminant uptake in reptiles. Like other vertebrates, reptiles are exposed to contaminants via dermal, inhalation, or ingestion exposure routes. The relative importance of each pathway will depend on a variety of factors, including the ecology and physiology of the organism, characteristics of the contaminant, and the physico-chemical environment in which the organism encounters the contaminant. In most cases, skin contact and ingestion will be the most important routes of exposure and will be the focus here. However, the importance of inhalation of certain contaminants (e.g., ozone[54]) should not be dismissed.

The importance of dermal exposure in reptiles may be underestimated for some compounds because of the common misperception that reptilian integument is relatively impermeable. Reptilian skin actually varies tremendously in permeability, primarily as a function of cutaneous lipid layers.[55–57] Many terrestrial reptiles, particularly those inhabiting arid regions, have relatively impermeable integument to prevent water loss.[58] This impermeability may also form a protective barrier to dermal exposure to some contaminants, but this relationship has not been adequately tested. In fact, pyrethrins and pyrethroids are known to be dermal toxicants in several snake species, at least one of which inhabits fairly arid habitats.[48–50,59] In contrast, some aquatic and semiaquatic reptiles are known to have permeable integument, probably to facilitate cutaneous respiration,[60,61] and may be more vulnerable to dermal contact to toxicants than their terrestrial counterparts. To date, there are no studies in reptiles that document tissue residues of a contaminant after controlled dermal exposure.

Ingestion likely results in the greatest exposure to contaminants in most reptiles, but because reptiles have widely divergent feeding ecologies, they experience equally diverse trophic exposure scenarios. Like other high trophic–level predators, carnivorous reptiles are likely at risk of contaminants that biomagnify, and a wealth of field surveys generally support this assertion (reviewed in references 1–5). Other trophic interactions, however, have received less attention. For example, many reptiles are opportunistic and will scavenge carrion,[62] placing them at risk of ingesting concentrations of contaminants that were lethal to prey. Secondary poisoning of other scavenging wildlife has been well documented (e.g., raptors[63]) but has yet to be studied in reptiles. Herbivorous reptiles, such as some turtles and lizards, are at risk of exposure not only to contaminants accumulated by plants, but also to contaminants such as metals[64] and pesticides that adhere to edible plant surfaces. Trophic transfer in herbivorous reptiles is generally less studied than transfer to predatory species. Although contaminated food is probably the most important ingestion source, soil ingestion may also be an important source of uptake for some reptiles. Incidental ingestion of substrate while foraging is likely widespread, and several species of reptiles are known to deliberately ingest soils, sand, or gravel. The purpose of this behavior is unclear, but it may occur to obtain limiting minerals or micronutrients, to macerate food in the gut, or to maintain intestinal microflora.[65–67] Regardless of the reason, such behavior likely increases the risk of exposure to soilborne contaminants.

Unfortunately, clear linkages between contaminant exposure and resulting tissue residues are rare in the literature. As discussed earlier, studies have demonstrated that laboratory feeding experiments are practical methods for addressing exposure, accumulation, and effects in snakes and lizards.[11,30,31] Clearly, additional experimental studies on a wider range of reptile species, contaminants, and exposure routes need to be performed to examine uptake processes. Field surveys that consider various reptiles that have different feeding ecologies but coinhabit a site would be valuable in assessing relative risk, especially if they are coupled with measures of environmental exposure and effects. Inference of environmental exposure conditions needs to be carefully considered given the complications outlined earlier (i.e., the comparison of field and laboratory studies on banded water snakes) but can be accomplished with analysis of gut contents to determine recent exposure and stable isotopes to better describe trophic position. Field transplant experiments of reptiles would further facilitate addressing such questions.

B. *Mechanisms of Contaminant Uptake in Embryos*

Embryonic reptiles are exposed to contaminants via two primary mechanisms: transfer from mother to offspring and absorption from the egg's surroundings (e.g., soil pore water). Most of what is currently known about embryonic exposure to contaminants comes from surveys of reptilian egg residues, but the actual route of exposure is sometimes not identified. In many cases, maternal transfer is probably the most important route of exposure, but studies should account for environmental contributions before reaching this conclusion.

Maternal transfer is initiated early in vitellogenesis when females begin synthesizing large quantities of lipoproteins that are critical for supplying nutrients, essential trace metals, and hormones to the developing embryo.[68] Unfortunately, these same lipoproteins may serve as transport molecules for environmental contaminants from mother to offspring. Compounds that tend to transfer from female to offspring include organic pollutants that associate with lipids (e.g., organochlorines), inorganic contaminants that are also essential nutrients or analogs of essential nutrients (e.g., Se and Sr as analogs for S and Ca), and contaminants that bind readily to egg proteins (e.g., Hg). Because the early stages of embryonic development direct the developmental trajectory of the individual, maternal transfer of such compounds represents an important vulnerability for developing offspring and ultimately for the mother's fitness. Indeed, in some cases, the quantities of contaminants transferred can be adequate to disrupt early development.[69,70] Chapter 6 deals with developmental effects of contaminants in more detail.

Because of their immense diversity, reptiles provide novel opportunities for evaluating the process of maternal transfer. Closely related reptiles display an enormous spectrum of reproductive strategies, ranging from the ancestral oviparous condition to viviparity, a strategy that has evolved at least 100 times in lizards alone.[71,72] Mechanisms of nutrient transport from mother to offspring vary significantly across this spectrum and within viviparous species with varying degrees of placental complexity and placentotrophy.[73] These different mechanisms probably have equally important implications for maternal transfer of contaminants. Within oviparous reptiles, ovulatory patterns may also influence the partitioning of contaminants among eggs.

Many reptiles produce a synchronous clutch of eggs, and concentrations of contaminants are fairly similar among reptile eggs within a clutch.[11,74,75] In contrast, birds often lay single eggs sequentially over a period of days to produce a clutch, a reproductive strategy that could result in contaminant apportionment among eggs that is influenced by laying order.[76] It is interesting that some reptiles, such as *Anolis* lizards, also repeatedly produce single eggs in succession over the course of the breeding season, a strategy that could influence maternal transfer of contaminants to eggs laid at different times in the laying sequence. Because these diverse life histories occur even within evolutionary lineages (e.g., within lizards), many novel opportunities to explore the consequences of reproductive strategy on maternal transfer and embryotoxicity exist in reptiles.

The importance of contaminant transfer from nesting substrate to the egg has not been adequately investigated. Reptiles generally deposit their eggs in soil, leaf litter, or other debris, where the potential exists for dissolved contaminants to traverse the eggshell.[77] However, the structure and calcification of the eggshell varies widely among reptiles and likely has important implications for contaminant transfer potential. For example, some turtles and most snakes and lizards have parchment-like eggs that are very permeable to water and perhaps dissolved contaminants. In squamates, water uptake during development can result in twofold to fourfold increases in egg mass.[78] In contrast, crocodilians and some geckos and turtles have hard-shelled eggs that are generally less permeable to water. Only two studies have explicitly examined the movement of contaminants from the surrounding incubation matrix into the egg. Nagle and colleagues[79] examined the potential transfer of As, Cd, Cr, Cu, and Se from coal fly ash used as nesting substrate by slider turtles (*Trachemys scripta*). The authors found that none of the elements were accumulated by turtles developing on this contaminated substrate. In contrast, Brasfield and colleagues[39] incubated lizard (*S. occidentalis*) eggs on perlite amended with various concentrations of Cd and found significant transfer of Cd into the egg at the three highest concentrations examined (148–14,800 μg Cd/g perlite). Transfer of contaminants to the egg is likely influenced by an array of factors, including characteristics of the species' egg, physico-chemical properties of the nest substrate and pore water, and characteristics of the contaminant. Additional studies that evaluate processes influencing transfer from the environment to the embryo should follow the lead of Nagle et al.,[79] Brasfield et al.[39,79] and quantify environmental concentration, egg and hatchling residues, and effects on embryonic development.

Because of the direct effects of contamination of eggs on reproductive success, studies examining maternal transfer and transfer from nesting substrate are important in reptile ecotoxicology. The use of reptilian eggs as bioindicators of environmental contamination has been fairly widespread, but the most useful studies are those that couple at least two of the key elements outlined in Figure 3.1. Studies are also needed that describe the relationships among environmental concentrations, maternal body burden, and egg concentration. However, quantification of maternal body burdens will not always be practical, especially for adult life stages of long-lived vertebrates such as turtles. Populations of slowly maturing, long-lived reptiles are usually intolerant of excessive harvesting of adults because adult mortality in nature is generally low.[80] However, harvest of eggs for residue analysis from species with these life history traits can still be useful because of the following:

1. Organisms with this life history strategy tend to be more resilient to harvest of young.
2. Concentrations of contaminants are often similar within a clutch, requiring only a subset of eggs be sacrificed.
3. The remainder of the clutch can be artificially incubated, or monitored during *in situ* incubation, to determine hatching success and malformation frequency.[69,74,79,81,82]
4. Healthy hatchlings can be released to their natal site at the end of the study.[74]

When describing the relationship between female and eggs is important,[83] laboratory models such as squamates may be most useful. For example, one laboratory study with snakes demonstrated that controlled trophic transfer, maternal transfer, and effects could simultaneously be determined using a laboratory food chain approach.[11]

C. *Factors Influencing Accumulation and Distribution among Tissues*

Work to date suggests that the distribution of contaminants among tissues in reptiles is generally similar to those in other vertebrate wildlife. Lipophilic compounds have been found in high levels in lipid stores of several reptiles.[84,85] Likewise, the liver, kidneys, and gonads appear to be important sites of accumulation of various hepatotoxic, renotoxic, and reproductively toxic compounds.[1–5] In addition, compounds that accumulate in bone (e.g., Pb and Sr) are also found in calcareous structures in reptiles, including bone and turtle shell.[86] However, because field surveys of residues usually do not relate residues to exposure conditions or effects, a number of significant knowledge gaps relating to tissue accumulation patterns will benefit from experimental approaches.

Reptiles have traits that may influence accumulation patterns in ways not experienced by traditionally studied wildlife species. First, reptiles have incredibly high conversion efficiencies (the amount of ingested energy converted to biomass). For example, the conversion efficiency of squamates is about 10 times higher than in birds and mammals of similar trophic level.[87] It is reasonable to predict that such high nutrient conversion efficiency would be associated with high rates of contaminant accumulation, yet this prediction has not been tested. Second, some reptiles ingest large meals (>30% of their body mass) at infrequent intervals (weeks or months between meals), a dietary pattern that would result in pulse exposure to contaminants not experienced by most other wildlife groups studied in toxicology. Finally, the enzymatic detoxification system of reptiles is thought to be less developed than in their endothermic counterparts. Available information suggests that reptiles have the major components of the vertebrate mixed function oxygenase system (e.g., cytochrome P450, NADPH-cytochrome C [P450] reductase), but the concentrations and activity of these components are sometimes much lower in reptiles than in other vertebrates.[88,89] Such observations could have important implications for biotransformation, accumulation, clearance, and toxicity of organic contaminants in reptiles. Clearly, the conversion efficiency, dietary patterns, and detoxification capacity of reptiles make them particularly interesting experimental subjects in studies of trophic transfer, yet controlled comparisons of contaminant accumulation in reptiles and other vertebrates have not been conducted.

Many reptiles have periods during their life history where stored contaminants could be remobilized, resulting in latent toxicity. For example, hibernation and aestivation are fairly common strategies for survival in reptiles in regions with climatic extremes. These key events represent periods when reptiles usually decrease activity and feeding and make a variety of physiologic adjustments to reduce metabolism and water loss. In some reptiles, such as freshwater turtles at the northern extremes of their range, periods of dormancy may account for half of their lives.[90] During these periods, reptiles usually rely on stored energy. Thus, contaminants associated with energy stores (e.g., lipophilic compounds sequestered in fat bodies) could be remobilized during what already tend to be precarious periods in the life history of these organisms. Likewise, many oviparous reptiles produce offspring with considerable yolk reserves important for posthatch survival and dispersal.[91,92] Some species, such as *Deirochelys reticularia* and *T. scripta,* even rely on stored reserves while they overwinter as hatchlings in the nest.[93] Because yolk stores can be significant sites of contaminant partitioning,[85] controlled studies that follow the remobilization of contaminants sequestered in yolk and track the success of hatchlings in the months posthatching would be valuable.

There are a variety of factors that commonly influence accumulation of contaminants in other wildlife that have received inadequate attention in reptiles. For example, accumulation of contaminants often differs between the sexes. Sexual differences in tissues residues observed in many field surveys of reptiles usually have been attributed to differences in behavior, feeding ecology, or reproduction,[45,94,95] the latter of which can result in elimination in females that maternally transfer contaminants to their eggs. However, most studies to date have sampled animals from the field with unknown exposure and reproductive histories. One laboratory study[31] has demonstrated that when dietary Se content, feeding behavior, and reproductive history are controlled for, differences in contaminant partitioning still arise between male and female western fence lizards (*S. occidentalis*). Although males and females accumulated the same amount of Se from their diet on a whole body basis, partitioning within organs differed between sexes. Females partitioned more Se (on a concentration and burden basis) to their gonads than did males, indicating that there are inherent differences in physiology between the sexes that influence toxicokinetics. It is reasonable to predict that such differences also exist under field conditions and may contribute to some phenomena observed in previous field surveys.

Age is another factor that can influence accumulation of contaminants, but it requires further investigation in reptiles. Some contaminants accumulate continually over an animal's lifetime, resulting in high tissue residues late in life.[83] However, estimates of age have seldom been related to accumulation of persistent contaminants in reptiles. Chronology techniques, such as growth rings in turtle scutes, can be used to relate age to accumulation, but the accuracy of these techniques needs to be thoroughly considered.[96,97] Many species also experience ontogenetic changes in physiology and ecology that should influence accumulation patterns. For example, slider turtles (*T. scripta*) are carnivorous as juveniles but primarily herbivorous as adults.[98] Because carnivorous turtles tend to accumulate

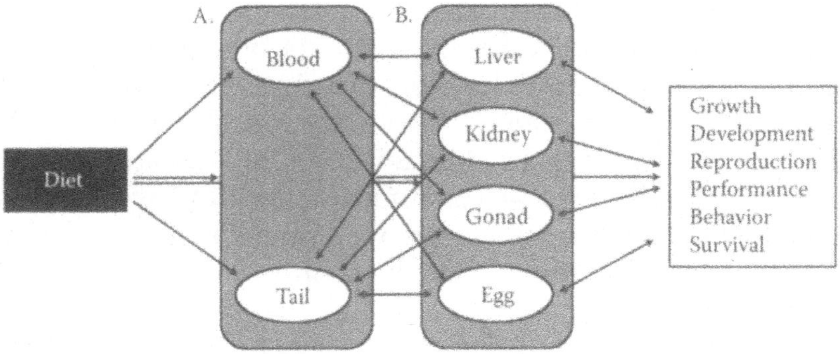

Figure 3.4 Conceptual figure describing how nondestructive tissues (A) can serve as a link among dietary concentrations, target tissue concentrations (B), and organism-level effects that may be ecologically important. Quantification of the functional relationships (*arrows*) among these factors will allow the use of tissue concentrations for predictive purposes.

higher concentrations of contaminants than do turtles that exhibit a higher degree of herbivory,[95,99,100] studies that examine the relationship between trophic level and accumulation could be achieved within species that exhibit ontogenetic shifts in diet. Similarly, ontogenetic shifts in enzymatic metabolism, such as age-dependent changes in P450 inducibility observed in snakes,[101] could influence the accumulation of contaminants as an individual ages.

Most tissue residue studies logically focus on accumulation of contaminants in tissues of toxicologic concern (e.g., liver, eggs), but sampling other tissues may also be of great importance, particularly within a conservation framework. For example, the use of nondestructive tissues (i.e., tissues that can be sampled from animals without killing them) has gained substantial attention in reptiles in recent years. The utility of chorioallantoic membranes (CAMs), skin, blood, tissue biopsy, scales, scutes, and claws have all been explored as indicators of exposure in reptiles.[31,32,36,102–112] Some tissues such as tail biopsy and CAMs show considerable promise as indicators of prior exposure. However, to be most useful, the functional relationships among nondestructive tissues, exposure conditions (e.g., concentrations in diet), target organ concentrations, and meaningful biological effects must be quantified (Figure 3.4). Future studies that apply mathematic methods to describe these functional relationships[111] will be extremely important in promoting the use of nondestructive tissues within a conservation framework. The use of CAMs in such mathematic models may have the highest probability of success because these tissues may be tightly associated with egg concentrations that influence embryonic development and ultimately the reproductive success of females.

D. Contaminant Transport in Ecologic Systems

The overall importance of reptiles in contaminant transport in food webs and ecologic systems is not understood. Because many reptiles can exploit resources with great efficiency (see earlier assimilation discussion) they may also serve as efficient transporters of pollutants into food chains. Niewiarowski[113] argued that the ability of reptiles to exploit invertebrate prey, which tend to be too energy poor for endotherms, make them crucial transporters of contaminants from low to high trophic levels. In some areas, reptiles can also occur at incredibly high densities and undoubtedly have considerable influence on carbon and energy flow. Garter snakes overwinter in dens that can contain as many as 10,000 individuals in Canada,[114] brown tree snakes occur at densities of several thousand per square kilometer on Guam,[51] *Anolis* lizards can reach densities of one per square meter in the tropics,[115] and *Sphaerodactylus* geckos achieve densities exceeding 50,000 per hectare.[116] In the United States the introduced brown anole (*Anolis sagrei*) is the most abundant terrestrial vertebrate in Florida, reaching densities of 10,000 lizards per hectare.[117] Given their high conversion efficiencies and ability to reach high densities, the role of reptiles in transporting contaminants through ecologic systems clearly deserves attention. Additionally, whereas many reptiles have fairly narrow home ranges,[106,107] other reptiles can use different habitat types for different stages of their life history and serve as significant contaminant transport vectors to otherwise uncontaminated regions. For example, sea turtles are known to be significant contributors of nutrients and energy into coastal dune ecosystems during the nesting season.[118] In fact, in some systems where nesting is intense (e.g., Tortuguero, Costa Rica, where ~80,000 green turtles deposit on average >100 eggs each[119]) turtles may be one of the most important biological transporters of energy and nutrients. Because turtles can maternally transfer high concentrations of contaminants to their eggs, the role of turtles in transporting contaminants from the ocean into coastal dune food webs should be evaluated.

Reptilian tissue residues may also be useful for understanding large-scale ecologic change and global patterns of contaminant transport. Because many turtles and crocodilians are long-lived (>30 years), archives of tissue residues from these reptiles could have great value in the future. Undoubtedly, the environment experienced by these animals now will be quite different in 30 years. Tissue residues archived from long-lived vertebrates can provide an integrative measure of environmental change, particularly when the same individuals, or at least individuals from the same area, are sampled repeatedly over long periods of time. The development of nondestructive tissue archives may be particularly useful in this regard. Because some reptiles inhabit remote arid and tropical regions of the planet,[120] they may also serve as excellent models for studying global transport of contaminants. For example, in some isolated tropical regions atmospheric input may currently represent the primary source of contamination (although point-source inputs are increasingly common). Long-lived top carnivores in these regions such as crocodilians offer the opportunity to study compounds such as Hg that are subject to atmospheric transport and readily biomagnify.[121] This approach has been successfully adopted with other long-lived vertebrates occupying remote regions such as marine mammals in the arctic.[122] Long-term monitoring of tissue residues will eventually be of great value but may require substantial dedication by scientists as a result of commonly inadequate support from funding agencies for such long-range initiatives.

V. Summary

We live in a time of enormous environmental change that rivals many large-scale ecologic transitions over the course of geologic history. In the face of such rapid change we must carefully select our research paths to be the most fruitful given limited resources. If the conservation of reptiles is our ultimate objective, I argue that it is imperative that we challenge ourselves and our colleagues to become more innovative as we press forward in the field of reptile ecotoxicology. With regard to uptake and accumulation of contaminants (the focus of this chapter), future studies should focus on *how* a contaminant is accumulated, *what* influences its accumulation and partitioning among tissues, and *why* accumulation matters to the organism. Thus, tissue residues will be most useful as the intermediary between the environmental setting in which exposure occurs and the resulting effects of contaminants.

Field surveys of tissue residues in reptiles have illustrated the utility of reptiles as indicators of contamination of habitats, but they have provided limited insight into the fundamental processes influencing uptake and accumulation of contaminants. Because of these knowledge gaps, most residue studies to date provide little information directly useful for developing conservation initiatives and risk assessments for reptiles exposed to contaminants. It is important to note that hypothesis-driven, experimental research is generally absent from the reptilian tissue residue literature. Future studies that adopt a pluralistic approach, in which field surveys are coupled with laboratory and field experiments, will provide a more holistic perspective on contaminant issues in reptiles. Moreover, interdisciplinary research is desperately needed in reptile ecotoxicology. Many of the most interesting questions in reptile ecotoxicology, as well as in most scientific fields today, lie at the boundaries among disciplines and require collaboration of scientists with different expertise.[123] By combining the expertise of chemists, toxicologists, physiologists, and ecologists, the linkages between tissue concentrations of environmental contaminants, effects that influence fitness, and population-level consequences of these effects can be realized. Such ambitious goals are not only exciting from a scientific perspective but are arguably required to make significant advances in the conservation of reptiles.

Finally, it is important to recognize that the study of reptiles can contribute to our fundamental understanding of problems in ecotoxicology and biogeochemistry. Perhaps the best existing example of this is the use of reptile models for understanding the influence of endocrine disruptors on sex determination.[124] However, there are also many ways in which reptile models could be useful for understanding processes directly related to contaminant uptake and accumulation. For example, reptile models may provide fundamental insight into biotransformations and the transport of contaminants through ecologic systems. Additionally, the diverse reproductive strategies of reptiles within closely related lineages permit a thorough examination of processes that regulate maternal transfer of contaminants. Thus, reptiles may serve as important models in the study of environmental pollution, just as they have historically offered insight as model species in the fields of ecology, evolution, physiology, and animal behavior.

Acknowledgments

This chapter is dedicated to my 2-year old daughter, Carly, whose innocent affection and curiosity for snakes and frogs gives me renewed hope that my work is worthwhile. B. Rothermel, D. Scott, D. Fletcher, C. Strojan, J. Unrine, K. Campbell, and an anonymous reviewer provided helpful comments on an early draft of the chapter. L. Zweifel, J. Derrick, and J. Mobley helped compile the literature. WAH was supported by the University of Georgia's Savannah River Ecology Laboratory. Funding was provided by the Environmental

Remediation Sciences Division of the Office of Biological and Environmental Research, U.S. Department of Energy through the Financial Assistance Award no. DE-FC09-96SR18546 to the University of Georgia Research Foundation.

Literature Cited

1. Campbell, K.R. and Campbell, T.S., Lizard contaminant data for ecological risk assessment, *Rev. Environ. Contam. Toxicol.*, 165, 39, 2000.
2. Campbell, K.R. and Campbell, T.S., The accumulation and effects of environmental contaminants on snakes: a review, *Environ. Mon. Ass.*, 70, 253, 2001.
3. Campbell, K.R., Ecotoxicology of crocodilians, *Appl. Herpetol.*, 1, 45, 2003.
4. Meyers-Schöne, L. and Walton, B.T., Turtles as monitors of chemical contaminants in the environment, *Rev. Environ. Contam. Toxicol.*, 135, 93, 1994.
5. Sparling, D., Linder, G., and Bishop, C., *Ecotoxicology of Amphibians and Reptiles*, SETAC Press, Pensacola, FL, 2002.
6. Bishop, C.A. and Gendron, A.D., Reptiles and amphibians: shy and sensitive vertebrates of the Great Lakes and St. Lawrence River. *Environ. Mon. Ass.*, 53, 225, 1998.
7. Gibbons, J.W. et al., The global decline of reptiles, déjà vu amphibians, *BioScience*, 50, 653, 2000.
8. Letnic, M.I. and Fox, B.J., The impact of industrial fluoride fallout on faunal succession following sand mining of dry sclerophyll forest at Tomago, NSW. — I. lizard recolonisation, *Biol. Conserv.*, 80, 63, 1997.
9. Luiselli, L. and Akani, G.C., An indirect assessment of the effects of oil pollution on the diversity and functioning of turtle communities in the Niger Delta, Nigeria, *Anim. Biol. Conserv.*, 26, 57, 2003.
10. Suter, G.W., *Ecological Risk Assessment*, Lewis Publishers, Chelsea, 1993.
11. Hopkins, W.A. et al., Trophic and maternal transfer of selenium in brown house snakes (*Lamprophis fuliginosus*), *Ecotoxicol. Environ. Saf.*, 58, 285, 2004.
12. Snodgrass, J.W. et al., Species-specific responses of developing anurans to coal combustion wastes, *Aquatic Toxicol.*, 66, 171, 2004.
13. Diamond, J., *Overview: Laboratory Experiments, Field Experiments, and Natural Experiments*, Harper and Row, Inc., New York, 1986.
14. Sadinski, W.J. and Dunson, W.A.,1992. A multilevel study of effects of low pH on amphibians of temporary ponds, *J. Herpetol.*, 26, 413, 1992.
15. Rowe, C.L. and Dunson, W.A., The value of simulated pond communities in mesocosms for studies of amphibian ecology and ecotoxicology, *J. Herpetol.*, 28, 346, 1994.
16. Willemsen, R.E. and Hailey, A., Effects of spraying the herbicides 2, 4-D and 2, 4, 5-T on a population of the tortoise *Testudo hermanni* in southern Greece, *Environ. Pollut.*, 113, 71, 2001.
17. Peveling, R. et al., Impact of locust control on harvester termites and endemic vertebrate predators in Madagascar, *J. Appl. Ecol.*, 40, 729, 2003.
18. Turner, F.B. and Lannom, J.R., Jr., Radiation doses sustained by lizards in a continuously irradiated natural enclosure, *Ecology*, 49, 548, 1968.
19. Turner, F.B. et al., A demographic analysis of continuously irradiated and nonirradiated populations of the lizard, *Uta stansburiana*, *Rad. Res.*, 38, 349, 1969.

20. Turner, F.B. et al., Radiation-induced sterility in natural populations of lizards (*Crotophytus wislizenii* and *Cnemidophorus tigris*), in Proc. of the Third National Symposium on Radioecology, Nelson, D.J. Ed., Radionuclides in Ecosystems, Oak Ridge, 1973, 1131.
21. Medica, P.A., Turner, F.B., and Smith, D.D., Effects of radiation on a fenced population of horned lizards (*Phrynosoma platyrhinos*) in southern Nevada, *J. Herpetol.*, 7, 79, 1973.
22. Turner, F.B. and Medica, P.A., Sterility among female lizards (*Uta stansburiana*) exposed to continuous irradiation, *Rad. Res.*, 70, 154, 1977.
23. Fleeger, J.W., Carman, K.R., and Nisbet, R.M., Indirect effects of contaminants in aquatic ecosystems, *Sci. Tot. Environ.*, 317, 207, 2003.
24. Metts, B.S., Hopkins, W.A., and Nester, J.P., Density-dependent effects of an insecticide on a pond-breeding salamander assemblage, *Fresh Biol.*, 2005.
25. Tuberville., T. and Gibbons, W., unpublished data.
26. Littrell, E.E., A study on the effects of Bolero 10G[R]on the mountain garter snake, *Thamnophis elegans elegans*, *Calif. Fish Game*, 69, 186, 1982.
27. Tinkle, D.W., Effects of radiation on the natality, density and breeding structure of a natural population of lizards, *Uta stansburiana*, *Health Phys.*, 11, 1595, 1965.
28. Dana, S.W., and Tinkle, D.W., Effects of X-irradiation on the testes of the lizard, *Uta stansburiana stejnegeri*, *Int. J. Radiat. Biol.*, 9, 67, 1965.
29. Hopkins, W.A., Rowe, C.L., and Congdon, J.D., Increased maintenance costs and trace element concentrations in banded water snakes, *Nerodia fasciata*, exposed to coal combustion wastes, *Environ. Toxicol. Chem.*, 18, 1258, 1999.
30. Hopkins, W.A. et al., Trace element accumulation and effects of chronic dietary exposure on banded water snakes (*Nerodia fasciata*), *Environ. Toxicol. Chem.*, 21, 906, 2002.
31. Hopkins, W.A. et al., Transfer of Se from prey to predators in a simulated terrestrial chain, *Environ. Pollut.*, 134, 447, 2005.
32. Hopkins, W.A. et al., Nondestructive indices of trace element exposure in squamate reptiles, *Environ. Pollut.*, 115, 1, 2001.
33. Rania, L.C. et al., Liver histopathology of the southern water snake, *Nerodia fasciata*, following chronic exposure to trace element-contaminated prey from a coal ash disposal site, *J. Herpetol.*, 37, 219, 2003.
34. Roe, J.H., Hopkins, W.A., and Jackson, B.P., Species- and stage-specific differences in trace element concentrations in amphibians: Implications for the disposal of coal-combustion wastes. *Environmental Pollution*. 136:353–363.
35. Hopkins., W.A., unpublished data.
36. Jackson, B.P., Hopkins, W.A., and Baionno, J.A., Laser Ablation-ICP-MS analysis of micro-dissected tissue: a conservation-minded approach to assessing contaminant exposure, *Environ. Sci. Technol.*, 37, 2511, 2003.
37. Yoshimaga, J. et al., Mercury concentration correlates with the nitrogen stable isotope ratio in the animal food of Papuans, *Ecotoxicol. Environ. Saf.*, 24, 37, 1992.
38. Talent, L.G. et al., Evaluation of western fence lizards (*Sceloporus cccidentalis*) and eastern fence lizards (*Sceloporus undulatus*) as laboratory reptile models for toxicological investigations, *Environ. Toxicol. Chem.*, 21, 899, 2002.
39. Brasfield, S.M. et al., Development of a terrestrial vertebrate model for assessing bioavailability of cadmium in the fence lizard (*Sceloporus undulatus*) and in ovo effects on hatchling size and thyroid function, *Chemosphere*, 54, 1643, 2004.

40. Rodda, G.H. and Fritts, T.H., The impact of the introduction of the colubrid snake *Boiga irregularis* on Guam's Lizards, *J. Herpetol.*, 25, 166, 1992.
41. Rodda, G.H., Fritts, T.H., and Chiszar, D., The disappearance of Guam's wildlife, *BioScience*, 47, 565, 1997.
42. Rodda, G.H. et al., An overview of the biology of the brown tree snake (*Boiga irregularis*), a costly introduced pest on pacific islands, in *Problem Snake Management: The Habu and the Brown Tree Snake*, Rodda, G. H., Sawai, Y., Chiszar, D., and Hiroshi, T., Eds., Cornell University Press, Ithaca, 1999, 534.
43. Butterfield, B.P., Meshaka, W.E., Jr., and Guyer, C., Nonindigenous amphibians and reptiles, in *Strangers in Paradise: Impact and Management of Nonindigenous Species in Florida*, Simberloff, D., Schmidz, D.C., and Brown, T.C., Eds., Island Press, Washington, D.C., 1997, 123.
44. Campbell, T.S. and Echternacht, A.C., Introduced species as moving targets: changes in body sizes of introduced lizards following experimental introductions and historical invasions, *Biol. Invasions*, 5, 193, 2003.
45. Burger, J., Campbell, K.R., and Campbell, T.S., Gender and spatial patterns in metal concentrations in brown anoles (*Anolis sagrei*) in southern Florida, USA, *Environ. Toxicol. Chem.*, 23, 712, 2004.
46. Campbell, K.R. and Campbell, T.S., submitted for publication in *Appl. Herpetol.*
47. Hopkins, W.A., Reptile toxicology: opportunities and challenges on the last frontier of vertebrate ecotoxicology, *Environ. Toxicol. Chem.*, 19, 2391, 2000.
48. Brooks, J.E., Savarie, P.J., and Johnston, J.J., The oral and dermal toxicity of selected chemicals to brown tree snakes (*Boiga irregularis*), *Wildl. Res.*, 25, 427, 1998.
49. Brooks, J.E., Savarie, P.J., and Bruggers, R.L., The toxicity of commercial insecticide aerosol formulations to brown tree snakes, *Snake*, 28, 23, 1998.
50. Brooks, J.E. et al., Toxicity of pyrethrin/pyrethroid fogger products to brown tree snakes (*Boiga irregularis*) in cargo containers, *Snake*, 28, 33, 1998.
51. Savarie, P.J. and Bruggers, R.I., Candidate repellents, oral and dermal toxicants, and fumigants for brown treesnake control, in *Problem Snake Management: The Habu and the Brown Tree Snake*, Rodda, G.H., Sawai, Y., Chizar, D., and Tanaka, H., Eds., Cornell University Press, Ithaca, 1999, 417.
52. Johnston, J.J. et al., Risk assessment of an acetaminophen baiting program for chemical control of brown tree snakes on Guam: evaluation of baits, snake residues, and potential primary and secondary hazards, *Environ. Sci. Technol.*, 36, 3827, 2002.
53. Primus, T.M. et al., Determination of propoxur residues in whole body brown tree snakes, *J. Agric. Food Chem.*, 46, 2647, 1998.
54. Dohm, M.R. et al., Effects of ozone on evaporative water loss and thermoregulatory behavior of marine toads (*Bufo marinus*), *Environ. Res.*, 86, 274, 2001.
55. Roberts, J.B. and Lillywhite, H.B., Lipid barrier to water exchange in reptile epidermis, *Science*, 207, 1077, 1980.
56. Roberts, J.B. and Lillywhite, H.B., Lipids and the permeability of epidermis from snakes, *J. Exp. Zool.*, 228, 1, 1983.
57. Palmer, B.D., Aspects of reptilian anatomy and physiology, in *Ecotoxicology of Amphibians and Reptiles*, Sparling, D.W. et al., Eds., SETAC Press, 2000, chap 3B.
58. Dmi'el, R., Perry, G., and Lazell, J., Evaporative water loss in nine insular populations of the lizard anolis cristatellus group in the British Virgin Islands, *Biotropica*, 29, 111, 1997.

59. Toriba, M., Senbo, S., and Kosuge, Y., New dermal toxicants and methods of application for venomous snakes, in *Problem Snake Management: The Habu and the Brown Tree Snake*, Rodda, G.H., Sawai, Y., Chizar, D., and Tanaka, H., Eds., Cornell University Press, Ithaca, 1999, 411.
60. Heatwole, H. and Seymour, R.S., Cutaneous oxygen uptake in three groups of aquatic snakes, *Aust. J. Zool.*, 26, 481, 1978.
61. Winne, C.T. et al., A comparison of evaporative water loss in two natricine snakes: *Nerodia fasciata* and *Seminatrix pygaea*, *J. Herpetol.*, 35, 129, 2001.
62. DeVault, T.L. and Krochmal, A.R., Scavenging by snakes: an examination of the literature, *Herpetologica*, 58, 429, 2002.
63. Elliott, J.E. et al., Poisoning of bald eagles and red tailed hawks by carbofuran and fensulfothion in the Fraser Delta of British Columbia, *J. Wild. Dis.*, 32, 486, 1996.
64. Punshon, T. et al., Mass loading of nickel and uranium on plant surfaces, *J. Enivron. Mon.*, 6, 153, 2004.
65. Sokol, O.M., Lithophagy and geophagy in reptiles, *J. Herpetol.*, 5, 69, 1971.
66. Sylber, C.K., Feeding habits of the lizards *Sauromalus varius* and *S. hispidus* in the gulf of California, *J. Herpetol.*, 22, 413, 1988.
67. Beyer, W.N., Connor, E.E., and Gerould, S., Estimates of soil ingestion by wildlife, *J. Wildl. Manage.*, 58, 375, 1994.
68. Specker, J. and Sullivan, C.V., Vitellogenesis in fishes: status and perspectives, in *Perspectives in Comparative Endocrinology*, Davey K.G., Peter, R.G., and Tobe, S.S., Eds., NRC Canada, Toronto, 1994, 304.
69. Bishop, C.A. et al., Environmental contamination and developmental abnormalities in eggs and hatchlings of the common snapping turtle (*Chelydra serpentina serpentina*) from the Great Lakes-St. Lawrence River basin (1989–91), *Environ. Pollut.*, 101, 143, 1998.
70. Guillette, L.J., Jr., Contaminant-associated endocrine disruption in reptiles, in *Ecotoxicology of Amphibians and Reptiles*, Sparling, D.W. et al., Eds., SETAC Press, 2000, chap 10B.
71. Shine, R., The evolution of viviparity in reptiles: an ecological analysis, in *Biology of the Reptilia*, Vol. 15, Gans, G. and Billet, F., Eds., Academic Press, New York, 1985, 605.
72. Tinkle, D.W. and Gibbons, J.W., The distribution and evolution of viviparity in reptiles, Misc. Publications # 154, Museum of Zoology, University of Michigan, 1977.
73. Thompson, M.B., Stewart, J.R., and Speake, B.K., Comparison of nutrient transport across the placenta of lizards differing in placental complexity, *Comp. Biochem. Physiol.*, 127, 469, 2000.
74. Roe, J.H. et al., Maternal transfer of selenium in *Alligator mississippiensis* nesting downstream from a coal burning power plant, *Environ. Toxicol. Chem.*, 23, 1969, 2004.
75. Heinz, G.H., Percival, H.F., and Jennings, M.L., Contaminants in American alligator eggs from Lake Apopka, Lake Griffin, and Lake Okeechobee, Florida, *Environ. Mon. Ass.*, 16, 277, 1991.
76. Bryan, L. et al., Maternal transfer of contaminants to eggs in common grackles (*Quiscalus quiscala*) nesting on coal fly ash basins, *Arch. Environ. Contam. Toxicol.*, 45, 273, 2003.
77. Wu, T.H. et al., Organochlorine contaminants in Morelet's crocodile (*Crocodylus moreletii*) eggs from Belize, *Chemosphere*, 40, 671, 2000.

78. Pough, F.H. et al., *Herpetology*, Pearson Prentice Hall, Upper Saddle River, NJ, 2004.
79. Nagle, R.D., Rowe, C.L., and Congdon, J.D., Accumulation and selective maternal transfer of contaminants in the turtle *Trachemys scripta* associated with coal ash deposition, *Arch. Environ. Contam. Toxicol.*, 40, 531, 2001.
80. Congdon, J.D. and Dunham, A.E., Contributions of long-term life history studies to conservation biology, in *Principles of Conservation Biology*, Sinauer, Sunderland, 1994, 181.
81. De Solla, S.R., Fletcher, M.L., Bishop, C.A., Relative contributions of organochlorine contaminants, parasitism, and predation to reproductive success of eastern spiny softshell turtles (*Apalone spiniferus spiniferus*) from Southern Ontario, Canada, *Ecotoxicology*, 12, 261, 2003.
82. Bishop, C.A. et al., The case for a cause-effect linkage between environmental contamination and development in eggs of the common snapping turtle (*Chelydra serpentina serpentina*) from Ontario, Canada, *J. Toxicol. Environ. Health*, 33, 521, 1991.
83. Hebert, C.E. et al., Organic contaminants in snapping turtles (*Chelydra serpentina*) populations from southern Ontario, Canada, *Arch. Environ. Contam. Toxicol.*, 24, 35, 1993.
84. Meeks, R.L., The accumulation of ^{36}Cl Ring-Labeled DDT in a freshwater marsh, *J. Wildl. Manage.*, 32, 376, 1968.
85. Bryan, A.M., Olafsson, P.G., and Stone, W.B., Disposition of low and high environmental concentrations of PCBs in snapping turtle tissues, *Bull. Environ. Contam. Toxicol.*, 38, 1000, 1987.
86. Hinton, T.G. and Scott, D.E., Radioecological techniques for herpetology, with an emphasis on freshwater turtles, in *Life History and Ecology of the Slider Turtle*, Gibbons, J.W., Ed., Smithsonian Institution Press, Washington, D.C., 1990, 267.
87. Pough, F.H., The advantages of ectothermy for tetrapods, *Am. Nat.*, 115, 92, 1980.
88. Schwenn, R.J. and Mannering, G.J., Hepatic cytochrome P450-dependent monooxygenase systems of the trout, frog, and snake, I. Components, *Comp. Biochem. Physiol.*, 71, 431, 1982.
89. Jewell, C.S.E. et al., The hepatic microsomal mixed-function oxygenase (MFO) system of *Alligator mississippiensis*: induction by 3-methylcholanthrene (MC), *Xenobiotica*, 19, 1181, 1989.
90. Ultsch, G.R., Ecology and physiology of hibernation and overwintering among freshwater fishes, turtles, and snakes, *Biol. Rev.*, 64, 435, 1989.
91. Congdon, J.D. and Gibbons, J.W., Egg components and reproductive characteristics of turtles: relationships to body size, *Herpetologica*, 41, 194, 1985.
92. Congdon, J.D. and Gibbons, J.W., Turtle eggs: their ecology and evolution, in *Life History and Ecology of the Slider Turtle*, Gibbons, J.W., Ed., Smithsonian Institution Press, Washington, D.C., 1990, 109.
93. Congdon, J.D., Gibbons, J.W., and Greene, J.L., Parental investment in the chicken turtle (*Deirochelys reticularia*), *Ecology*, 64, 419, 1983.
94. Albers, P.H., Sileo, L., and Mulhern, B.M., Effects of environmental contaminants on snapping turtles of a tidal wetland, *Arch. Environ. Contam. Toxicol.*, 15, 39, 1986.
95. Meyers-Schöne, L. et al., Comparison of two freshwater turtle species as monitors of radionuclide and chemical contamination: DNA damage and residue analysis, *Environ. Toxicol. Chem.*, 12, 1487, 1993.

96. Litzgus, J.D. and Brooks, R.J., Testing the validity of counts of plastral scute rings in spotted turtles, *Clemmys guttata*, *Copeia*, 1998, 222, 1998.
97. Milnes, M.R. et al., Plasma steroid concentrations in relation to size and age in juvenile alligators from two Florida lakes, *Comp. Biochem. Physiol.*, 131, 923, 2002.
98. Clark, D.B. and Gibbons, J.W., Dietary shift in the turtle *Pseudymys scripta* (Schoepff) from youth to maturity, *Copeia*, 1969, 704, 1969.
99. Holcomb, C.M. and Parker, W.S., Mirex residues in eggs and livers of two long-lived reptiles (*Chrysemys scripta* and *Terrapene carolina*) in Mississippi. 1970–1977, *Bull. Environ. Contam. Toxicol.*, 23, 369, 1979.
100. Clark, D.R. and Krynitsky, A.J., Organochlorine residues in eggs of loggerhead and green sea turtles nesting at Merritt Island, Florida — July and August 1976, *Pest. Monit. J.*, 14, 7, 1980.
101. Hecker, M., Murphy, M.B., Giesy, J.P., Hopkins, W.A., *In Press*. Induction of cytochrome P4501A in primary hepatocyte cultures of the African brown house snake (*Lamprophis fuliginosus*): An approach to model sensitivity of reptiles to TCDD and selected PCBs. *Environ. Toxicol. Chem.*
102. Kaur, S., Lead in the scales of cobras and wall snakes from rural and urban areas of Punjab, India, *Sci. Tot. Environ.*, 77, 289, 1988.
103. Cobb, G.P. and Wood, P.D., PCB concentrations in eggs and chorioallantoic membranes of loggerhead sea turtles (*Caretta caretta*) from the Cape Romain National Wildlife Refuge, *Chemosphere*, 34, 539, 1997.
104. Jagoe, C.H. et al., Mercury in alligators (*Alligator mississippiensis*) in the southeastern United States, *Sci. Tot. Environ.*, 213, 255, 1998.
105. Twining, J.R. et al., Osteoderms of estuarine crocodiles record their enhanced Pb exposure in Kakadu National Park, *Environ. Sci. Technol.*, 33, 4396, 1999.
106. Jeffree, R.A., Markich, S.J., and Twining, J.R., Element concentrations in the flesh and osteoderms of estuarine crocodiles (*Crocodylus porosus*) from the Alligator Rivers Region, Northern Australia: biotic and geographic effects, *Arch. Environ. Toxicol.*, 40, 236, 2001.
107. Markich, S.J., Jeffree, R.A., and Harch, B.D., Catchment-specific element signatures in estuarine crocodiles (*Crocodylus porosus*) from the Alligator Rivers Region, northern Australia, *Sci. Tot. Environ.*, 287, 83, 2002.
108. Cobb, G.P., Wood, P.D., and O'Quinn, M., Polychlorinated biphenyls in eggs and chorioallantoic membranes of American alligators (*Alligator mississippiensis*) from coastal South Carolina, *Environ. Toxicol. Chem.*, 16, 1456, 1997.
109. Cobb, G.P. et al., Using chorioallantoic membranes for non-lethal assessment of persistent organic pollutant exposure and effect in oviparous wildlife, *Ecotoxicology*, 12, 31, 2003.
110. Rainwater, T.R. et al., Organochlorine pesticides and mercury in cottonmouths (*Agkistrodon piscivorus*) from northeastern Texas, USA, *Environ. Toxicol. Chem.*, 24, 665, 2005.
111. Hopkins, W.A. et al., Functional relationships among selenium concentrations in the diet, target tissues, and nondestructive tissue samples of two species of snakes, *Environ. Toxicol. Chem.*, 24, 344, 2005.
112. Burger, J. et al., Metals and metalloids in tissues of American alligators in three Florida lakes, *Arch. Environ. Contam. Toxicol.*, 38, 501, 2000.
113. Niewiarowski, P.H., Aspects of reptile ecology, in *Ecotoxicology of Amphibians and Reptiles*, Sparling, D.W. et al., Eds., SETAC Press, 2000, chap. 4B.

114. Aleksiuk, M., Cold-induced aggregative behavior in the red-sided garter snake (*Thamnophis sirtalis parietalis*), *Herpetologica*, 33, 98, 1977.
115. Roughgarden, J., Anolis lizards of the Caribbean: ecology, evolution, and plate tectonics, in *Oxford Series in Ecology and Evolution*, Oxford University Press, New York, 1995.
116. Rodda, G.H., Perry, G., and Rondeau, R.J., The densest terrestrial vertebrate, in *Abstracts of the Society of for the Study of Amphibians and Reptiles*, Pennsylvania State University, State College, 1999, 195.
117. Campbell, T.S., Analysis of the effects of an exotic lizard (*Anolis sagrei*) on a native lizard (*Anolis carolinensis*) in Florida, using islands as experimental units, Ph.D. Dissertation, University of Tennessee, Knoxville, 2000.
118. Bouchard, S.S. and Bjorndal, K.A., Sea turtles as biological transporters of nutrients and energy from marine to terrestrial ecosystems, *Ecology*, 81, 2305, 2000.
119. Bjorndal, K.A. et al., Twenty six years of green turtle nesting at Tortuguero, Costa Rica: an encouraging trend, *Conserv. Biol.*, 13, 126, 1999.
120. Van der Valk, H.C.H.G., Community structure and dynamics in desert ecosystems: potential implications for insecticide risk assessment, *Arch. Environ. Contam. Toxicol.*, 32, 11, 1997.
121. Rainwater, T.R. et al., Mercury in Morelet's crocodile eggs from northern Belize, *Arch. Environ. Contam. Toxicol.*, 42, 319, 2002.
122. Vos, J.G. et al., *Toxicology of Marine Mammals*, Taylor and Francis, New York, 2002.
123. Collins, J.P., May you live in interesting times: using multidisciplinary and interdisciplinary programs to cope with change in the life sciences, *Bioscience*, 52, 75, 2002.
124. Crain, D.A. and Guillette, L.J., Jr., Reptiles as model of contaminant-induced endocrine disruption, *Anim. Reprod. Sci.*, 53, 77, 1998.

chapter 4

Tools for Assessing Contaminant Exposure and Effects in Reptiles

Carys L. Mitchelmore, Christopher L. Rowe, and Allen R. Place

Table of Contents

I. Introduction

A. Overview

In the 20th and 21st centuries an increasing amount and diversity of anthropogenic inorganic and organic chemicals have been released into the environment as a result of urbanization and industrial and agricultural activities. Examples include heavy metals; polychlorinated biphenyls (PCBs); organochlorine pesticides (OCPs); polycyclic aromatic hydrocarbons (PAHs); polybrominated diphenyl ethers (PBDEs); and, more recently, an array of pharmaceuticals and personal care products. The ultimate sink for many of these anthropogenic contaminants is the aquatic environment, either through direct discharge or as a result of hydrologic and atmospheric processes.[1] Many of these contaminants, particularly lipophilic chemicals, are readily accumulated by resident organisms through a variety of mechanisms such as direct contact (water, sediment) or dietary uptake. Indeed, contaminant accumulation has been reported in a diverse array of taxa, including reptiles. Accumulation-based studies are informative with respect to particular contaminants, but they fail to capture information regarding the thousands of environmental contaminants and their metabolites that are likely to be present in the environment, many of which have yet to be identified. In addition, accumulation of contaminants does not necessarily indicate a detrimental effect to the organism.[2] For accurate environmental risk assessments, relationships have to be established among contaminant exposure, accumulation, and biological and ecologic effects. The need to detect and assess impacts of contaminants, particularly low concentrations of complex mixtures, has led to the study and development of biomarkers to assess adverse biological responses to anthropogenic chemicals.[3,4]

B. *Tools to Assess Contaminant Exposure and Effects (Biomarkers)*

The search for biomarkers reflecting adverse biological responses to contaminants has been a major focus in environmental toxicology over the past decade. Many definitions exist for the term "biomarker." This definition has evolved over recent years along with an increased understanding of specific molecular, biochemical, and cellular responses that provide reliable means of quantifying potential biological effects. Formally, biomarkers are defined as "measurements in body fluids, cells or tissues indicating biochemical or cellular modifications due to the presence and magnitude of toxicants, or of host response."[5] A broader definition that accommodates whole animal studies would also include "measurements on whole animals" and "indicate in physiological, behavioral or energetic terms."[6] A contaminant may trigger a whole cascade of biological responses, each of which may, in theory, serve as a biomarker.[7] Thus there are biomarkers that can be used at various levels of organization, including molecular, biochemical, cellular, and individual levels.[4] Biomarkers are therefore used to describe effects at the individual level or below. However, the term "bioindicator" is specific to the individual level and defines the presence or absence of a species as a metric of environmental conditions.[8]

According to the National Research Council (NRC)[5] and the World Health Organization (WHO)[9] as detailed by Van der Oost and colleagues,[2] biomarkers can be subdivided into three main classes:

1. Biomarkers of *exposure:* indicating the presence of an exogenous substance or its metabolite or an interaction between a xenobiotic agent and a target molecule or cell
2. Biomarkers of *effect:* including measurable biochemical, physiologic, or other alterations within tissues or body fluids of an organism that are associated with an established or possible impairment or disease
3. Biomarkers of *susceptibility:* indicating the inherent or acquired ability of an organism to respond to the challenge of exposure to a specific xenobiotic substance, including genetic factors and changes in receptors that alter the susceptibility of an organism to that exposure

Currently there is debate concerning whether the bioaccumulation of persistent environmental contaminants in tissues may be considered to be true biomarkers. Whereas accumulation has been considered an exposure biomarker by NRC[5] and WHO,[9] definitions by Van Gestel and Van Brummelen[8] exclude contaminant body burdens from their definition of biomarkers. For the purpose of this review, body burdens will not be considered as biomarkers; they will be considered to be *bioaccumulation markers* as defined by Van der Oost et al.[2] However, the presence of *metabolites* of accumulated contaminants can be considered as biomarkers of exposure. A common example is the measurement of fluorescent aromatic compounds (FACs) in the bile of organisms that have metabolized accumulated PAHs.[10] Biomarkers can also be classified in terms of "specific" or "general" response — that is, some of the responses observed may be contaminant specific (see references 11 and 12) and identify a particular contaminants class (e.g., brain acetylcholinase [AChE] enzyme inhibition by organophosphate pesticides or production of the female protein vitellogenin [Vtg] in males by endocrine-disrupting chemicals [EDCs], although this now represents an array of chemical mixtures) or represent a general stress response (e.g., induction of heat shock proteins or DNA strand breaks).

The initial response to a contaminant occurs at the molecular level (e.g., via induction of detoxification enzymes or proteins). Therefore, a primary role of many molecular and biochemical biomarkers in environmental research is to serve as early-warning signals of contaminant exposure or effect.[4] These biomarkers represent early responses in the overall sequence of events from exposure through ecosystem-level effects and can be applied as reactive tools, allowing possible remedial actions to be implemented to reduce environmental risk. In general, molecular and biochemical processes are more highly conserved and thus more universal than processes operating at higher levels of biological organization. As a result, molecular and biochemical biomarkers have been the primary area of research dedicated to identifying early-warning biomarkers that may be universal across taxa.[4] However, as a result of extensive research it is clear that in some cases there are differences in biochemical function and regulation that limit the application of the biomarker across a diverse taxa. For example, cytochrome P4501A is present in all vertebrates, but in reptiles and invertebrates it is present at lower levels than in other groups and its inducibility, forms, and functions are unclear at present.[4,13] Species-specific differences in cytochrome P4501A responses among reptile species have been observed; three species of turtle have been shown to differ in P4501A induction[14] (see later in this chapter). These differences highlight the need for extensive research, characterization, and development of potential biomarkers in reptiles. It should also be noted that biomarker development and characterization in laboratory studies must always be validated with field research to avoid underestimations or overestimations of effects.

The most compelling reason for using biomarkers is that they can provide information on the biological effects of pollutants rather than a mere quantification of their environmental concentrations. Generally, biomarkers are considered to be intermediates between pollution sources and higher-level effects (i.e., at the population or community system levels).[15] A further advantage of biomarker research is that it may provide essential information concerning the potential mechanisms involved in toxic effects of contaminants. These mechanistic studies may aid in elucidating the fate of accumulated contaminants and could be important in determining toxicologic risks for higher trophic–level species. For example, from a human health perspective it is imperative that we understand the potential risks of consuming contaminated reptile meat. Therefore, mechanistic studies can investigate the possibility that detoxification mechanisms in reptiles transform benign parent contaminants into potentially carcinogenic metabolites.

Extensive biomarker research and application have been carried out in ecotoxicologic studies, particularly those focused on commercial fish species. Data generated from this field have formed a basis for biomarker development in reptile species, and clear parallels exist with the biomarkers most routinely used in these two groups (e.g., dominated by data on the cytochrome P450 system; see section B.1). To evaluate the strengths and weaknesses of reptile biomarkers throughout this chapter, we will adopt the six criteria based on those established for fish biomarkers.[2] Discussions will include the following points:

1. The assay used to quantify the biomarker should be reliable, relatively inexpensive, and easy to perform.
2. The biomarker response should be sensitive to low levels of contaminant exposure and early effects to serve as an early-warning parameter.
3. The baseline data of the biomarker should be well defined to distinguish between natural variability (noise) and contaminant-induced stress (signal).
4. The impacts of confounding factors, such as gender and seasonality, on the biomarker should be well established.

5. The underlying mechanism of the relationships between biomarker response and contaminant exposure (dosage and time) should be established.
6. The toxicologic significance of the biomarker (e.g., the relationships between biomarker expression and acute and chronic impact on the organism) should be established.

In comparison with biomarker research in fish species, very few biomarkers have been developed and applied to reptile species, and most of these criteria listed earlier are poorly established for biomarker applications to reptiles. In reptiles, as for most vertebrates, the liver is the organ most commonly involved in the detoxification of foreign compounds, and many biomarker responses have been characterized in this organ. However, currently there is debate on the use of these destructive endpoints with an associated interest in establishing tools that are nondestructive in nature (e.g., see reference 16).

C. *Assessing Contaminant Exposure and Effects in Reptiles — Why?*

The increasing concerns about global declines in reptilian populations have initiated investigations to evaluate the effects of contaminants on these species.[17,18] There are many aspects of the life history and physiology of reptiles that make them of particular risk to environmental contaminants. Their high trophic–level position, diets, and especially long life spans expose them to chronic and potentially very high concentrations of a wide variety of contaminants. Tissues, such as the liver and fat of reptiles, contain a high lipid concentration and therefore can potentially accumulate very high levels of lipophilic contaminants. In addition, during the process of vitellogenesis it has been shown that maternal transfer of accumulated pollutants can occur, posing a substantial risk to the developing embryos (see Chapter 3). The control of their reproductive development by temperature (temperature-dependent sex determination, or TSD; see Chapter 6) makes them exquisitely sensitive to contaminant effects, particularly to chemicals that have estrogenic activity (EDCs; see Chapter 6). However, a disproportionate amount of ecotoxicologic research has been directed toward reptilian species compared with other taxa. Of the families of reptiles, by far the largest data set exists for testudines. Many research studies have been directed toward studying contaminant accumulation in reptiles, and little emphasis has been placed on actual effects on individuals and populations. Research on contaminant accumulation is outside the scope of this chapter; the reader may find excellent reviews on this topic in recent literature.[18,19]

The life history and physiologic traits of reptiles discussed earlier make them a useful group of vertebrates for quantifying ecologic markers of stressors.[20] Many reptiles have a relatively small home range and are relatively sedentary,[21–23] thus they experience chronic exposure to local contaminants. In addition these upper trophic–level species can provide important information for ecologic risk assessments of contaminants that magnify through the food web. Currently, five reptilian species are listed in the U.S. Environmental Protection Agency's wildlife exposure factors handbook as potential indicator species,[24] and toxicologic research to date has primarily focused on these species. These five species of primary interest are the painted turtle (*Chrysemys picta*), eastern box turtle (*Terrapene carolina*), common snapping turtle (*Chelydra serpentina*), racer (*Coluber constrictor*), and northern water snake (*Nerodia sipedon*). Basic ecologic information is available for a large number of reptiles, and it is often a lack of sufficient toxicity data in the literature that precludes the use of reptiles in ecologic risk assessments.[20]

A need for further research on the mechanisms of toxicity of contaminants on reptile species was presented by Hall and Henry.[25] They stated that many species are endangered and that an emphasis should be placed on the potential of chemical contaminants to affect their survival, a statement that was frequently reiterated in a review.[18] Unlike amphibian research where a model species is widely used (e.g., the African clawed frog, *Xenopus laevis*), there has been no such standard test species for assessing contaminant toxicity to reptiles. There is also a need to examine variation and species sensitivity among reptiles to chemical toxicants. Meyers-Schöne[20] stated that further work in reptilian toxicology was recommended to prevent the underestimation of risk to a class of organisms that have been largely ignored in the ecologic risk assessment arena despite their obvious sensitivity based on their unique life history and physiologic traits. Specific needs are a better understanding of lethal and sublethal contaminant effects at all stages in the life cycle and identification of mechanisms of toxicity. The author recommended a particular emphasis should be placed on the effects of EDCs and the interactions of contaminants with other biological stressors (e.g., paratism, predation, and disease) at the population level (see Chapter 10).

The purpose of this chapter is to summarize and critically assess the tools (biomarkers) that have been used to assess contaminant exposure and effects in reptiles to date. In addition, we discuss methods (particularly molecular; see Section II.A) in development that could serve as future tools for assessing contaminant exposure in reptile species.

II. Current Status of Reptilian Biomarker Research

Reptile biomarker studies to date have followed in the footsteps of the well-developed biomarker studies in other vertebrate species, such as fish. For example, there is an emphasis on detoxification enzymes and proteins, such as the cytochrome P450s (see Section B.1). However, as a result of the clear demonstration of the effects of EDCs on reptiles, a substantial emphasis has also been placed on determining mechanisms of toxic action using biomarker endpoints and in developing early-warning biomarkers predictive of such reproductive perturbations by environmental contaminants. Therefore, many studies have investigated the changes in steroid hormones and reproductive proteins of the endocrine system (see Section B.4) in addition to using developmental indices, such as altered sex ratios (see Chapter 6 and Sections E and F.3).

We present this chapter hierarchically based on the levels of organization within which specific biomarkers may be applied. We have attempted to discuss all field-based studies using biomarkers as tools to assess contaminant exposure and effects. In addition, to address the utility and possible limitations of these biomarkers we also refer to laboratory-based studies throughout.

A. Molecular Biomarkers

It is likely that environmental toxicants that affect reproduction, metabolism, or development act either by disrupting the synthesis or degradation of endogenous hormones or by directly activating or deactivating hormone receptor–mediated gene pathways. For example, it is now well recognized that exposure to estradiol (E_2) results in the synthesis of specific proteins required for reproduction. Several genes that encode proteins induced by this process are the estrogen receptor (ER), Vtg, the egg yolk precursor proteins, and choriogenins, which are required for making the egg membrane. Exposure to E_2 or to an E_2 mimic increases the number of estrogen receptors[26,27] as well as induces the synthesis of Vtg in the liver.[28–30] Vtg is normally present at high levels only in females undergoing oogenesis. In males the gene for Vtg is normally suppressed; however, Vtg can be induced in males by exposure to E_2 or estrogen mimics. In addition to affecting the expression of Vtg and ER, E_2 induces an array of genes, some by direct interaction with the ER and others by alternative routes. The cascade of genes that are induced is tissue specific. For example, Vtg is produced only in the liver, yet we know that E_2 targets other tissues besides the liver, such as the gonads and the brain. Although the gene for Vtg is present in these other tissues, it is not induced by E_2. Other hormones such as androgens and thyroid hormones activate their own tissue-specific cascades of genes. The identity of many of those genes is limited, especially in reptiles. The two (and perhaps three) ERs, α and β (and possibly γ) in vertebrates,[31,32] alone or in combination, may control different subsets of genes. New research indicates that E_2 mimics may bind differently to these receptors, acting as agonists in one case and antagonists in another.[33] This complicates the issue of determining the risks of environmental exposure when only using one or a few biomarkers.

It is likely that many toxicants will have their own specific gene-expression profiles because they may bind with low affinity to more than one receptor, resulting in a complex gene activation pattern, and this may be species specific. It is clear that competition for ligands and transacting factors plays a large role in the molecular events that take place. If a compound can bind to both the ER and androgen receptor (AR) or progesterone receptor, are both pathways induced? Or, does one pathway predominate (see reference 26 for example)? What happens with mixtures that are the norm for environmental exposures? Do the specific compounds in mixtures interact with each other or compete? The use of global, open-ended gene-expression profiling experiments to determine the pathways that are affected has become a common approach with mammals and more recently with fish.

Unfortunately, if one were to search the available DNA databases for reptile-specific biomarker genes, one would find very few candidates at this writing. We expect this to change dramatically over the next 5 years. By using data from the whole genome projects for birds and fish, many homologous candidate genes will be found and used to isolate the reptile-specific genes. In 2001 the National Science Foundation (NSF) launched a new grant competition designed to help boost genetic research on organisms that are not models of human biology and therefore not normally supported by biomedical science. A handful of organisms were selected, with an attempt to represent as many branches of the Tree of Life as possible. During the selection process, favor was given to those species for which little resources currently exist. Fortunately, reptiles were included in this funding to build libraries of bacterial artificial chromosomes (BACs) for the painted turtle (*Chrysemys picta*), tuatara (*Sphenodon punctatus*), South American worm lizard (*Amphisbaena alba*), the American alligator (*Alligator missisippiensis*), and the emu (*Dromaius novaelhollandie*) as an outgroup. These BAC libraries will be essential for production of reptilian-specific DNA arrays and gene isolation. Because so little molecular data exist for reptiles, we will describe the approaches that can be used when such gene markers are available and cite examples when available for reptiles.

In this review we will examine several common approaches for quantifying gene expression using both targeted approaches such as Northern blotting and quantitative reverse transcriptase-polymerase chain reaction (RT-PCR) and open-ended approaches, such as differential display RT-PCR (DD RT-PCR), subtractive hybridizations, and gene arrays that have yet to be applied to reptiles. Each of these systems has advantages and disadvantages and strengths and weaknesses. Working with both targeted and open-ended approaches gives a clearer view of the molecular mechanisms deployed by particular exposures, which are important for obtaining a clear understanding of the risks to exposure of toxicants.

1. *mRNA Isolation*

Measurement of gene expression begins with the isolation of ribonucleic acid (RNA) transcripts from the target organ. Obtaining high-quality, intact RNA is the first and often the most critical step in performing many fundamental molecular biology experiments, including Northern analysis, nuclease protection assays, RT-PCR, RNA mapping, *in vitro* translation, and complementary deoxyribonucleic acid (cDNA) library construction. During tissue disruption for RNA isolation, it is crucial that the denaturant be in contact with the cellular contents at the moment that the cells are disrupted. This can be problematic when tissues or cells are hard (e.g., bone, roots); when they contain capsules or walls (e.g., yeast, Gram-positive bacteria); or when samples are numerous, making rapid processing difficult. A common solution to these problems is to freeze the tissue or cells in liquid nitrogen or on dry ice. The samples must then be ground with a mortar and pestle into a fine powder, which is added to the denaturant. However, products have become available (e.g., *RNAlater*, Ambion Inc.) that allow the researcher to postpone RNA isolation for days, weeks, or even months after tissue collection without sacrificing the integrity of the RNA. Dissected tissue or collected cells are simply dropped into the *RNAlater* solution at room temperature. The solution permeates the cells, stabilizing the RNA. Once isolated the RNA needs to be stored in RNAase-free buffers containing chelating agents (e.g., ethylenediamine tetraacetic acid, or EDTA) and must be at low pHs (<7.0).

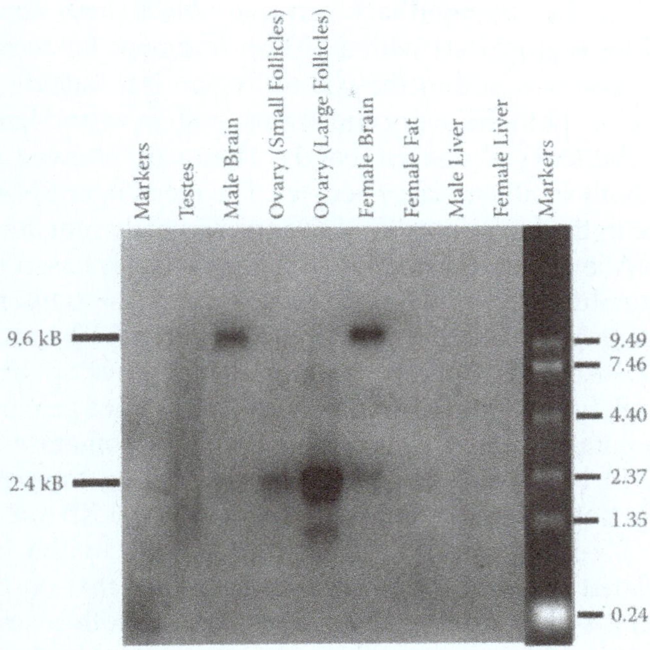

Figure 4.1 Northern blot of messenger ribonucleic acid (mRNA) isolated from tissues of the diamondback terrapin (*Malaclemys terrapin*) probed with a P450 aromatase probe. (Courtesy of Jeyasuria and Place, unpublished.)

2. *Northern Blotting*

The Northern blotting procedure can be used to determine the relative abundance of gene transcripts, their size, and the number of potential splice variants. This last piece of information is extremely important given that at least 74% of the human multiexon genes are alternatively spliced,[34] and there is no reason to expect much different levels in reptiles. RNA samples are separated on a denaturing agarose gel by electrophoresis and then transferred to a nylon or nitrocellulose membrane. The RNA is then probed with cDNA that is either radiolabeled or digoxigenin-labeled and that is complementary to the sequence of interest. To determine the relative abundance of the genes of interest, a specific probe to an invariant housekeeping gene (such as ß-actin, 18S ribosomal RNA) is also hybridized to the membranes either at the same time or sequentially.

A good example of this approach is the work of Custodia-Loria and colleagues[27] on the regulation of progesterone and estrogen receptors in the turtle, *Chrysemys picta*. Not only could transcript levels of different isoforms be quantified but also compared between sexes in response to endogenous estradiol and progesterone treatments. Another example is illustrated in Figure 4.1 for the expression of P450 aromatase in the diamondback terrapin, *Malaclemys terrapin*. The northern blot was probed with a 433-bp fragment for terrapin P450 aromatase that was within the coding region but outside the heme binding region. Both the ovary and brains (both male and female) gave signals in the form of discrete bands. The ovary showed a band at 2.4 Kb for both small and large oocytes. The messenger RNA (mRNA) abundance in the larger oocytes was approximately nine times higher than mRNA levels in the smaller (1–3 mm) oocytes based on densitometric quantification of the bands. Two aromatase transcripts were observed in both male and female brains, a large 9.6-Kb transcript and a smaller 2.4-Kb transcript. There is one smaller transcript in the ovary that is about 1.4 Kb, but this band probably does not produce a functional aromatase protein considering that the aromatase transcript itself is 1512 base pairs in length. The aromatase transcripts were observed in both male and female brains, a large 9.6-Kb transcript and a smaller 2.4-Kb transcript. The larger 9.6-Kb in the brain is a 5′ untranslated region (UTR) unspliced transcript that on processing produces a 2.4-Kb mature mRNA in the brain of both sexes.[35]

Slot and dot blots are simple variations of the Northern blot. In these procedures the RNA is not prefractionated on gels; instead, it is spotted directly onto membranes. Another variation is the ribonuclease protection assay, in which a radiolabelled probe is allowed to bind in solution to its target to form a double-stranded RNA/DNA hybrid. After digestion with a single-strand–specific ribonuclease, the protected fragment is fractionated by acrylamide gel and its radioactivity is measured to determine the amount of mRNA present in the original sample. This has largely been supplanted by real-time quantitative PCR assay described later.

These assays have been used for three decades to measure expression levels of mRNAs. However, they normally are used for measuring only a handful of genes and they have largely been superseded by newer methods, such as real-time PCR (described later in this chapter), that are more quantitative and have higher throughput. They remain, however, an important screen that must be run when examining a new species or tissue. Northern blots are also an important tool to validate gene-induction profiles suggested by any differential display or subtractive method as well as for gene arrays.

3. *Quantitative Real-Time Polymerase Chain Reaction*

Quantitative real-time PCR (Q-PCR) is a fairly new technology that emerged in the early 1990s[36,37] as a more accurate and sensitive tool than Northern blotting to measure gene expression. Q-PCR is a major advancement from its predecessor, competitive (also called semi-quantitative or comparative) RT-PCR (see Figure 4.2 and reference 38 as an example), and takes advantage of the exonuclease activity of Taq polymerase to more accurately quantitate levels of gene transcripts. The Q-PCR reaction is monitored in real time by fluorescence either with the incorporation of the SYBR green dye that fluoresces only when it is intercalated into double-stranded DNA or by a fluorescent probe that is complementary in sequence to the cDNA of interest. The fluorescent probe hybridizes with the target sequence between the gene-specific forward and reverse primers. The probe contains a reporter dye on one end and a quencher dye on the other that inhibits a fluorescent signal when the probe is intact.

Figure 4.3 is a representative amplification plot for snapping turtle ß-actin and Cyp17a mRNA in the brain of a single mature female. The fluorescence level (ΔRn) increases with each PCR cycle as the amplified product increases. During the log-linear phase, the increasing fluorescence signal is directly proportional to the initial amount of target mRNA in the sample. Expression levels of a gene can then be determined relative to other genes or can be quantified by using a standard curve. To minimize variability in RNA quantitation and integrity, samples run by Q-PCR are normalized either to a stable housekeeping gene (e.g., ß-actin) or to 18S ribosomal RNA. It is always important to determine whether expression of the housekeeping gene used to normalize the data is in fact invariant because some commonly used housekeeping genes can be differentially regulated. For example, we have found that β-actin mRNA levels change in response to estrogen exposure in terrapin hatchlings (unpublished data).

The Q-PCR assay is a fairly quick and efficient method to quantify mRNA. This assay has several advantages over ribonuclease protection assays, Northern blot or slot-blot hybridization, and competitive RT-PCR for measuring gene expression. These advantages include the sensitivity of the assay (theoretically to a single copy of the mRNA of interest); the small amount of total RNA required (100 ng or less); the elimination of the need for post-PCR processing and radioisotope labeling; and the simplicity of the assay, which facilitates processing large numbers of samples. The Q-PCR assay can also be designed to measure several genes at once (multiplexing), thus making it a high-throughput assay. The assay is able to measure differences in gene expression over 7–8 log values. A few disadvantages exist with the Q-PCR assay. At least a portion of the gene sequence must be known to which primers and a probe can be designed. Also, the assay

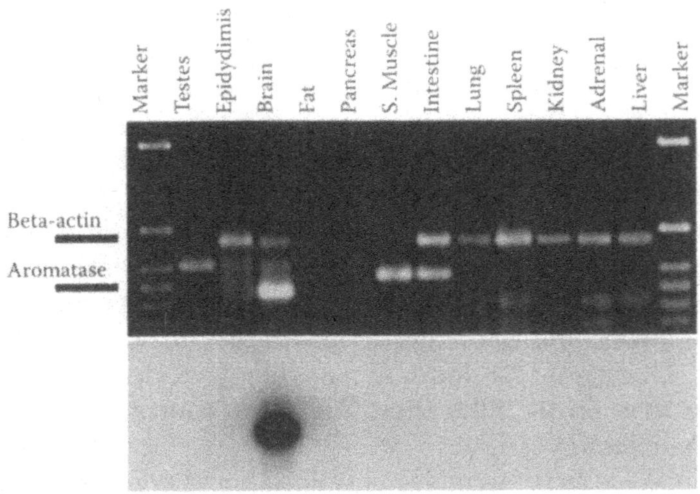

(a)

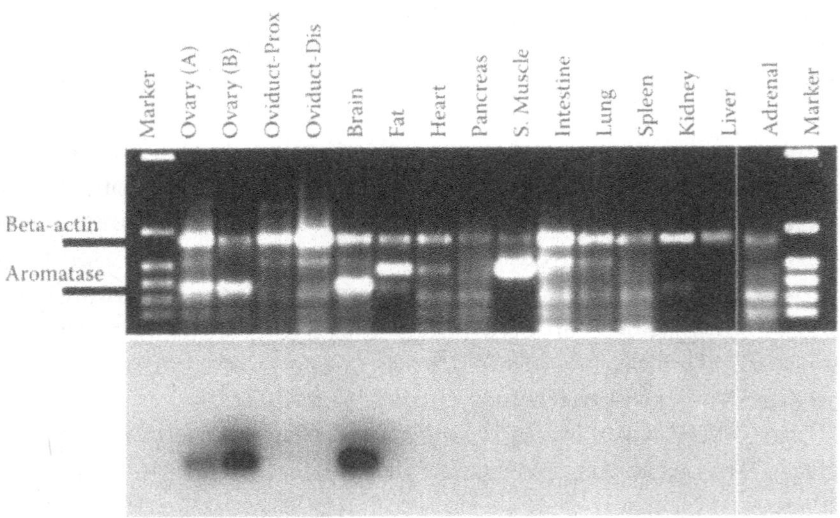

(b)

Figure 4.2 Semiquantitative reverse transcriptase-polymerase chain reaction (RT-PCR) using a combination of four primers: two are specific for ß-actin and two are specific for terrapin P-450 arom. Some bands were not of expected size, prompting blotting and probing with an internal oligonucleotide (5'-AATGAGGGGACCAATTCC-3') specific for terrapin P-450arom. Gel (a) shows RT-PCR for male tissues and gel (b) shows RT-PCR for female tissues. (Courtesy of Jeyasuria and Place, unpublished.) Both gels were probed, and the results are shown below the SYBR Green I stained gels. The size markers (PGEM Markers, Promega Corp.) seen in the gel are from the top 1198, 676, 517, 460, 396, 350, and 222 base pairs. The expected size is 630 base pairs for the ß-actin PCR product and 433 base pairs for the aromatase.[35]

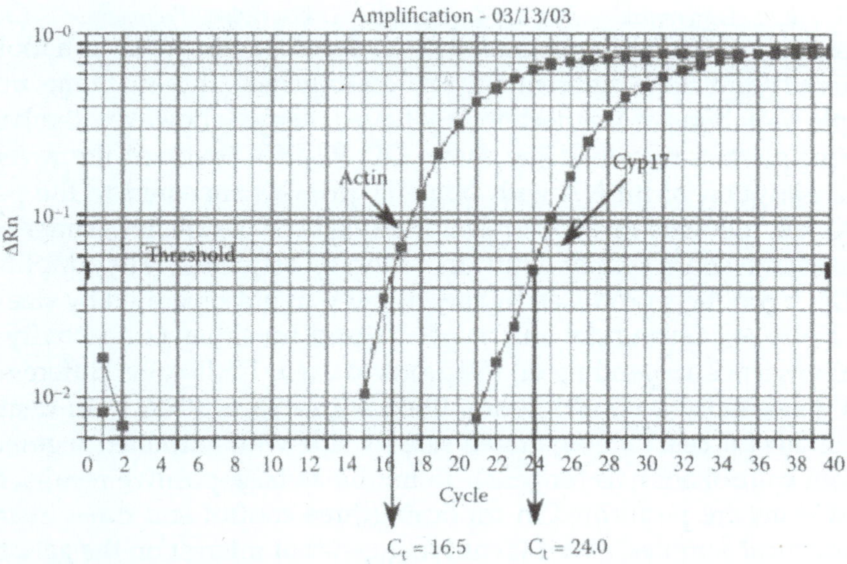

Figure 4.3 Quantitative reverse transcriptase-polymerase chain reaction (RT-PCR) using a combination of four primers: two are specific for ß-actin and two are specific for snapping turtle *Cyp17a*. Twenty nanograms of total RNA from the brain of a mature adult female were used. The amplification plots represent duplicate samples probed for ß-actin and *Cyp17a*. The difference between the C_t for ß-actin and C_t for *Cyp17a* represents ($\Delta C_t = -7.5$), the fold difference in transcript abundance ($2^{-\Delta Ct} = 181$ fold greater ß-actin transcript abundance). (Courtesy of Goto-Kazeto and Place, unpublished.)

will not provide information about the size of the target mRNA because only a short (approximately 100 bp) amplicon is amplified and measured. High-quality total RNA is required for accurate quantitation with little to no DNA contamination. If the gene structure is known, designing primers or a probe so that intron and extron boundaries are included ensures that genomic DNA contamination will not bias the results. Another consideration is the relatively high cost for hardware, probes, and other reagents.

4. *Open-Ended Methods*

Although directed methods to measure gene expression are very useful for known gene transcripts, these methodologies are not useful for exploratory research. For example, if the goal is to identify the full set of genes that are regulated by specific hormones or environmental contaminants, a more global, open-ended technology should be used. The most commonly used open-ended procedures include DD RT-PCR, subtractive hybridizations, and gene arrays.

4.1. Differential Display Reverse Transcriptase-Polymerase Chain Reaction. DD RT-PCR technology was developed in 1992[39] as a tool to identify and clone differentially expressed genes. Since that time, multiple variations of this technology have emerged; however, the basic principle has remained the same. DD RT-PCR involves the reverse transcription of mRNA with oligo-dT primers anchored to the poly (A) tail, followed by a PCR reaction using both the anchor primer and a second short primer with an arbitrary sequence. The amplified cDNA products produced by the primer pair are separated by size on a DNA sequencing gel and are visualized by either radioactivity or fluorescence, depending on the protocol used. To discover differences in gene transcription, PCR-amplified segments of RNA from treated animals (or cells) are separated side by side with amplified segments from control animals (or cells). To minimize false-positive results, the reactions are performed in triplicate (three control and three treated biological samples). After identifying genes of interest on the gels, the corresponding gel bands are cut out, reamplified with the same primer pair, cloned, and sequenced for further study. Visualization and identification of many mRNA species are possible by using different combinations of anchor primers and arbitrary primers.

A good example of this approach with reptiles is the work of Herbst and colleagues[40] on a comparison of normal and fibropapilloma fibroblasts from the green turtles (*Chelonia mydas*). Two overexpressed transcripts (β-hexosaminidase and chain termination factor) and one underexpressed transcript (thrombospondin) were observed in this comparison. Similarly, Trudeau and colleagues[41] used DD RT-PCR to show that octylphenol upregulated amyloid precursor like protein (APLP-2) in the hypothalamus of the snapping turtle (*Chelydra serpentina*).

The main advantage of DD RT-PCR technology, compared with other open-ended procedures such as subtractive hybridizations, is that this procedure can be used to identify differentially regulated up- and down-regulated genes within a specific tissue of interest. A refinement of this procedure, called suppressive subtractive hybridization (SSH), now enables a researcher to obtain both high- and low-abundance transcripts to be screened (Clontech, Palo Alto, CA).[42]

4.2. Gene Arrays. Gene arrays (also referred to as microarrays, macroarrays, gene chips, or DNA arrays) are a high-throughput technology that can be used to simultaneously measure the expression of hundreds to thousands of genes. In this procedure, RNA is reverse transcribed with fluorescently or radiolabeled markers (called the target) and is subsequently hybridized to DNA sequences (called probes) that are attached to a solid support matrix.[43] In essence DNA arrays are reverse Northern blots, from which the terminology for "target" and "probe" was derived. The most commonly used arrays can be divided into two groups: oligonucleotide arrays and cDNA arrays. To construct oligonucleotide arrays, probes can be designed and synthesized separately and then attached to the matrix or they can be synthesized *in situ* (directly on the chip). In contrast, cDNA clones obtained from differential display, subtractive hybridizations, or other methods are then attached to the matrix. A robotic workstation usually is used to make the oligonucleotide or cDNA arrays to ensure spot reproducibility and equal deposition of minute volumes. Gene chips can also be constructed manually by using a hand-held spotter. In both methods, the probes are spotted onto glass slides or nylon membranes. Expression analysis using glass slide arrays usually uses competitive hybridization of two different fluorescently labeled targets (e.g., Cy3-dNTP and Cy5-dNTP dyes). For membrane arrays, control and treated samples are hybridized individually on separate membranes using radioactive or chemiluminescent targets.

Gene arrays offer the advantage of simultaneously monitoring hundreds to thousands of genes at once, enabling one to examine whole metabolic pathways affected by a particular contaminant. However, a disadvantage is that arrays may not be sensitive enough to detect rare gene transcripts. Additionally, the function of many genes spotted on an array may be unknown. A major disadvantage for those working with reptiles is that there are no current arrays available, but that should change in the near future. Using arrays designed for the chicken may provide a stopgap approach until species-specific arrays become available for reptiles.

B. *Biochemical Biomarkers*

The biomarkers that have been investigated most extensively in ecotoxicology are enzymes that are involved in substrate metabolism, including biotransformation and antioxidant enzymes. All organisms possess at least some of these biotransformation enzymes whose functions involve many endogenous processes (e.g., steroid metabolism). Of particular interest in this discussion are their roles in converting lipophilic organic xenobiotics into more water-soluble and thus excretable metabolites (i.e., detoxification processes). Various contaminants have been shown to inhibit or enhance many endogenous enzyme activities, some of which may be toxicant specific[11,12] and hence their application and wide use as biomarkers. Biotransformation enzymes consist of both phase I and phase II metabolism reactions, where phase I enzymes introduce a functional group (-OH, $-NH_2$, -COOH, etc.) to the xenobiotic on which phase II reactions can act on (e.g., conjugate or attach a large water-soluble moiety). The fate and toxicity of an accumulated contaminant may be affected by biotransformation, which may be beneficial (detoxification) or harmful (bioactivation or toxification) to the organism. The responses of biotransformation enzymes to xenobiotics include alterations (induction and inhibition) in their levels and activities. In reptiles, as for most vertebrates, the liver is the organ most commonly involved in the detoxification of foreign compounds, and the majority of the studies detailed later in this chapter use enzymes that are isolated from this tissue. The following section details the main biochemical biomarkers that have been used in reptiles to date and discusses their utility and limitations as tools for assessing contaminant exposure.

1. *Phase I Biotransformation; Cytochrome P450 Enzyme System*

One of the most extensively studied exposure biomarkers used in environmental biomonitoring studies is that of the hepatic microsomal cytochrome P450 system. Cytochrome P450s, which are a multi-gene family consisting of many different isoforms, are the terminal components of the mixed-function oxygenase (MFO) enzyme system and are of central importance in phase I metabolism reactions. They are essential for many endogenous reactions, such as steroid hormone pathways, and are capable of biotransforming many chemicals, including xenobiotic substrates. Cytochromes P450 catalyze monooxygenation reactions with enzymes that provide the reducing equivalents (cytochrome b5, cytochrome P450 reductase, and other cytochrome c reductases). Cytochrome P450 reductase mediates the transfer of reducing equivalents from NAD(P)H to cytochromes P450, although it can also reduce certain substrates directly. Cytochrome P450s can result in a contaminant's detoxification (removal and excretion) or in its toxification (i.e., metabolites are produced that are more toxic than the original parent compound). For example, cytochrome P450s can activate some PAHs to form mutagenic metabolites; therefore, increased synthesis of PAHs has consequences for carcinogenicity in the organism and even has secondary implications in human health risk assessments in commercial species that are used as a human food source.

Some cytochromes, in particular cytochrome P4501A, have been shown to be induced by many organic metabolites such as certain PAHs, dioxins, and PCBs via binding of the compound to a soluble protein known as the Ah (aromatic hydrocarbon) receptor.[44] It is the strong, dose-dependent inducibility of some P450 isoenzymes that forms the basis for its use as a sensitive biomarker of exposure to certain classes of organic contaminants.[3] Responses have been reported as changes in total cytochrome P450 content, although generally more sensitive and contaminant-specific measures of levels or activities of selected isoenzymes have been used.[3] In addition to serving as biomarkers of contaminant exposure, the response of certain P450 isoforms to contaminants can have implications for higher-level effects (e.g., development; see Section E). For example, sex determination can be affected by contaminants that may induce or inhibit the cytochrome P450s responsible for maintaining normal steroid hormone levels such as aromatase (cytochrome Cyp19).

Reptiles have been shown to possess all the components of the MFO system (for a review see reference 13). A summary of studies investigating the cytochrome P450 response in reptile species is given in Tables 4.1 and 4.2. Table 4.1 details the response of P450s in reptiles sampled from contaminated and reference sites. Table 4.2 lists laboratory-based studies using known P450 inducers (e.g., Aroclor 1254, 3-methylcholanthrene [3MC], and phenobarbital [PB]). These laboratory studies provide essential data as to the characteristics of the P450 system and the utility of using P450 responses as a tool (biomarker) to assess contaminant exposure. In comparison with mammalian species (e.g., rat) the basal quantity of reptilian P450s is much lower (i.e., often <40% of rat total P450 content[13,45–49]). P450 activity, isoform content, and induction patterns (and extent) also appear to be different between reptiles and mammals.[13,50] In parallel with P450 content, reptiles tend to have significantly lower microsomal NAD(P)H reductase activities than do mammalian species.[48,49,51] The relatively low content of the MFO system in comparison with other species (i.e., mammals) may be a factor that confers an increased sensitivity of reptiles to some contaminants.[48,49] Stafford and colleagues[52] attempted to correlate MFO activity and organochlorine residues with occurrence of five species of snakes from sites of heavy and low insecticide use in central Texas. Cottonmouths (*Agkistrodon piscivorus*), copperheads (*Agkistrodon contortrix*), and blotched water snakes (*Nerodia erythrogaster transversa*) demonstrated a high oxidase activity and were present in the highly contaminated site. In contrast, diamondback water snakes (*Nerodia rhombifera*) had low activity for the detoxifying enzymes and were absent from the highly contaminated site. These findings led the authors to conclude that the snake species with the highest activity of detoxification enzymes may be better able to survive in contaminated ecosystems.

Table 4.1 Cytochrome P450 Levels and Activities in Reptiles Collected from Contaminated Field Sites[1]

Species	Study conditions	Results and Findings	Reference
Marsh turtle (*Mauremys caspica rivulata*)	Petrochemical polluted site	No P4501A induction at contaminated site	14
Snapping turtle (*Chelydra serpentina serpentina*)	Multiple contaminant sites (chemistry in eggs measured), EROD and CYP1A protein	Induction of EROD (x8) and CYP1A (x50) protein at contaminated site, low levels cf. other vertebrates	54
Painted turtle (*Chrysemys picta picta*)	Superfund site, EROD and CYP1A protein	Induced at contaminated sites; seasonal, sex differences noted. No linear correlation with EROD and CYP1A protein, low levels. cf. other vertebrates	55
American alligator *mississippiensis*	3 site gradient of mixed contaminants, CYP1A (using EROD and MROD)	Inhibition of EROD activity at contaminated sites, MROD no differences, negative relationship between EROD, MROD, and body size at least contaminated site	56

Abbreviations: EROD, ethoxyresorufin-*O*-deethylase; MROD, methoxyresorufin-*O*-demethylase.
[1]Excluding data on CYP45019 (aromatase).

The majority of cytochrome P450 studies in reptiles to date are of those involved in hormone biosynthesis (e.g., CYP19/aromatase) or are of the classic biomarker CYP1A subfamily (form induced by PAHs). To quantify the induction and response of CYP450s to xenobiotics, various biochemical assays have been used with isolated microsomal fractions, usually from liver tissue. One method often used to quantify CYP450 induction is the determination of enzyme-specific activity using a substrate (specific for certain CYP450 isoforms), such as the *O*-dealkylation of ethoxyresorufin in the EROD (ethoxyresorufin-*O*-deethylase) assay, indicative of CYP1A1 induction. Because EROD activity appears to be highly inducible in most vertebrate species, including reptiles, it has become the benchmark MFO activity for studies of contaminant exposure, although other substrates have been used for other cytochrome P450 activities.[13]

TABLE 4.2 Cytochrome P450 Levels and Activities in Reptiles After Laboratory Exposures to Contaminants or Known Inducers[1]

Species	Study conditions	Results and Findings	Reference
Snake (*Thamnophis* sp.)	3MC, PB injections, total P450, enzyme activities (various substrates)	Low induction by 3MC (<x2) for all activities, PB no induction	51
Mauremys caspica rivulata	ARO	No induction of P4501A	45
Alligator mississippiensis	3MC, PB, CLO *in vivo*, enzyme activites (EROD, MROD, PROD)	No induction with CLO (EROD/PROD), induction of MROD by 3MC (x15.6) and PB (x2)	48, 49
M. caspica rivulata (adrenal microsomes)	ARO. P450 activity (progesterone 21-hydroxylation)	Induction (x3)	164
A. mississippiensis	3MC, protein levels	Induction of CYP1A1 and 1A2. Metabolism of BaP to mutagens	165
Chrysemys picta picta	BNF, TCB, ARO, enzyme activities, protein levels	No EROD induction, but protein induction	53
M. caspica rivulata	*In vivo* injection with ARO, CYP1A protein	Fivefold induction	166
C. picta picta; *C. picta elegans*; *Mauremys caspica rivulata*	*In vivo* injection ARO (or TCB except *M. caspica rivulata*), total P450, P4501A protein	No effect ARO on total P450, species differences in ARO P4501A protein induction. *C. picta picta* and *elegans* both 4–5x induction, *M. caspica rivulata* very low induction	14
Corn snake (*Elaphe guttata emoryi*)	3MC or PB, total P450 and b5 reductase, enzyme activities, CYP1A2/2B1 protein	No effect on total P450 or b5 reductase. 3MC-induced AHH, EROD, PROD other activities (e.g., ECOD) not affected, 3MC induced CYP1A2 protein	65
A. mississippiensis	3MC, PB *in vivo*, enzyme activites (EROD, MROD, PROD, BROD), CO binding (total CYP) and protein levels (2B and 1A)	Induction rates slower and less cf mammals. Enzyme and protein inductions	167

TABLE 4.2 (continued) Cytochrome P450 Levels and Activities in Reptiles After Laboratory Exposures to Contaminants or Known Inducers[1]

Species	Study conditions	Results and Findings	Reference
A. mississippiensis	Antibiotic treatment, total P450 content, enzyme activities and various protein isoforms	Total P450, BROD decrease with some antibiotics. Little or no induction of proteins	168
Chrysemys picta picta	In vivo exposure to TCB, PB, ARO, BNF, activities (EROD, MROD, AHH, ECOD), proteins (1A/2B/3A)	Induction of various specific activities and protein	169
Alligator mississippiensis	PB and/or 3MC, ARO, TCB, CLO,;10 CYP antibodies	Low levels cf. rat. Induction of various isoforms with 3MC, PB, ARO, TCB (less responsive to ARO and TCB), no response with CLO	170
Chrysemys picta picta	*In vitro* microsome exposures to TCB to assess TCB oxidation, CYP1a protein, EROD activity, and ROS production measured	Correlation between EROD activity and TCB oxidation	171

Abbreviations: 3MC, 3-methycholanthrene; PB, phenobarbital; ARO, aroclor 1254; CLO, clofibrate; EROD, ethoxyresorufin-*O*-deethylase; MROD, methoxyresorufin-*O*-demethylase; BNF, ß-naphthoflavone; TCB, 3,3′,4,4′-tetrachlorobiphenyl.

[1]Excluding data on CYP45019 (aromatase).

It should be noted that certain P450 activities representing classic activities found in other phyla appear to be absent in reptile species.[13] For example, aniline hydroxylase activity (indicative of CYP4502E) observed in the rat was not detected in alligator tissue.[48,49] In addition to lower induction levels, classic model inducers of P450s (e.g., 3MC, βNF, and PB) used in other species have shown varied responses (including no response) in reptiles, and the responses are often species specific (see Table 4.2). For example, Yawetz and colleagues[14] compared induction of cytochrome P4501A protein by Arochlor 1254 in liver microsomes of three species of freshwater turtles. A highly significant four- to fivefold increase in P4501A occurred in the yellow-bellied pond slider (*Chrysemys picta picta*) and red-eared slider (*Chrysemys picta elegans*), whereas a very low but significant induction was observed in the marsh turtle (*Mauremys caspica rivulata*). This inducibility, however, was not demonstrated in *M. caspica rivulata* that were collected from a contaminated site (petrochemical polluted waste oxidation pond). Similarly, in a prior study Yawetz and colleagues[45] showed no induction of P4501A in the *M. caspica rivulata* after dosing with Arochlor 1254. However, it should be noted that the turtles used for this induction study were collected from a sewage canal. It is possible that the cytochrome P450 system may already have been stimulated in the field, causing these turtles to be unresponsive to further chemical challenges. In summarizing these studies, Yawetz and colleagues[14] concluded that the cytochrome P4501A system of the *M. caspica rivulata* is insensitive to induction by PCBs. These results highlight the need to determine the sensitivity and characteristics of cytochrome P450 induction in individual species before it is used as a possible biomarker of exposure in field situations (see reference 13 for a complete summary).

Further complications in the use of CYP450 as contaminant biomarkers in the field result from the observation that certain contaminants can inhibit and mask catalytic responses, leading to underestimations of effects in organisms exposed to unknown contaminant mixtures.[2] For example, in highly polluted sites, inhibition of EROD activity occurred, but an elevation of P4501A protein was detected.[4] Similarly, in contrast to the induction of P4501A protein observed in *C. picta picta* by Yawetz and colleagues,[14] the same authors in 1993 detailed a lack of induction of CYP1A-type activities (i.e., EROD activity) with Aroclor in this species, although CYP1A protein was induced.[53] This protein induction with a lack of induced activity possibly indicates that the inducible isoform exhibits a different substrate specificity in other vertebrate species and highlights the need for further characterization of cytochrome P450 activity and regulation in reptile species.

Coupling enzyme activity levels with measures of P450 protein or mRNA levels has been suggested as a means to fully determine CYP450 responses. Determination of the level of CYP1A protein using either Western blotting or ELISA (enzyme-linked immunosorbent assays) techniques have been used to overcome this possible problem. Often a limiting step is that a cross-reactive antibody is required for these measurements. All reptile studies to date have used a variety of commercial P450 antibodies raised against specific CYP450 isoforms in other species (i.e., fish), which have been shown to be cross-reactive with reptile proteins. Current analyses are also being directed toward determination of changes at the very early-warning molecular level, such as up-regulation of the gene (mRNA) coding for cytochrome P450 (e.g., P45019, see Section A). Yet there remains the question as to whether this gene up-regulation manifests itself as increased protein levels and ultimately catalytic activity. To fully investigate this phenomenon in reptiles, analyses of cytochrome P450 response should be carried out at these different response levels to fully characterize this system to determine its utility as a biomarker.

The utility of using CYP450 induction as a biomarker of contaminant exposure in the field was demonstrated by Bishop and colleagues[54] who investigated cytochrome P450 responses in hatchling snapping turtles (*Chelydra serpentina*) from a contaminated site (Lynde Creek, Lake Ontario) and a less contaminated reference site (Algonquin Provincial Park, Ontario). Both EROD activity and cytochrome P4501A protein content were significantly higher in the livers of hatchling snapping turtles from Lynde Creek.[54] It was suggested that the high levels of PCBs, PCDD/F, and organochlorine pesticides in snapping turtle eggs from Lynde Creek were the cause of this induction. In addition to some of the limitations discussed earlier concerning the use of cytochrome P450 responses as a biomarker of contaminant exposure (particularly species-specific differences), further complications may include seasonal, sex, age, and ontogenetic influences, which have been described for other vertebrate species. As poikilotherms, reptiles, may be particularly susceptible to seasonal changes in environmental conditions.[13] Enzyme activities may show temperature optima of a broad or narrow range. For example, the optimum temperature for B[a]P hydroxylase activities in spectacle caiman (*Caiman sclerops*) is 20°C, whereas an optimum of 30°C for ECOD (ethoxycoumarin-*O*-deethylase) and EROD activities was observed.[46] In addition, gender differences and life cycle changes may influence the MFO system, and future studies should address these issues. A field study by Rie and colleagues[55] demonstrated seasonal and sex-related differences in EROD activity and cytochrome P4501A protein content in adult painted turtles (*Chrysemys picta picta*). However, despite these complications statistically significant inductions in enzyme activity and protein levels were observed in the contaminated vs. reference sites. A study by Gunderson and colleagues[56] assessed the response of CYP1A in juvenile alligators (*Alligator mississippiensis*) collected from three sites with varying contamination in the Kissimmee-Everglades drainage in South Florida. A significant negative relationship between body size and EROD and MROD (methoxyresorufin-*O*-demethylase) activity was observed in animals collected from the site of lowest contamination.

The formation of estrogens from androgens is catalyzed in all vertebrates by the "aromatase" complex, which consists of a membrane-bound P450 enzyme, P450 aromatase, CYP19 (which binds the androgen substrate and inserts an oxygen into the molecule), and a flavoprotein (NADPH-cytochrome P450 reductase). Reducing equivalents from NADPH are transferred via the flavoprotein to P450 aromatase. The overall reaction involves a hydroxylation of the C2 on the A ring and two hydroxylations on the C-19 methyl group, thus spontaneously aromatizing the A ring. During this aromatization formic acid and water are released. Based on this reaction scheme, the standard activity assay for aromatase involves measuring the release of tritiated water from the 1-β position using [$1^1\beta$-^3H] androstenedione with nicotinamide adenine dinucleotide phosphate (NADPH) (or a NADPH-regenerating system) as a cofactor. Extracts, microsomes, cells, and whole fetal organs can be used. The assay is extremely sensitive, being able to measure subfemtomole levels of activity. Assays should always be done in the presence or absence of specific P-450arom inhibitors (e.g., 1 μM 4-hydroxyandrostenedione [4OH-A4[], CGS16949A [fadrozale], or CGS20267 [letrazole]) to ensure the specificity of the assay.

The usefulness of this biomarker is that it tends to be a highly accurate indicator of feminization in all vertebrates. During development in reptiles, aromatase transcript and protein tend to be higher in developing ovaries than in developing testes.[57–61] In some adults species little to no aromatase activity is found in the testes.[62] Lower but measurable activities are also found in the brain and pituitary and show seasonal variation depending on the sex and reproductive status.

Further laboratory- and field-based studies in reptile species clearly are required to fully characterize the cytochrome P450 response in reptiles. Similarities and differences in responses have been demonstrated with respect to other vertebrate taxa, as well as among closely related reptilian species. The majority of studies have used destructive sampling procedures (hepatic tissues), although it may be possible in the future to perform tissue biopsies similar to methods that are being developed for marine mammals using skin biopsies.[63]

2. *Phase II Enzymes*

Enzymes involved in phase II reactions catalyze the conjugation of endogenous substrates, xenobiotics, or their phase I metabolites with an endogenous ligand (i.e., polar groups such as glutathione [GSH] and glucuronic acid [GA]). These enzymes can play an important role in homeostasis and in the detoxification and clearance of many xenobiotic compounds. To date, only a few studies have investigated the response of phase II enzymes to xenobiotics in reptile species, and therefore they have received very little attention as to their role as biomarkers of contaminant stress. The majority of studies in reptile species have investigated these enzymes with respect to their responses to anoxia and hypoxia and overwintering (freezing). We include these studies in this section to assess the utility of these enzymes as possible biomarkers of contaminant stress. However, we limit our discussion to two of the major phase II enzymes: glutathione *S*-transferases (GSTs) and uridine diphosphate glucuronosyl transferases (UDPGTs). Conjugation of electrophilic contaminants (often metabolites of phase I reactions) with GSH is catalyzed by cytosolic GSTs, which represent a multigene and multifunctional family.[64] GSTs play an important role in many endogenous processes (i.e., intracellular transport of steroid hormones) and protecting against contaminant-driven oxidative damage and carcinogenesis. UDPGTs catalyze the transfer (conjugation) of glucuronic acid to compounds (e.g., nucleophilic xenobiotics) forming glucuronides.[64] Multiple isoenzymes of UDPGT exist, some of which show very specific substrate binding, whereas others display a degree of overlapping specificity. UDPGTs are involved in various endogenous processes (e.g., the glucuronidation of steroids) and detoxify a variety of xenobiotics.

The toxicity of many contaminants can be modulated by the induction of GSTs and UDPGTs, and induction of these enzymes by various contaminants has been described, particularly in mammalian and fish species. Like the phase I CYP1A genes, the Ah gene battery also comprises phase II genes including GST and UDPGT. The mechanism of induction for these enzymes therefore is probably regulated via the Ah-receptor, although induction responses of phase II enzymes are less pronounced compared with phase I systems.[64] For example, various UDPGT isoenzymes have been shown to be induced in the rat by 3MC and ßNF (i.e., the phenol UDPGT) and also PB (i.e., steroid UDPGT[64]). However, Bani and colleagues[65] exposed the corn snake (*Elaphe guttata emoryi*) to 3MC and PB and did not observe any changes in the activities of GSTs or UDPGTs. GSTs (and associated enzymes and cofactors) play a major role in alleviating conditions of oxidative stress.

Many reptilian species have been shown to be able to tolerate, under natural conditions, situations that pose a large potential for oxidative stress. These include anoxia (i.e., some turtles) to severe hypoxia in organs of freeze-tolerant snakes or diving turtles (for a review see reference 66). In an *in vitro* laboratory study using the freeze-tolerant red-sided garter snakes (*Thamnophis sirtalis parietalis*) Hermes-Lima and Storey[67] demonstrated a decrease in GST levels (50%) when snake muscle homogenates were exposed to H_2O_2 and Fe (reactive oxygen species [ROS] generation). The same authors also demonstrated that after freezing (5 h at –2.5°C) levels of GST were reduced significantly. Anoxic conditions (10 h at 5°C), however, demonstrated no significant response in GST activity compared with controls, although other antioxidant responses (see later) were altered.[68] They noted that their control values (no exposures) for GST were of similar levels to those determined in other nonmammalian vertebrates, although they were lower than for mammalian species (see reference 66). Willmore and Storey[69,70] demonstrated an inhibition of GST activities in turtle (*Trachemys scripta elegans*) liver and heart (25% and 42%, respectively) after anoxic conditions. Although not investigating chemical contaminant-driven responses, this suggests that GSTs are responsive to stress and highlight the need for further investigations as to contaminant response. These studies also highlight the fact that natural and seasonal responses (i.e., overwintering and diving) may detract from the use of GST as a biomarker of contaminant stress and may depend on the organism and timing of the study.

One field-based contaminant study investigating the GST response in the painted turtle (*Chrysemys picta*) was carried out by Rie and colleagues.[55] Hepatic cytosolic GST activity, measured by 1,2-chlorodinitrobenzene conjugation, showed a seasonal pattern in females but not in males. No differences in activity were observed between males and females. Statistically significant elevations in GST activity were demonstrated in females from the contaminated site compared with the reference site. This elevated activity at the impacted site indicates a response to contaminants superimposed on a normal seasonal pattern. A study by Gunderson and colleagues[56] demonstrated no differences in GST activities in juvenile alligators (*Alligator mississippiensis*) collected from three sites with varying contamination. The authors did observe a significant negative relationship between GST activities and body size at the site with the lowest contamination. However, no relationship between body size and GST activity was found in animals at the other more contaminated sites, suggesting that contaminants present at these sites acted to alter this relationship.

Clearly, more laboratory- and field-based studies are required to fully characterize phase II responses and to determine their utility as biomarkers. However, the literature that exists for fish species is often conflicting in terms of their utility as biomarkers (see reference 2 for a review). For example, although increases in hepatic GST activity have been observed in fish species after laboratory exposures to contaminants, such as PAHs and PCBs, many other studies did not demonstrate any significant alterations or describe an inhibitory response.[2] Studies aimed at detecting chemically induced activities of GSTs and UDGTPs in field-collected specimens from polluted sites reported mixed results ranging from strong inductions to no response to significant decreases in activity.[2] The authors noted that, considering the importance of these enzyme systems in contaminant detoxification reactions, more research was required to characterize the systems. Additional information may elucidate specific isoenzymes that have a more sensitive and selective response to pollutants. Results of the previously mentioned studies could also be applied to research for reptile species where phase II responses to contaminant exposure (laboratory or field based) are limited. The strong involvement of GST in alleviating oxidative stress conditions as a result of natural stressors (i.e., overwintering and diving) may hamper the use of this system as a biomarker of contaminant stress.

3. *Antioxidant Enzymes and Factors*

Many contaminants (and their metabolites) are toxic via oxidative stress mechanisms, which result in the production of a variety of cytotoxic ROS. Conditions of elevated oxidative stress have been found to result in numerous deleterious conditions, including the development of neoplasia and a variety of conditions linked with aging. ROS are produced as by-products of many endogenous processes, including aerobic metabolism.[71] ROS include the superoxide anion radical ($O_2^{\bullet -}$), hydrogen peroxide (H_2O_2; not a radical species but involved in ROS generation pathways), and the hydroxyl radical ($OH^{\bullet}$). These very potent oxidants are capable of reacting with many critical cellular targets, leading to consequences such as enzyme inactivations, DNA damage, and ultimately cell death.[72] Contaminants have been shown to evoke conditions of oxidative stress (elevated ROS) via a number of processes, including via redox cycling. These redox-active contaminants include aromatic diols and quinones (e.g., B[a]P quinone), nitroaromatics (e.g., nitrofurantoin), and certain transition metal chelates. The process of redox cycling involves the enzymatic reduction (e.g., 1 electron reduction via cytochrome P450 reductase) of the xenobiotic to yield a xenobiotic radical species. In the presence of molecular oxygen, this xenobiotic radical donates its unshared electron, yielding the superoxide anion radical ($O_2^{\bullet -}$), and in doing so reforms the parent compound. This cyclic process can then repeat itself if all cofactors are available, thereby leading to conditions of oxidative stress.

Organisms have developed a variety of enzymatic and nonenzymatic antioxidant systems to detoxify these toxic products and alleviate injury. Organisms may adapt to elevated levels of ROS by up-regulating these responses. However, if conditions of oxidative stress occur for long periods or at such high levels (i.e., contaminant-driven redox cycling), then endogenous detoxification pathways may become overwhelmed. Biomarkers of oxidative stress conditions have been developed for numerous endpoints, including adaptive responses such as increased activities of antioxidant enzymes and nonenzymatic compounds or manifestations of ROS-mediated toxic effect such as oxidations of lipids, proteins, and nucleic acids. Biomarkers that have been used in vertebrates have included responses of antioxidant enzymes such as superoxide dismutase (SOD), catalase (CAT), glutathione peroxidase (GPx), and glutathione reductase (GR), whose functions are to detoxify radical species into nonreactive molecules. In addition, numerous low-molecular–weight antioxidants, particularly GSH, have been shown to have biomarker utility.[2]

Information regarding the application of antioxidants as biomarkers of contaminant exposure in reptiles is limited. Rozhaja and colleagues[73,74] and Elezaj and colleagues[75] examined blood samples of the Greek tortoise from habitats severely affected by heavy metals and found that blood CAT and GPx activities were significantly lower in these turtles compared with controls.[75] As mentioned earlier, these enzymes are often used as indicators of oxidative stress (either up- or down-regulations); however, they may also be affected by heavy metals (hence indicators of metal intoxication). Both of these enzymes have metal cofactors (Fe and Se) by which other metals may interfere and inhibit these enzymes.[12,76]

The majority of studies regarding these antioxidant enzyme systems have focused on adaptations and alterations of these systems with respect to "natural" stressors, physiology, and behavior (i.e., altered oxidative status as a result of overwintering and diving). Many species of reptiles are able to survive for hours to months without oxygen, such as during hibernation. For example, species of *Trachemys* and *Chrysemys* can endure oxygen deprivation for many months when submerged in cold waters. However, resurfacing from diving or underwater hibernation potentially results in enhanced ROS generation. Similarly, in freeze-tolerant species, thawing may also expose animals to oxidative stress as a result of oxygen reintroduction into the tissues. Hermes-Lima and Zenteno-Savin[66] reviewed the antioxidant responses of reptiles in relation to changes in oxygen availability and physiologic oxidative stress. Induction of total SOD was observed in garter snakes (*Thamnophis sirtalis*) on exposure to anoxia for 10 h (at 5°C) in muscle and liver tissues (59% and 118% elevations, respectively). Similarly, levels of GSH were increased in liver tissues in response to anoxic conditions.[68] Antioxidant enzyme responses were also observed after freezing exposures in the same snakes (e.g., increase in CAT activity in muscle tissue). Several studies have detailed alterations in a variety of antioxidant enzymes and cofactors in turtle species in response to anoxia and subsequent reoxygenation.[66,70] Herme-Lima and Zenteno-Savin[66] described relatively high antioxidant enzyme activities in control animals when compared with other nonmammalian vertebrates, possibly reflecting adaptations of these species to these seasonal and behavioral changes in oxygen availability. Other complicating factors that could mask any contaminant-driven antioxidant response include factors such as mixed exposures (inhibitory and inducing contaminants), diet, environmental variables (e.g., temperature, hypoxia), and age and sex of the organism. For example, Elsey and Lance[77] demonstrated elevations in the levels of plasma and erythrocyte GPx in immature alligators in response to a diet supplemented with selenite. Jena and colleagues[78] demonstrated that the inhibition of CAT activity by aluminum in brain homogenates from the male garden lizard was age dependent.

Clearly, the contaminant-dependent responses (induction and inhibition) and the sensitivity of the antioxidant systems to the natural variables discussed earlier may complicate and limit the utility of these endpoints as biomarkers of contaminant exposure in the field. It is interesting to note that Van der Oost and colleagues[2] summarized that many antioxidant endpoints appeared unsuitable as biomarkers for ecologic risk assessment in fish as a result of the mixed responses observed in both laboratory- and field-based contaminant exposures. Because of the lack of contaminant-related exposure experiments and the observation that, for at least some reptile species, the response of these antioxidants are mediated by natural variables, it appears that these antioxidant systems may not be suitable as biomarkers of contaminant exposure. More research is required to determine the potential utility of these activities and responses in reptiles as biomarkers of contaminant exposure.

4. *Steroid Hormone and Protein Biomarkers*

Compared with other biomarker deployments, one of the most commonly used endpoints of contaminant exposure in reptilian research has focused on the effects of EDCs on circulating blood (plasma or serum) steroid hormone levels (e.g., E_2, testosterone [T], etc.) and the female yolk protein Vtg. These biomarkers have been used as early-warning biomarkers because changes in the normal levels of these circulating hormones and proteins can disrupt the endocrine system and have been shown to impair immune, neurologic, behavioral, and reproductive functioning. In addition these biomarkers use blood samples that can be collected from specimens by using non-destructive and minimally invasive procedures. The endocrine system in reptiles has been discussed in detail in Chapter 6 and elsewhere.[18,79] Here we provide only a brief introduction of studies demonstrating the utility of a variety of biomarker endpoints as tools to assess contaminant exposure.

Many studies have been undertaken to assess alterations in circulating blood steroid hormone levels in response to contaminants in a variety of reptilian species. Usually these assays use detection via antibodies by either radioimmunoassay, Western blotting, or ELISA-based techniques. For example, Guillette and colleagues[80] found elevated levels of plasma E_2 in female alligators and depressed levels of T in male juvenile alligators collected from the pesticide-contaminated Lake Apopka. In contrast to this study, de Solla and colleagues[81] examined levels of E_2 and T in adult male snapping turtles captured from both uncontaminated sites and sites contaminated with a mixture of organic compounds and concluded that no correlation existed between hormone and contaminant levels. The observation that the turtles collected from the contaminated site contained large body burdens of mixtures of organochlorines and showed evidence of decreased sexual dimorphism led the authors to suggest that disruption occurred during sexual differentiation during embryonic developments. However, studies[82–84] have demonstrated both seasonal variation and handling stress in levels of E_2 and T.

Descriptions on the effects of contaminants on the thyroid system resulting in alterations in levels of circulation thyroid hormone levels and their ratios (triiodothyronine, T3, and thyroxine, T4) have been discussed at length in reviews.[18] For example, studies of thyroid hormone levels in alligators from Lake Apopka have been carried out by several workers.[85–87] Brasfield and colleagues[88] exposed lizard eggs to Cd-spiked substrate during development and demonstrated that in addition to Cd accumulation, exposed lizards exhibited a decrease in their T3:T4 ratio at the highest surviving dose of Cd. However, no differences were observed in whole-body thyroid hormone levels or in other physiologic traits. There are many examples of positive and negative correlations of steroid hormone levels with contaminant levels in reptiles species, which are discussed in further detail in Chapter 6 and in reference 18. Further data are required to assess the utility of these biomarkers to act as tools of contaminant exposure and effect, particularly regarding species differences, complex mixtures, and observations of natural (sex, season) and handling stress variations. For example, Gunderson and colleagues[89] in studying the response of plasma T4 in juvenile alligators at sites along a contaminant gradient demonstrated both temporal and spatial variation in response. The authors concluded that mean plasma T4 concentrations did not match sediment contaminant data.[89]

One biomarker that has been extensively used as a specific indicator of exposure to EDCs is that of (Vtg) protein production in males (see Chapter 6). As stated earlier (see Section II.A) the synthesis of Vtg, a precursor of yolk proteins, is affected by E_2. Vtg synthesis does not normally occur in males but can be induced when exposed to EDCs, such as alkylphenols. Production of Vtg in juvenile or male specimens has been extensively used as a sensitive biomarker of exposure to estrogenic compounds. Detection methods use Western blotting or ELISA techniques using a variety of anti-Vtg antibodies that have shown to be cross-reactive with the reptile species of choice. For quantitation of the levels of Vtg protein (and for use in competitive ELISAs) it is necessary to include a Vtg standard (i.e., isolated Vtg from the species of choice), although semiquantitative methods using absolute value comparisons have been used.[90] Palmer and Palmer[30] demonstrated that this marker could be used in reptiles to test the estrogenicity of a variety of environmental contaminants. They observed that treatment of adult freshwater turtles with o,p′-DDT would induce Vtg synthesis in males in a dose-dependent manner.

Many studies (see Chapter 6 and Sparling and colleagues[18]) have indicated that induction of Vtg in males may be a useful nondestructive biomarker of contaminant-derived estrogenic activity in reptiles. However, other studies have also demonstrated a lack of Vtg induction in males and juveniles after exposure to contaminants. For example, Shelby and Mendonça[91] collected plasma samples from male yellow-blotched map turtles captured at a historically polluted site (including 2,3,7,8-tetrachlorodibenzo *p*-Dioxin [TCDD] and other contaminants from a wood pulp mill) and a reference site. No detectable levels of Vtg were found in males from either site. The authors concluded that the current impact of contaminants on reproductive parameters was limited in this population. However, the authors focused on historical exposures (1984–1995) because the focus of the research was to demonstrate the potential for long-term effects. Indeed, liver tissues in specimens collected from both sites have been shown to have low concentrations of PCBs and TCDD at the time (2000) of study.[92] A lack of Vtg induction was also demonstrated by Irwin and colleagues[90] in the painted turtle *Chrysemys picta* collected from a farm (contaminated with steroids similar to those found in sewage treatment plants). In addition, they found that they could not induce Vtg levels in males after laboratory water-borne exposures of E_2 for 28 days. However, in other studies, male painted turtles have been shown to respond to E_2 injections by increasing plasma Vtg levels.[29,30] It is interesting to note that in the field-collected specimens Irwin and colleagues[90] found an induction in levels in female specimens, which the authors speculated could ultimately change their reproductive fitness. The use of Vtg induction in females is limited because of their seasonal variation in response to reproductive changes, hence the focus on the induction of this protein in juvenile and male specimens. The lack of induction observed even by high doses of E_2[90] warrants further studies into the utility of this endpoint as a biomarker of exposure to EDCs. Observations of negative responses in field-collected specimens should be complemented with laboratory-based induction studies in the same species.

5. *Esterases*

The characteristic response of esterases to organophosphate (OPs) insecticides has led to their widespread use as biomarkers of exposure. Esterases consist of a large group of enzymes (e.g., "A" esterases that hydrolyze OPs and "B" esterases that are inhibited by OPs). The inhibition of brain "B" esterases (i.e., AChE) has been used in many biomarker studies but is limited in application because destructive sampling is required. Brain AChE was shown in laboratory studies to be inhibited by the OPs parathion and trichlorphon in the lizard (*Gallotia galloti*[93,94]). A field study by McLean and colleagues[95] failed to detect a significant inhibition of AChE after the application of OPs. McLean and colleagues[95] collected anoles (*Anolis c. coelestinus*) in areas that had received (24–36 h prior) an aerial dosing of malathion. Lizards collected in the treated area had levels of brain AChE activity similar to levels measured in specimens collected before application, leading the authors to speculate that the lush forest had intercepted the spray and prevented direct exposure of the lizards. Brain AChE inhibition after parathion exposure has also been measured in turtles. Yawetz and colleagues[45] studied various responses of PCB-treated or untreated male Caspian terrapins (*Mauremys caspica rivulata*) to dietary parathion exposure. A high LD50 for parathion in these specimens was observed compared with other vertebrate species. The authors speculated that this was a result of the relatively low rate of AChE inhibition observed in this species.

As stated earlier, it is advantageous to use biomarkers that are nondestructive in nature to monitor the exposure and effects of contaminants on reptile species. Type "A" esterases in serum (plasma) are lower in activity in reptile species compared with mammals.[96] Therefore, the utility of using serum "B" esterases (i.e., butyrylcholinesterase, or BChE) has been investigated by a variety of workers.[93,94,97] For example, Sanchez and colleagues[94] carried out a field study to validate the use of serum B esterases as a biomarker of exposure to organophosphorus insecticides in the lizard (*Gallotia galloti*). Serum butyrylcholinesterase (BChE) in addition to carboxylesterase (CbE) activities were measured in 213 lizards 23 days after an aerial spraying event (parathion). In addition, a control group of 39 lizards were sampled before the spraying. A significant inhibition (>40% BChE and >50% CbE) in activities was observed in the lizards in response to exposure to the organophosphorus insecticide. It is interesting that the authors noted no relationships between activity levels and sex or other biometric parameters. However, in a field study by Fossi and colleagues,[93] blood esterase activities did not show any changes 24 and 48 h after an area was sprayed with trichlorphon, demonstrating the relative "nontoxicity" of trichlorphon for lizards at the average concentrations used in agriculture.

Laboratory studies by the same researchers have aimed to characterize this biomarker response (i.e., response to dose, time to effect and recovery [persistence], and correlations with brain and liver endpoints[97,98]). Single or repeated oral doses of parathion were given to the lizards. Brain, serum, and liver microsomal esterase activities (plus liver monooxygenase activities; see Section B.1) were measured, and the authors concluded that there was a good correlation (inhibitions) with the well-known brain AChE (destructive) biomarker and that of the serum (nondestructive) "B" esterases (albeit nonlinear), which was similar to previous studies using the OP trichlorophon where levels of BChE in serum exhibited the highest percentage inhibition compared with brain AChE, liver microsomal CbE, and EROD measures.[93] Serum esterase activities recovered extremely slowly, particularly after consecutive treatments. However, this slow recovery (compared with other vertebrate species), coupled with the high sensitivity of this lizard to OPs, led the authors to conclude that this measurement was ideal for monitoring OP exposure to nontarget organisms in the field. This biomarker approach was deemed especially useful for evaluating exposure, considering that rapidly degradable compounds, such as OPs, are not easily detectable by chemical analyses.[93]

6. *Additional Plasma and Serum Endpoints*

A variety of additional serum or blood plasma endpoints have been used as very general biomarkers (indicators) of overall health of the organism. For example, Overmann and Krajicek[99] collected blood samples from snapping turtles and measured the activity of aminolevulinic acid dehydratase (ALAD), in addition to other endpoints, such as hematocrit, hemoglobin, plasma glucose, osmolarity, and chlorine ion content. The majority of these measures showed no effect relating to contamination at the capture location. However, they did find that d-ALAD was dramatically lower in turtles collected in areas of lead contamination in comparison with those taken from the reference sites. This enzyme activity in the blood of turtles from the medium and most contaminated sites was reduced about 57% to 60% and 93% to 94%, respectively, when compared with turtles collected from the least contaminated site. Lead exposure in other organisms has also demonstrated decreased ALAD levels. Indeed blood ALAD activity has been considered to be the most sensitive biological indicator of lead exposure, although this gives no indication of effect.[100]

The data by Overmann and Krajicek[99] contrast with a previous study by Albers and colleagues[101] who found no significant differences in blood ALAD activity levels, albumin, glucose, and other blood endpoints in snapping turtles collected from brackish water sites contaminated with PCBs and trace metals. In fact, the authors found that there were elevated levels of protein, albumin, and plasma glucose in turtles from uncontaminated sites compared with contaminated sites. The authors said this finding represented a function of age rather than contaminant exposure. For example, rapidly growing animals need high protein requirements, and growth hormones have been related to an elevated blood glucose level in reptiles.[102] Furthermore, elevated plasma glucose levels may also have been caused by handling stress.[101] These results demonstrate one of the requirements for the successful field deployment of biomarkers — that they are rigorously characterized so that factors such as seasonality, sex, and age do not confound results.

Lovelette and Wright[103] demonstrated a long-term effect of Pb on reducing ALAD activity in pond slider turtles (*Trachemys scripta*) compared with controls (to 73% decline). In addition, a small decline (10%) over time in hemoglobin levels was observed. Rozhaja and colleagues[73,74] showed a pronounced decrease in total serum proteins and lipid content and an increase in total cholesterol, number of reticulocytes, and serum glutamic oxaloacetic transaminase (SGOT), whereas no changes in total red blood cell numbers were observed in controls. Glutamic oxaloacetic transaminase (GOT) is released as a result of destruction of cells and cell components. However, data were not supported by tissue–metal concentrations. An additional study that also measured ALAD activity in hatchling snapping turtles collected from a highly industrialized urban site did not give any indication of Pb poisoning, although tissue residue levels were not provided.[104]

Phillips and Wells[105] examined the effect of DDT on the adenosine triphosphatase (ATPase) activity of five species of turtles after *in vitro* exposure. Of the species studied, the snapping turtle and eastern spiny softshell turtle (*Apalone spinifera spinifera*) showed the greatest inhibition of ATPase activity. In a similar study, map turtles also demonstrated an inhibition of ATPase activity after an *in vitro* exposure to aldrin and dieldrin.[106] It was suggested that this inhibition of ATPase by DDT, aldrin, and dieldrin in some species may be sufficient to reduce organ functions if uptake from the environment by turtles were comparable with levels used in the two studies.[105,106]

C. *Cellular Biomarkers*

The majority of cellular level biomarkers that have been used in reptilian toxicology studies are focused on responses of various cells of the immune (e.g., white blood cell counts) and reproductive systems (e.g., sperm, follicle, germ cell numbers). These studies are reviewed extensively in Chapters 6 and 8, and will not be discussed in this section. In the aquatic toxicology literature cellular endpoints, such as the extent of DNA damage, have often been used in routine biomonitoring programs.[2,4] In contrast with other vertebrate species (e.g., fish), assessment of the exposure to environmental genotoxicants (chemicals, radiation, etc.) in reptiles is limited and is discussed at length in Chapter 9. Many techniques are available to assess the extent of DNA damage in cells and include analyses of DNA adducts (chemicals attached to DNA), DNA strand breakage (single- or double-stranded breaks; see Figure 4.4), apoptosis, induction of micronuclei, chromosome aberrations, and alterations in DNA bases. These have been used as biomarkers of intermediate severity, which may precede (if not effectively repaired) more serious endpoints, such as carcinogenesis and the incidence of neoplasia and heritable mutations. As highlighted in Chapter 9, genotoxicity studies in reptiles have mainly focused on exposure to radionucleotides.[107] Techniques

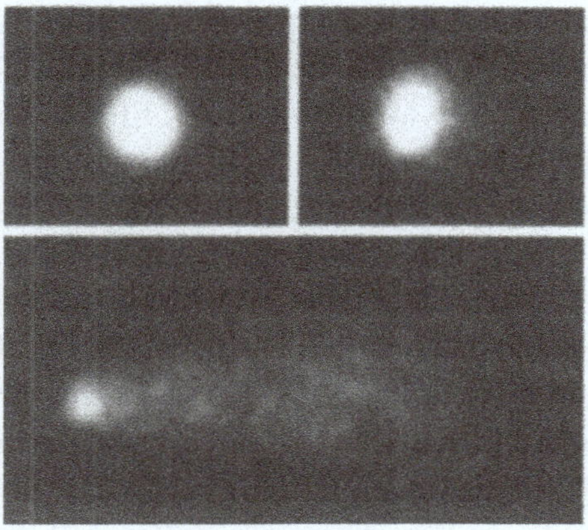

Figure 4.4 Assessment of DNA damage induced in diamondback terrapin (*Malaclemys terrapin*) red bloods cells by hydrogen peroxide using the single-celled gel electrophoresis or COMET assay. (A) control (buffer alone); (B) 25μM H_2O_2, and (C) 150μM H_2O_2. (Courtesy of Mitchelmore and Spinicchia, unpublished.)

commonly used in reptilian studies include flow cytometry, micronucleus assays, and quantitation of DNA strand breaks.

Quantification of DNA strand breaks by the single-celled gel electrophoresis assay (or COMET assay) has been described as a useful nonspecific general biomarker of exposure to and effects of genotoxic compounds.[108] This assay has been extensively used in various aquatic vertebrate and invertebrate species and can be applied to any nucleated cell type.[108,109] This assay is particularly useful for assessing genotoxic exposure in organisms that have nucleated red blood cells. Therefore, this nondestructive, minimally invasive assay can be used to assess DNA damage using less than 5 μl of blood. We are unaware of any studies using this assay to assess genotoxic insult in any reptile species. The response of diamondback terrapin red blood cells (*Malaclemys terrapin*) to *in vitro* exposures to a known direct (not requiring metabolism) damaging agent (hydrogen peroxide) is depicted in Figure 4.4 (Mitchelmore and Spinicchia, unpublished data). Further studies are under way to assess the utility of this biomarker to detect exposure and effects of genotoxic chemicals in reptile species by using nondestructive collections of red blood cells. Potentially confounding factors of seasonality, sex, age, and other environmental variables (e.g., oxygen saturation, temperature, etc.) will need to be addressed to fully characterize this potential biomarker.

D. Tissue and Organ System Biomarkers

Histologic examinations have often been used to detect morphologic changes, reproductive anomalies, and disease (pathologic) states (see reference 110 for a review). Histologic biomarkers, as opposed to many of the biochemical biomarkers that have previously been discussed (see Section B), are higher-level responses after chemical and cellular interactions, which are generally indicative of irreversible damage. Although not used as early-warning biomarkers, often these endpoints are used to overcome sex, seasonal (natural variations), and handling-stress alterations that can complicate many biochemical biomarkers (see Section B). In addition, these endpoints have been used to demonstrate relationships between early-warning exposure and effects biomarkers with the occurrence of the final disease state (e.g., FACs, DNA adducts, and neoplasia correlations in hepatic samples from English sole from Puget sound[111]). These types of studies help characterize the early-warning biomarkers and provide mechanistic links of these biomarkers with later diseased states. Although exposure to toxins may result in histologic and pathologic changes, many diseases can also be brought about by infectious agents and physical and nutritional factors. However, it is also possible that these events occur as a secondary effect of contaminant exposure (i.e., immunosuppression by chemical toxins leading to a reduced resistance to infective agents). It is not within the scope of this chapter to describe these disease states (for a review see reference 110). To date, contaminant-linked histologic changes in reptile species have focused mainly on the reproductive and endocrine glands or on those glands important for normal immune responses (e.g., thymus and spleen). Reproductive organ alterations in response to contaminants are discussed at great length elsewhere (see Chapter 6) and in previous reviews (e.g., see reference 79). Our aim is not to repeat these reviews but to provide a general overview on some of the types of histologic alterations (endpoints) that have been investigated in studies of contaminant effects in reptiles.

1. *Gross Indices*

Gross indices have often been used to indicate contaminant effects. Morphologic parameters that are often examined in field studies include the liver somatic index (LSI), which is the ratio between the weight of the liver and the total body weight of the reptile — that is, 100 × liver weight (g)/body weight (g). This is used to identify possible liver diseases. The condition factor (CF; body weight/length^3) is used to assess the organisms general condition. The gonadal somatic index (GSI; ratio between the weight of the gonad and the total body weight) is used to assess the reproductive health of the individual. The condition of the liver, gonads, and whole body may provide general nonspecific information on potential pollution effects. Although general and somewhat insensitive as a result of being influenced by natural factors (e.g., season, disease, nutritional level), these condition indices may serve as initial screening biomarkers to indicate exposure and effects or to provide information on energy reserves.[2] Overmann and Krajicek[99] examined common snapping turtles collected from sites differing in levels of lead (Pb). Despite high tissue residue levels in turtles collected from severely contaminated sites, no changes in gross condition (CF, LSI) were observed between animals collected from less contaminated sites. Similarly, Hopkins and colleagues[112] did not find differences in CF or GSI among juvenile banded water snakes (*Nerodia fasciata*) despite their accumulation of high concentrations of pollutants as a result of being fed for 2 years with coal-ash–exposed prey items.

Further data are required, particularly with respect to exposures to organic contaminants, to assess the utility of gross indices to serve as biomarkers of overall health in reptile species. It is possible that these endpoints may not have the sensitivity to act as useful indicators of contaminant impact.

2. *Histologic Endpoints*

Altered immune functioning (immunomodulation) as a result of exposure to environmental contaminants has been demonstrated in many wildlife and aquatic species and linked to increased incidences of disease states. Indeed, contaminant-driven immunosuppression has been postulated as a possible factor responsible for the significant declines in many wildlife populations, including reptiles. Among the most important immune organs of vertebrates are the thymus and the spleen. Although the immune system of reptiles has not been extensively studied, it has been shown that these organs are similar in their structure and function to that of other vertebrate species (see Chapter 8 and reference 113). Histologic analyses have been used as biomarkers because morphologic differences in spleen and thymus have been shown to be closely related to functional immune differences.[114] Hormonal regulation of the immune system has been demonstrated in reptilian thymic and splenic tissues (see reference 113 for examples). For example, exogenous doses of T contributed toward the involution of the thymus and depletion of the white pulp from the spleen of the lizard *Chalcides ocellatus*.[115] A study by Rooney and colleagues[113] aimed to determine if juvenile American alligators exposed in the field to an array of contaminants would exhibit morphologic alterations in thymic or splenic tissues. In comparison with two reference sites, significantly smaller thymic ratios (medula:cortex) were observed in alligators from the pesticide-contaminated site (Lake Apopka). This enlarged thymic cortex suggested a change in T-lymphocyte maturation, although the authors noted that further studies were required to examine the functional significance of these observations. Involution of the thymus and depletion of splenic white pulp often occurs seasonally (e.g., in turtles[114]). Observations of seasonal changes to these organs must be critically assessed and characterized and may limit the use of these endpoints as biomarkers (and those of general immune status) of contaminant stress.

Many studies have demonstrated altered gonad morphology and contaminant exposure links (see Chapter 6). For example, Guillette and colleagues[80] demonstrated a variety of gonadal alterations in juvenile alligators collected from the pesticide-contaminated Lake Apopka site, such as males with poorly organized testes, females with ovaries containing polyovular follicles and many multinucleated oocytes, and evidences of clitero-phallic abnormalities (e.g., reduced penis size in males). These studies have been discussed at length elsewhere in this book (see Chapter 6) and in previous reports (e.g., see reference 79).

Some emerging persistent contaminants (e.g., PBDEs) have been shown to affect the thyroid system, and studies have been directed to assess perturbations to this system. For example, Hewitt and colleagues[87] found differences in thyroid morphology in animals collected from a contaminated site, including increased epithelial cell widths and significantly less colloid present in their follicles compared with a reference site.

Lind and colleagues[116] investigated the effects of EDCs on bone tissue in female juvenile American alligators (*Alligator mississippiensis*). Bone tissue homeostasis is controlled by estrogens (i.e., decreasing estrogenic levels are associated with age-related bone loss) and therefore may be a target for disruption by EDCs. Indeed, significant differences in bone composition were observed between the contaminated site (Lake Apopka) and reference site (Lake Woodruff) alligators. These alterations included an increase in total bone mineral density (BMD) in alligators collected from Lake Apopka; however, the effects on juvenile males or adults were not determined.[116] This study demonstrated that effects of EDCs can affect many other tissues that are controlled by hormones, although the significance of these alterations on the overall health of the affected individuals and utility as biomarkers of contaminant impact has not yet been determined.

The appearance of a variety of neoplasms and lesions has been demonstrated in reptile species (for reviews see references 110, 117, and 118); however, the exact etiologic agents responsible for these alterations have not been well studied. Contaminants may cause these directly (i.e., genotoxic mechanisms) or indirectly as a result of immune deficiency (immunosuppression). For example, the increase in shell lesions in many U.S. turtles[119] and the occurrence of a variety of bacterial and fungal infections in the upper respiratory tract and inner ear were demonstrated in eastern box turtles (*Terrapene carolina*) collected from a contaminated site (Long Island, NY[120]). Many cases of neoplastic disease have been demonstrated in reptiles[121–123]; however, we are not aware of any studies showing contaminants as the etiologic agents. Exposure to environmental contaminants acting via impairment of the immune system (hence increasing susceptibility to infectious agents) has been considered as a possible cause of Green turtle (*Chelonia mydas*) fibropapillomas (GTFP; see reference 124 and Chapter 9), although this relationship has not been proved.

E. Developmental Indicators

In ovo exposure of embryos to contaminants can bring about disruptions in developmental processes that ultimately may affect traits of the individual later in life. Here we provide only a brief introduction to examinations of developmental traits for determining contaminant exposure and effect; this topic is thoroughly covered in Chapters 3 and 6.

There are primarily two sources of contaminants to a developing reptile embryo: maternal transfer or accumulation directly from the environment. Maternal transfer of contaminants can occur when lipophilic compounds found in storage sites or in the diet of the female are incorporated into Vtg, the lipoprotein precursor of yolk, and ultimately are partitioned into the yolk. An example is the transfer of PCBs to eggs of turtles; in some highly contaminated systems such as the upper Hudson River, New York, maternal transfer can result in yolks with [PCB_{Total}] in the order of 29 ppm ww (Table 4.3). Other contaminants, such as some trace elements, can mimic and replace specific elements of macromolecules. These contaminants can be incorporated into the egg during oogenesis when, for example, the female distributes proteins into the developing oocyte. Examples of this pathway of transfer are the maternal contribution of selenium to eggs in the slider turtle (*Trachemys scripta*[125]) and in the American alligator (*Alligator mississippiensis*[126]). Because of its physico-chemical properties, selenium can act as a sulfur mimic, replacing sulfur in the

Table 4.3 Average Concentrations of PCBs in Several Tissues of Snapping Turtles in the Upper Hudson River and Sites Outside the Hudson River (Data are from Stone et al.[172] Only eggs derived from females in the Hudson River were analyzed.)

Tissue	Site	N	Average $[PCB]_{Total}$ (ppm ww)
Fat	Hudson River	11	2990.6
Fat	Elsewhere	9	464.2
Liver	Hudson River	23	66.1
Liver	Elsewhere	8	7.8
Skeletal muscle	Hudson River	23	4.2
Skeletal muscle	Elsewhere	6	0.4
Eggs	Hudson River	6	28.9

constitution of sulfur-rich proteins. Protein contributions to the oocyte by the female can thus contribute selenium to the developing offspring in place of the sulfur portions of some proteins.

In some situations, contaminants may also enter the eggs of oviparous reptiles after the eggs have been deposited in the environment by the female. Most reptile eggs will gain a considerable mass in water during development, with this water being derived from the nest substrate. Therefore, hydrophilic contaminants present in the nest substrates could passively enter the eggs in solution. However, despite the somewhat porous nature of many reptilian eggs, the overall importance of embryonic water exchange with the nest substrate for delivering significant doses of xenobiotics is questionable. For example, Nagle and colleagues[125] incubated slider (*T. scripta*) eggs for the entire embryonic period in nests constructed of coal ash, a substrate that is extremely rich in water-soluble trace elements.[12] Subsequent analyses of hatchlings for contaminant body burdens did not reveal elevation of any trace elements in hatchlings derived from the ash nests compared with reference hatchings; only through maternal contribution did embryos receive contaminants.[125]

Regardless of the source of the contaminants, several developmental traits can be monitored to determine contaminant effects. The most obvious effect of developmental changes after contaminant exposure is the presence of gross abnormalities in embryos and hatchlings. For example, Bishop and colleagues[54,128] found correlations between [PCB], [PCDD], and [PCDF] and developmental deformities of the limbs, skull, and carapace of hatchling snapping turtles collected from several sites in Ontario, Canada. In some of these same sites, de Solla and colleagues[81] found that a sexually dimorphic trait (precloacal length) was skewed toward feminine morphology in adult male snapping turtles collected from the most contaminated site, suggesting an endocrine-mediated mode of action.

Because of environmental sex determination in crocodilians and many turtles, endocrine disrupting compounds have been hypothesized to have the potential to counteract environmental influences on sex determination.[129,130] In studies with red ear slider turtles (*Trachemys scripta elegans*), mixtures of organic contaminants or their metabolites applied topically to eggs were found to result in morphologically female embryos and hatchlings at temperatures normally inducing development of males.[131,132] However, dosing of snapping turtle (*C. serpentina*) and green sea turtle (*Chelonia mydas*) eggs with DDE [1,1′-(dichloroethenylidiene) bis 4-chlorobenzene] at environmentally relevant concentrations failed to elicit effects on sex determination.[133,134]

Development of somatic and sexual structures in crocodilians has also been shown to be influenced by environmental contaminants. Milnes and colleagues[135] reported that hatchling American alligators from Lake Apopka, Florida, which is contaminated with DDE and numerous other agricultural chemicals, displayed distinct morphologies compared with hatchlings from less contaminated sites. In the same sites, secondary sexual development was altered in Lake Apopka males; reductions in phallus size in juvenile and adult alligators in Lake Apopka compared with reference lakes were reported by Guillette and colleagues[136,137] and Pickford and colleagues.[138] These differences in phallus size among resident populations of alligators generally reflected decreased circulating androgens in animals from the contaminated site, suggesting that the activity of an antiandrogenic xenobiotic compound may be the source of the differences.[126–138] Guillette and colleagues[137] suggested that morphologic and hormonal abnormalities observed in the highly contaminated Lake Apopka population may extend into other populations with a less severe history of contamination. This observation by Guillette and colleagues[137] suggests that circulating androgen concentrations and phallus size may provide useful indicators of contaminant effects even in systems that are not characterized by extreme chemical contamination. The study by Pickford and colleagues[138] found that female as well as male alligators in Lake Apopka may be affected by contaminants. Whereas male phallus size showed the same trend as previous studies,[136,137] juvenile females in Lake Apopka experienced increased concentrations of circulating dihydrotestosterone compared with females from reference sites.[138]

In a laboratory study, Stoker and colleagues[139] examined the effects of a single xenoestrogenic compound (bisphenol A) on sexual development of caiman (*Caiman latirostris*). Embryonic exposure to a high concentration of bisphenol A (9 mg of 140 ppm solution added topically per egg) resulted in sex reversal of males; animals hatched at male-producing temperature formed ovaries, but the ovaries were less well organized than those of normal females.[139] When a lower dose of bisphenol A was applied (90 ng/egg of 1.4 ppm solution), evidence of sex reversal was not found, although males displayed abnormalities of seminiferous tubules.[139] As mentioned earlier, topical application of nonylphenol to terrapin eggs resulted in complete sex reversion at male-producing temperatures.[61]

Aside from morphologic effects of contaminants on developing reptiles, neurologic development may also be impaired by some compounds. Having been studied primarily in birds and mammals, compounds modifying neurologic development have primarily been associated with changes in learning and behavior. Behavioral responses of reptiles to contaminants will be addressed in the following section.

F. *Individual Responses*

1. *Bioenergetics: Metabolic Pathways and Regulation of Growth and Reproduction*

Regardless of the taxa in question, there is a fundamental priority in allocation of energy assimilated from the environment; only the assimilated energy that exceeds that dedicated to costs of maintenance can be allocated to production pathways (growth, reproduction). Thus, growth and reproductive potential are ultimately governed by the investments in support of basic physiologic processes and activities necessary for short-term survival, which may account for up to 80% of the annual energy budget for some reptiles.[140] This relationship between maintenance metabolism and production forms the basis for the use of standard metabolic rate (SMR; a measure of maintenance costs estimated empirically by respiration rate) as an indicator of significant contaminant effects on individuals.[141–144]

As with any physiologic process, "normal" SMR is a range of values that have provided optimal fitness within the historical environmental and physiologic conditions experienced by the species in question. Outside this range of values, fitness is suboptimal as a result of energetic constraints presented to the individual. In some situations, chronic sublethal exposure to contaminants has been shown to result in elevated SMR in reptiles[145] and other taxa.[146,147] Given no compensatory increase in feeding or assimilation, individuals having SMRs elevated above normal would be expected to experience fitness costs associated with reduced growth or reproductive potential as a result of decreased energetic contributions to the production budget.[143,148] For example, a subtle inverse relationship between SMR and growth rate has been observed in hatchling snapping turtles, which naturally varied in SMR,[149] as well as other taxa having experienced elevated SMR after chronic exposure to contaminants.[147,148] Note that growth or reproductive responses to elevation in SMR may be expected even if the elevations in SMR are not sustained over long periods. In many natural situations resources are limiting, and thus individuals are unable to compensate for elevated SMR via increasing consumption and assimilation of energy. As a result, even temporary elevations in SMR would be predicted to translate into reduced production, cumulatively reducing the slope of the growth trajectory.[147]

There have also been observations of reduced SMR after contaminant exposure. Hatchling slider turtles (*Trachemys scripta*) having received selenium via maternal transfer had significantly decreased SMR compared with controls.[125] Although growth was not measured in the study,[125] one would predict that growth would be decreased in this situation as well because a reduction in SMR below optimal values translates into an overall slowing of physiologic processes, including assimilation of energy.[148] Thus, SMR, whether increased or decreased significantly from the normal range, can be a useful indicator for potential effects of contaminants on fitness traits of individuals, which ultimately translate into population-level features such as demographics or population genetics.[150,151]

2. *Growth*

Growth is an integrated response to numerous physiologic processes, which influence the production budget of the individual. Thus, growth is a cumulative process integrating those processes, which have positive and negative effects on the production budget. Growth rates are therefore often used as an index of overall individual health because an individual that has positive growth rates displays an ability to satisfy underlying survival (maintenance) costs while also allocating energy to the production of new tissues.

Most reptiles display indeterminate growth, accruing new biomass throughout their lifetimes. However, in most reptiles growth rates decline after the rapidly growing juvenile period, and whereas the growth rate remains positive, it can become quite low in old individuals. Thus, the growth curve approaches an asymptote, although the response trajectory does not quite reach a horizontal path (e.g., does not reach a fixed size). Because growth rate depends on the initial size of the individual during the interval of interest, especially during the rapidly growing phase, environmental factors that lead to a period of reduced growth can be particularly important during the juvenile period. It is during the rapidly growing juvenile period that the individual is growing toward a threshold beyond which further rapid growth is less necessary for survival — for example, when a size is attained at which gape limited predators can be avoided. Therefore, growth reductions before reaching the inflection of the growth trajectory will result in a longer period of risk of size-dependent mortality.

As discussed previously, growth reductions can result from effects of contaminants on underlying bioenergetic processes such as energy assimilation and metabolic expenditures.[146–148] Although these processes can be directly modified by contaminants, they may also be modified indirectly. For example, contaminants with behavioral effects on foraging activities may result in reduced feeding and thus assimilation. Morphologic abnormalities resulting from contaminant exposure could have impacts on feeding and assimilation, depending on the type and severity of the malformation(s). For example, in a study with a larval amphibian that experienced abnormalities of the oral region after chronic exposure to mixed contaminants, abnormal individuals displayed significant reductions in grazing ability and concomitantly negative growth rates.[152,153] Thus, reductions in growth rates can serve as a biomarker of stressful environmental conditions, although the mechanisms responsible for the reductions may not be immediately evident.

3. *Reproduction*

Reproductive success is the most basic index of an individual's fitness and potential to contribute to the population as a whole. Thus, significant, negative effects of contaminants on fecundity, age at reproductive maturity, clutch size, or offspring performance would be expected to have fitness consequences for the affected individuals. Moreover, if reproductive performance is hindered in a large proportion of individuals in a given population, modified population dynamics could result.

There have been no reports of contaminants having direct, reproductively toxic effects on reptiles at concentrations representative of those in natural systems. Numerous studies, primarily with crocodilians and turtles, have demonstrated impairment of traits related to reproduction, such as contaminant-induced anomalies of sex structures and endocrine disruption or sex determination. However, there have not yet been definitive studies demonstrating that reproductive success or population dynamics can be regulated by contaminant-induced changes in these suborganismal responses. One might expect effects such as modified circulating hormone concentrations[81,86,130,138,154,155] and sex reversals[129,131,132,139] to influence reproductive success and population traits. However, making the link connecting subindividual responses, lifetime reproductive fitness, and population change is an arduous exercise; without an understanding of population demographics, the incidence and frequency of anomalies in the population, and the ultimate reproductive effects of the anomalies, reproductively mediated population-level effects of suborganismal traits cannot be assumed.

There are several reproductive traits that theoretically could be, or that have been shown in reptiles or other taxa to be, modified by contaminants in such a way that fitness of the individual could be compromised. Transfer of contaminants from the female to her offspring has been well studied in reptiles (see Chapters 3 and 6). Although there is yet no strong evidence to indicate that maternal transfer of contaminants in wild populations of reptiles significantly influences offspring performance, in other taxa (especially birds) effects of maternally derived contaminants include decreased hatching success and reproductive success of the second generation.[156,157]

Effects of contaminants on time to reproduction or clutch size can also modify the lifetime reproductive fitness of an individual. Such effects would be expected to be most drastic for shorter-lived reptiles producing relatively small clutches (many lizards and snakes) compared with longer-lived species with high annual reproductive output (crocodilians, many turtles). For long-lived species that may reproduce iteroparously over decades, a delay in reproduction by one or a few seasons may essentially be ameliorated over the very long ensuing reproductive period. However, a species that survives and reproduces over a period of less than a decade could lose more than 10% of its lifetime reproductive capacity from each season in which reproduction is delayed.

A basic life history feature of all animals is a species-specific relationship between the number of offspring produced per breeding event and the per capita parental investment in the offspring.[158] Natural selection has acted to define the relationship between offspring number and offspring size such that it is optimized for the environmental conditions historically experienced by the species. Therefore, a divergence from the normal clutch size and energetic investment per offspring can translate to a reproductive strategy that does not incur the optimum for fitness as defined by evolutionary history. By examining clutch size and offspring quality (energy content, size), potential environmental constraints on reproduction may be addressed. In other words, given suboptimal conditions for production, whether resulting from anthropogenic activities or natural phenomena, a female is faced with the physiologic options of reducing the number of offspring produced or reducing the per capita energetic investment in the normal number of offspring (e.g., a few large offspring or many small offspring). Although the female's "choice" will reflect that which was most successful under suboptimal conditions in the evolutionary past, it remains a suboptimal strategy compared with what it could be under less stressful conditions. Thus, offspring number and size comparisons among sites or treatments can provide an index of stress experienced by adult females before reproduction. However, care must be taken in attempting to identify factors such as contaminants that may be responsible for changes in reproductive strategy. Because these strategies are ultimately driven by energy assimilation and allocation by the female, any host of environmental factors could play a role in shaping them.

4. *Behavior*

As with growth, behavior is an integrative process, reflecting numerous physiologic processes. Thus, behavioral changes after contaminant exposure may reflect changes in neurologic, endocrinologic, and other processes operating at the suborganismal level. If the behaviors that are modified by contaminants are those known to have strong influences on survival or reproduction, such endpoints may have ecologic significance.

Reptiles display numerous behaviors that may be used as indicators of significant contaminant effects. Most reptiles display basking behaviors, courtship displays, foraging strategies, and predator escape and avoidance behaviors. In controlled situations, these behaviors can be monitored for indications of changes as a result of contaminant exposure, especially when the contaminants in question are thought to exert toxicity via neurologic pathways.

Bain and colleagues[159] examined thermoregulation and prey capture behaviors in Australian central bearded dragons (*Pogona vitticeps*) after ingestion of fenitrothion, a cholinesterase inhibitor. Thermal preference did not differ between controls and treated animals. However, for the group receiving the highest dose of fenitrothion, there was a slight (but statistically insignificant) increase in the number of strikes required to capture prey when tested 24 h postdosing.

In another study using lizards, Peveling and Demba[160] tested the effects of oral dosing of fringe-toed lizards (*Acanthodactylus dumerili*) with the mycoinsecticide *Metarhizium anisoplia* var. *acridum* or the phenylpyrazole insecticide Fipronil. Feeding activities were reduced by both insecticides, and Fipronil additionally brought about reduced food consumption. The effects of the insecticides on feeding behaviors translated into reductions in relative liver mass (*M. anisoplia*) and body weight (Fipronil).

In turtles, a behavior that can have direct relevance to survival is the individual's righting response, the time required to return to a prone position after being placed on its carapace.[161] Especially in species such as snapping turtles (*Chelydra serpentina*), which have a greatly reduced plastron size, the speed at which an individual can right itself after being overturned by a predator can determine the likelihood of surviving the attack. An example of applying this behavior is provided by Burger and colleagues[162] who examined righting response in hatchling slider turtles (*T. scripta*) several months after having received injections of lead solutions. Righting response time was correlated with lead dose, indicating that lead may have indirect effects on survival via behavioral modifications.

Hopkins and colleagues[163] compared swimming behaviors in neonates of two species of water snakes (black swamp snakes, *Seminatrix pygaea* and diamondback water snakes, *Nerodia rhombifer*) after topical application of Carbaryl, a cholinesterase-inhibiting insecticide. The highest exposure concentration brought about reduced swimming speeds in both species. *Seminatrix pygaea* was more sensitive than *N. rhombifer*, suffering the greatest reduction in swim velocity and incurring the longest recovery times postdosing. The authors suggest that the permeability of *S. pygaea*'s integument, which exceeds that of most water snakes, allowed greater concentrations of the insecticide to cross the integument, compared with *N. rhombifer*.

III. Summary and Conclusions

Reptiles provide excellent models for examining contaminated systems. Although reptilian physiology and ecology have received considerable study, many underlying toxicologic mechanisms and responses remain to be established and fully characterized. As a result, many of the tools required for mechanistic examination of toxicologic responses currently must be extrapolated from studies in other taxa. A current reliance on extrapolation is particularly demonstrated by the lack of reptile-specific information on molecular and cellular toxicologic mechanisms. As evidenced by the content presented here, hypotheses regarding toxicologic mechanisms at the molecular and cellular levels must, in many instances, be based on models developed for fish toxicology. A prime example of the need for verification of results based on other animals is the dearth of information regarding genes that code for commonly used cellular and subcellular biomarkers in reptiles. Issues such as the conflicting evidence that exists regarding induction and mechanisms of, for example, phase II enzymes in fish also hinder extrapolation from fish models to reptile models. Although we might expect that many suborganismal traits may be highly conserved across taxa, there remains a requirement for empirical work focused specifically on reptilian physiology.

Indicators of pollutant effects on reptiles may be found at many levels of biological organization, from molecules to physiologic and ecologic traits of individuals. Work at these higher levels of organization often encompasses metabolic, behavioral, and reproductive effects. Because processes operating at the individual level are integrative of additive, synergistic, and competing processes occurring at the subindividual level, it is extremely difficult to identify those specific cellular and subcellular processes that may be most biologically labile (e.g., effective in transcending levels of organization from subindividual to individual traits). Certainly this difficulty is not limited to reptiles — it exists in practically all toxicologic and physiologic models. Yet, if we are to interpret subindividual biomarkers as indicators of contaminant effects rather than signals of exposure, research is required in which induction of specific subindividual responses can be correlated with overall effects on overall physiologic health.

Literature Cited

1. Stegeman, J.J. and Hahn, M.E., Biochemistry and molecular biology of monooxygenase: current perspective on forms, functions, and regulation of cytochrome P450 in aquatic species, in *Aquatic Toxicology; Molecular, Biochemical and Cellular Perspectives*, Malins, D.C. and Ostrander, G.K., Eds., Lewis Publishers, CRC Press, Boca Raton, FL, 87, 1994.
2. Van der Oost, R., Beyer, J., and Vermeulen, N.P.E., Fish bioaccumulation and biomarkers in environmental risk assessment: a review, *Environ. Toxicol. Pharmacol.*, 13, 57, 2003.
3. Bucheli, T.D. and Fent, K., Induction of cytochrome P450 as a biomarker for environmental contamination in aquatic ecosystems, *Crit. Rev. Environ. Sci. Technol.*, 25, 201, 1995.
4. Livingstone, D.R., Biotechnology and pollution monitoring—use of molecular biomarkers in the aquatic environment, *J. Chem. Technol. Biotechnol.*, 57, 195, 1993.

5. NRC: Committee on Biological Markers of the National Research Council, Biological markers in environmental health research, *Environ. Health Perspect.*, 74, 3, 1987.
6. Widdows, J. and Donkin, P., Role of physiological energetics in ecotoxicology, *Comp. Biochem. Physiol.*, 100, 69, 1991.
7. McCarthy, J.F., Halbrook, R.S., and Shugart, L.R., *Conceptual strategy for design, implementation, and validation of a biomarker-based biomonitoring capability,* Publ. no. 3072, RNL/TM-11783. Environ. Sciences Div., Oak Ridge National Laboratory, TN, 1991.
8. Van Gestel, C.A.M. and Van Brummelen, T.C., Incorporation of the biomarker concept in ecotoxicology calls for a redefinition of terms, *Ecotoxicology*, 5, 217, 1996.
9. WHO International Programme on Chemical Safety (IPCS), *Biomarkers and risk assessment: concepts and principles. Environmental Health Criteria 155*, World Health Organization, Geneva, 1993.
10. Huggett, R.J. et al., Biomarkers in fish from Prince William Sound and the Gulf of Alaska: 1999–2000, *Environ. Sci. Technol.*, 37, 4043, 2003.
11. Merian, E., Ed., *Metals and their compounds in the environment: occurrence, analysis, and biological relevance*, VCH Verlagsgesellschaft, Weinheim, Germany, 1991, 1438.
12. Niesink, R.J.M, de Vries, J., and Hollinger, M.A., *Toxicology: Principles and Applications*, CRC Press, Boca Raton, FL, 1996, 1284.
13. Ertl, R.P. and Winston, G.W., The microsomal mixed function oxidase system of amphibians and reptiles: components, activities and induction, *Comp. Biochem. Physiol.*, 121, 85, 1998.
14. Yawetz, A., Benedek-Segal, M., and Woodin, B., Cytochrome P4501A immunoassay in freshwater turtles and exposure to PCBs and environmental pollutants, *Environ. Toxicol. Chem.*, 16, 1802, 1997
15. Suter, G.W., II, Use of biomarkers in ecological risk assessment, in *Biomarkers of Environmental Contamination*, McCarthy, J.F. and Shugart, L.R., Eds., Lewis Publishers, Boca Raton, FL, 1990, 419.
16. Hopkins, W.A. et al., Nondestructive indices of trace element exposure in squamate reptiles, *Environ. Pollut.*, 15, 1, 2001.
17. Spotila, J.R., Dunham, A.E., Leslie, A.J., Steyermark, A.C., Plotkin, P.T., Paladino, F.V. 1996. Worldwide population decline of *Dermochelys Coriacea*: Are leatherback turtles going extinct? *Chelonian Conserv. Biol.*, 2, 209–222.
18. Sparling, D.W., Linder, G., and Bishop, C., Eds., *Ecotoxicology of Amphibians and Reptiles*, SETAC Press, Pensacola, FL, 2000.
19. Meyers-Schöne, L. and Walton, B.T., Turtles as monitors of chemical contaminants in the environment, *Rev. Environ. Contam. Toxicol.*, 135, 93, 1994.
20. Meyers-Schöne, L., Ecological risk assessment of reptiles, in *Ecotoxicology of Amphibians and Reptiles*, Sparling, D.W., Linder, G., and Bishop, C., Eds., SETAC Press, Pensacola, FL, 2000, 793.
21. Hirth, H.F. et al., Dispersal of snakes from a hibernaculum in Northwestern Utah, *Ecology*, 50, 332, 1969.
22. Goodwin, T.M. and Marion, R.W., Aspects of the nesting ecology of American alligators (*Alligator mississippiensis*) in north central Florida, *Herpetologica*, 34, 43, 1978.
23. Berry, J.F. and Shine, R., Sexual size dimorphism and sexual selection in turtles (order Testudines), *Oecologia*, 44, 185, 1980.

24. [USEPA] U.S. Environmental Protection Agency, *Wildlife Exposure Factors Handbook*, Office of Research and Development, Washington, DC, EPA/600/R-93/187, 1993.
25. Hall, R.J. and Henry, P.F.P., Assessing effects of pesticides on amphibians and reptiles: status and needs, *Herpetol. J.*, 2, 65, 1992.
26. Custodia-Lora, N., Novillo, A., and Callard, I.A., Effect of gonadal steroids on progesterone receptor, estrogen receptor, and vitellogenin expression in male turtles (*Chrysemys picta*), *J. Exp. Zoolog.*, 301, 15, 2004a.
27. Custodia-Lora, N., Novillo, A., and Callard, I.P., Regulation of hepatic progesterone and estrogen receptors in the female turtle, *Chrysemys picta*: relationship to vitellogenesis, *Gen. Comp. Endocrinol.*, 136, 232, 2004b.
28. Pakdel, F. et al., In vivo estrogen induction of hepatic estrogen-receptor messenger-RNA and correlation with vitellogenin messenger-RNA in Rainbow trout. *Molec. Cell. Endocrinol.*, 75, 3, 205, 1991.
29. Ho, S.M., Danko, O., and Callard, I.P., Effect of exogenous estradiol-17_ on plasma vitellogenin levels in male and female *Chrysemys* and its modulation by testosterone and progesterone, *Gen. Comp. Endocrinol.*, 43, 413, 1981.
30. Palmer, B. and Palmer, S., Vitellogenin induction by xenobiotic estrogens in the Red-eared turtle and African clawed frog, *Environ. Health Perspect.*, 103, 19, 1995.
31. Hawkins, M.B. et al., Identification of a third distinct estrogen receptor and reclassification of estrogen receptors in teleosts. *Proc. Natl. Acad. Sci. USA*, 97, 20, 10751, 2000.
32. Katsu, Y. et al., Molecular cloning of the estrogen and progesterone receptors of the American alligator. *Gen. Comp. Endocrinol.*, 136, 122, 2004.
33. Paech, K. et al., Differential ligand activation of estrogen receptors ER alpha and ER beta at API sites. *Science*, 277, 5331, 1508, 1997.
34. Johnson, J.M. et al., Genome-wide survey of human alternative pre-mRNA splicing with exon junction microarrays. *Science*, 302, 5653, 2141, 2003.
35. Jeyasuria, P., *The Embryonic Brain Gonadal Axis in the Temperature-Dependent Sex Determination of* Malaclemys terrapin, *A Role for P450 aromatase*, Univ. of Maryland Eastern Shore, 1997.
36. Higuchi, R. et al., Kinetic PCR analysis — real-time monitoring of DNA amplification reactions. *Bio-technology*, 11, 9, 1026, 1993.
37. Heid, C.A. et al., Real time quantitative PCR. *Genome Research*, 6, 10, 986, 1996.
38. Jeyasuria, P. and Place, A.R., Temperature-dependent aromatase expression in developing diamondback terrapin (*Malaclemys terrapin*) embryos, *J. Steroid Biochem. Molec. Biol.*, 61, 415, 1997.
39. Liang P., Pardee A.B. *Science*, 257, 967–971, 1992.
40. Herbst, L.H. et al., Differential gene expression associated with tumorigenicity of cultured green turtle fibropapilloma-derived fibroblasts, *Cancer Genet. Cytogenet.*, 129, 35, 2001.
41. Trudeau, V.L. et al., Octylphenol (OP) alters the expression of members of the amyloid protein family in the hypothalamus of the snapping turtle, *Chelydra serpentina serpentine*, *Environ. Health Perspect.*, 110, 269, 2002.
42. Diatchenko, L. et al. *Proc. Natl. Acad. Sci. USA*, 93, 12, 6025–6030, 1996.
43. Schena, M. et al., Microarrays: biotechnology's discovery platform for functional genomics. *Trends Biotech.*, 16, 7, 301, 1998.
44. Hahn, M.E., Aryl hydrocarbon receptors: diversity and evolution, *Chem. Biol. Interact.*, 141, 131, 2002.

45. Yawetz, A., Sidis, I., and Gasith, A., Metabolism of parathion and brain cholinesterase inhibition in Aroclor 1254-treated and untreated Caspian terrapin (*Mauremys caspica rivulata*, Emydidae, Chelonia) in comparison with two species of wild birds, *Comp. Biochem. Physiol.*, 75C, 377, 1983.
46. Andersson, T. and Nilsson, E., Characterization of cytochrome *P*-450-dependent activities in hagfish, dogfish, perch and spectacle caiman, *Comp. Biochem. Physiol.*, 94B, 99, 1989.
47. Haasch, M.L. et al., Cloned rainbow trout liver $P_{(1)}$450 complementary DNR as a potential environmental monitor, *Toxicol. Appl. Pharmacol.*, 98, 362, 1989.
48. Jewell, C.S.E. et al., The hepatic microsomal mixed-function oxygenase (MFO) system of *Alligator mississippiensis*: induction by 3-methylcholanthrene (MC), *Xenobiotica*, 19, 1181–1200, 1989a.
49. Jewell, C.S.E. et al., Induction of the hepatic microsomal mixed-function oxygenase (MFO) system of *Alligator mississippiensis* by 3-methylcholanthrene (3-MC), *Mar. Environ. Res.*, 28, 73–79, 1989b.
50. Walker, C.H. and Ronis, M.J., The monooxygenases of birds, reptiles and amphibians, *Xenobiotica*, 19, 111, 1989.
51. Schwen, R.J. and Mannering, G.J., Hepatic cytochrome *P*-450-dependent monooxygenase systems of the trout, frog and snake — III. Induction, *Comp. Biochem. Physiol.*, 71, 445, 1982.
52. Stafford, D.P., Plapp, F.W., and Fleet, R.R., Snakes as indicators of environmental contamination: relation of detoxifying enzymes and pesticide residues to species occurrence in three aquatic ecosystems, *Arch. Environ. Contam. Toxicol.*, 5, 15, 1976.
53. Yawetz, A. et al., Induction, fractionation, and localization of cytochrome P450 isozymes in the liver of the freshwater turtle, *Chrysemys picta picta*, *Mar. Environ. Res.*, 35, 205, 1993.
54. Bishop, C.A. et al., Environmental contamination and developmental abnormalities in eggs and hatchlings of the common snapping turtle (*Chelydra serpentina serpentina*) from the Great Lakes–St. Lawrence River Basin (1989–1991), *Environ. Pollut.*, 101, 143, 1998.
55. Rie, M.T. et al., Hepatic biotransformation enzymes in a sentinel species, the painted turtle (*Chrysemys picta*), from Cape Cod, Massachusetts: seasonal-, sex- and location related differences, *Biomarkers*, 5, 382, 2000.
56. Gunderson, M.P., Oberdorster, E., and Guillette, L.J., Jr., Phase I and II liver enzyme activities in juvenile alligators (*Alligator mississippiensis*) collected from three sites in Kissimmee–Everglades drainage, Florida (USA), *Comp. Biochem. Physiol.*, 139C, 39, 2004.
57. Desvages, G., Girondot, M., and Pieau, C., Sensitive stages for the effects of temperature on gonadal aromatase activity in embryos of the marine turtle *Dermochelys coriacea*, *Gen. Comp. Endocrin.*, 92, 54, 1993.
58. Desvages, G. and Pieau, C., Aromatase activity in gonads of turtle embryos as a function of the incubation temperature of eggs, *J. Steroid Biochem. Molec. Biol.*, 41, 851, 1992.
59. Jeyasuria, P., Roosenburg, W.M., and Place, A.R., Role of P-450 aromatase in sex determination of the diamondback terrapin, *Malaclemys terrapin*, *J. Exp. Zool.*, 270, 95, 1994.
60. Gabriel, W.N. et al., Alligator aromatase cDNA sequence and its expression in embryos at male and female incubation temperatures, *J. Exp. Zool.*, 290, 439, 2001

61. Place, A.R. et al., Expression of P450(arom) in *Malaclemys terrapin* and *Chelydra serpentina*: a tale of two sites, *J. Exp. Zool.*, 290, 673, 2001.
62. Lance, V.A. et al., Does alligator testis produce estradiol? A comparison of ovarian and testicular aromatase, *Biol. Reprod.*, 69, 1201, 2003.
63. Fossi, M.C. and Marsili, L., The use of non-destructive biomarkers in the study of marine mammals, *Biomarkers*, 2, 205, 1997.
64. George, S.G., Enzymology and molecular biology of phase II xenobiotic-conjugating enzymes in fish, in *Aquatic Toxicology Molecular, Biochemical and Cellular Perspectives*, Malins, D.C. and Ostrander, G.K., Eds., Lewis Publishers, Boca Raton, FL, 1994.
65. Bani, M.H. et al., Modulation of snake hepatic cytochrome P450 by 3-methylcholanthrene and phenobarbital, *Comp. Biochem. Physiol.*, C, 119, 2, 143, 1998.
66. Hermes-Lima, M. and Zenteno-Savin, T., Animal response to drastic changes in oxygen availability and physiological oxidative stress, *Comp. Biochem. Physiol.*, 133, 537, 2002.
67. Hermes-Lima, M. and Storey, K.B., In vitro oxidative inactivation of glutathione *S*-transferase from a freeze tolerant reptile, *Molec. Cell. Biochem.*, 124, 149, 1993a.
68. Hermes-Lima, M. and Storey, K.B., Role of antioxidants in the tolerance of freezing and anoxia by garter snakes, *Am. J. Physiol.*, 265, R646, 1993b.
69. Willmore, W.G. and Storey, K.B., Antioxidant systems and anoxia tolerance in a freshwater turtle, *Trachemys scripta elegans*, *Molec. Cell. Biochem.*, 170, 177, 1997a.
70. Willmore, W.G. and Storey, K.B., Glutathione systems and anoxia tolerance in turtles, *Am. J. Physiol.*, 273, R219, 1997b.
71. Halliwell, B. and Gutteridge, J.M.C., *Free Radicals in Biology and Medicine*, 3rd ed., Oxford University Press, Oxford, UK, 1999.
72. Winston, G.W. and Di Giulio, R.T., Prooxidant and antioxidant mechanisms in aquatic organisms, *Aquat. Toxicol.*, 19, 137, 1991.
73. Rozhaja, D.A. et al., Some biochemical characteristics of the blood sera of the turtle *Testudo hermanni* Gmel from the immediate surrounding of lead and zinc foundry in Zvecan Yugoslavia, *Acta. Biol. Med. Exper.*, 5, 43, 1980.
74. Rozhaja, D.A. et al., The effect of lead metallurgy pollutants on blood reticulocyte number and serum glutamate oxal-acetate transaminase activity of the turtles *Testudo hermanni* Gmel, *Acta. Biol. Med. Exper.*, 8, 25, 1983.
75. Elezaj, I., Halili, F., and Rozhaja, D.A., Changes in blood catalase and peroxidase activities of land turtles (*Testudo hermanni* Gmal) living under conditions of industrial lead contamination. *Acta. Biol. Med. Exper.*, 8, 29, 1983.
76. Beyersmann, D., The significance of interactions in metal essentiality and toxicity, in *Metals and Their Compounds in the Environment. Occurrence, Analysis and Biological Relevance*, Merian, E., Ed., VCH Verlagsgesellschaft, Weinheim, Germany: 491, 1991.
77. Elsey, R.M. and Lance, V., Effect of diet on blood selenium and glutathione peroxidase activity in the alligator, *Comp. Biochem. Physiol.*, 76B, 831, 1983.
78. Jena, B.S., Nayak, S.B., and Patnaik, B.K., Age-related effect of aluminum on the catalase activities of the brains of two species of poikiloremic vertebrates, *Gerontology*, 48, 1, 34–38.
79. Guillette, L.J., Jr., Contaminant-associated endocrine disruption in reptiles, in *Ecotoxicology of Amphibians and Reptiles*, Sparling, D.W., Linder, G, and Bishop, C., Eds., SETAC Press, Pensacola, FL, 2000, 595.

80. Guillette, L.J., Jr. et al., Developmental abnormalities of the gonad and abnormal sex hormone concentrations in juvenile alligators from contaminated and control lakes in Florida, *Environ. Health Perspect.*, 102, 680, 1994.
81. de Solla, S.R. et al., Impact of organochlorine contamination on levels of sex hormones and external morphology of common snapping turtles (*Chelydra serpentina serpentina*) in Ontario, Canada, *Environ. Health Perspect.*, 106, 253, 1998.
82. Gunderson, M.P. et al., Effect of acute stress on plasma B-corticosterone, estradiol-17B and testosterone concentrations in juvenile American alligators collected from three sites within the Kissimmee Everglades drainage basin in Florida (USA), *Comp. Biochem. Physiol.*, 135C, 365, 2002.
83. Shelby, J.A. et al., Seasonal variations in reproductive steroids of male and female yellow-blotched map turtles *Grapteymys flavimaculcta, Gen. Comp. Endocrinol.*, 119, 143, 2000.
84. Rooney, A.A. et al., Seasonal variation in plasma sex steroid concentration in juvenile American alligators, *Gen. Comp. Endocrinol.*, 135, 25, 2004.
85. Gross, D.A. et al., Characterization of potential contaminant-induced clinical manifestation in neonatal alligators from contaminated and control lakes in central Florida. Society of Environmental Toxicology and Chemistry 17th Annual Meeting; 17–21 Nov.; SETAC, Washington, DC, Pensacola, FL, 1996, 213.
86. Crain, D.A. et al., Sex-steroid and thyroid hormone concentrations in juvenile alligators (*Alligator mississippiensis*) from contaminated and reference lakes in Florida, *Environ. Toxicol. Chem.*, 17, 446, 1998.
87. Hewitt, E.A. et al., Thyroid status in juvenile alligators (*Alligator mississippiensis*) from contaminate and reference sites on Lake Okeechobee, FL, *Chemosphere*, 47, 1129, 2002.
88. Brasfield, S.M., Bradham, K., Wells J.B., et al. Development of a terrestrial vertebrate model for assessing bioavailability of cadmium in the fence lizard (*Sceloporus undulatus*) and in ovo effects on hatchling size and thyroid function, *Chemosphere*, 54, 1643, 2004.
89. Gunderson, M.P. et al., Temporal and spatial variation in plasma thyroxine (T-4) concentrations in juvenile alligators collected from Lake Okeechobee and the northern Everglades, Florida, USA, *Environ. Toxicol. Chem.*, 21, 5, 914, 2002.
90. Irwin, L.K., Gray, S., and Oberdörster, E., Vitellogenin induction in painted turtle, *Chrysemys picta*, as a biomarker of exposure to environmental levels of estradiol, *Aquat. Toxicol.*, 55, 49, 2001.
91. Shelby, J.A. and Mendonça, M.T., Comparison of reproductive parameters in male yellow-blotched map turtles (*Graptemys flavimaculata*) from a historically contaminated sited and a reference site, *Comp. Biochem. Physiol.*, 129C, 233, 2001.
92. Kannan, K., Ueda, M., Shelby, J.A. et al., Polychlorinated dibenzo-*p*-dioxins (PCDDs), dibenzofurans (PCDFs), biphenyls (PCBs) and organochlorine pesticides in yellow-blotched map turtle from the Pascagoula River Basin in Mississippi, USA, *Arch. Environ. Contam. Toxicol.*, 38, 362, 2000
93. Fossi, M.C. et al., The lizard *Gallotia galloti* as a bioindicator of organophosphorous contamination in the Canary Islands, *Environ. Pollut.*, 87, 289, 1995.
94. Sanchez, J.C., Fossi, M.C., and Focardi, S., Serum "B" esterases as a nondestructive biomarker in the lizard *Gallotia galloti* experimentally treated with parathion, *Environ. Toxicol. Chem.*, 16, 1954, 1997a.

95. McLean, R.G., Spillane, J.T., and Miles, J.W., A prospective study of the effects of ultralow volume (ULV) aerial application of malathion on epidemic *Plasmodium falciparum* malaria, *Am. J. Trop. Med. Hyg.*, 24, 193, 1975.
96. Hall, R.J. and Clark, D.R., Jr., Responses of the iguanis lizard *Anolis carolinensis* to four organophosphorus pesticides, *Environ. Pollut.*, 28, 45, 1982.
97. Sanchez, J.C., Fossi, M.C., and Focardi, S. Serum "B" esterases as a non-destructive biomarker for monitoring the exposure of reptiles to organophosphorous insecticides, *Ecotox. Environ. Saf.*, 38, 45, 1997b.
98. Sanchez-Hernandez, J.C. and Moreno-Sanchez, B., Lizard cholinesterases as biomarkers of pesticide exposure: enzymological characterization, *Environ. Toxicol. Chem.*, 21, 2319, 2002.
99. Overmann, S.R. and Krajicek, J.J., Snapping turtles (*Chelydra serpentina*) as biomonitors of lead contamination of the Big River in Missouri's Old Lead Belt, *Environ. Contam. Toxicol.*, 14, 689, 1995.
100. Ewers, U., Schlipköter, H.-W., Lead, in *Metals and Their Compounds in the Environment. Occurrence, Analysis and Biological Relevance*, Merian, E., Ed., VCH Verlagsgesellschaft, Weinheim, Germany, 971, 1991.
101. Albers, P.H., Sileo, L., and Mulhern, B.M., Effects of environmental contaminants on snapping turtles of a tidal wetland, *Arch. Environ. Contam. Toxicol.*, 15, 39, 1986.
102. Licht, L.E., 1976.
103. Lovelette, C.A. and Wright, E., In vivo effects of Pb^{+2} upon 5-aminolevulinate dehydratase in *T. scripta*. Abstract papers. *Am. Chem. Soc.*, 211, 90, 1996.
104. Bishop, C.A. et al., Biochemical indicators of contaminant exposure in passerines colonial waterbirds and snapping turtles of the Great Lakes Basin, *Mar. Environ. Res.*, 42, 273, 1996.
105. Phillips, J.B. and Wells, M.R., Adenosine triphosphatase activity in liver, intestinal mucosa, cloacal bladder, and kidney tissue of five turtle species following in vitro treatment with 1,1,1-trichloro-2,2-bis(p-chlorophenyl)ethane (DDT), *J. Agric. Food Chem.*, 22, 404, 1974.
106. Wells, M.R., Phillips, J.B., and Murphy, G.G., ATPase activity in tissues of the map turtle, *Graptemys geographica* following in vitro treatment with aldrin and dieldrin, *Bull. Environ. Contam. Toxicol.*, 11, 572, 1974.
107. Meyers-Schöne, L. et al., Comparison of two freshwater turtles as monitors of radionuclide and chemical contamination: DNA damage and residue analysis, *Environ. Toxicol. Chem.*, 12, 1487, 1993.
108. Mitchelmore, C.L. and Chipman J.K., DNA strand breakage in aquatic organisms and the potential value of the comet assay in environmental monitoring. *Mutat. Res.*, 399, 135, 1998.
109. Mitchelmore, C.L. et al., Evidence for cytochrome P450 catalysis and free radical involvement in the production of DNA strand breaks by benzo[a]pyrene and nitroaromatics in mussel (*Mytilus edulis* L.) digestive gland cells. *Aquat. Toxicol.*, 41, 3, 193, 1998.
110. Crawshaw, G.J., Diseases and pathology of amphibians and reptiles, in *Ecotoxicology of Amphibians and Reptiles*, Sparling, D.W., Linder, G., and Bishop, C., Eds., SETAC Press, Pennsacola, FL, 199, 2000.
111. Malins, D.C. et al., Toxic chemicals in marine sediment and biota from Mukilteo, Washington: relationships with hepatic neoplasms and other hepatic lesions in English sole (*Parophrys vetulus*). *J. Natl. Cancer Inst.*, 74, 487, 1985.

112. Hopkins, W.A. et al., Trace element accumulation and effects of chronic dietary exposure on banded water snakes (*Nerodia fasciata*), *Environ. Toxicol. Chem.*, 21, 906, 2002.
113. Rooney, A.A., Bermudez, D.S. and Guillette Jr., L.J. 2003. Altered histology of the thymus and spleen in contaminant-exposed juvenile American alligators. *J. of Morphology*, 256, 349–359.
114. Leceta, J. and Zapata, A.G., Seasonal changes in the thymus and spleen of the turtle, *Mauremys caspica*. A morphometrical, light microscopical study, *Dev. Comp. Immunol.*, 9, 653–668, 1985.
115. El Deeb et al., Neuroimmunomodulation in reptiles. III. Morphological and immunological changes in the spleen of juvenile lizards following steroid hormones treatment, *J. Egypt Ger. Soc. Zool.*, 12, 489, 1993.
116. Lind, P.M. et al., Abnormal bone composition in female juvenile American alligators from a pesticide-polluted lake (Lake Apopka, Florida), *Environ. Health Perspect.*, 112, 359, 2004.
117. Frye, F.L., *Biomedical and Surgical Aspects of Captive Reptile Husbandry*, 2nd ed., Krieger, Melbourne, 1991.
118. Done, L., Neoplasia, in *Reptile Medicine and Surgery*, Mader, D.R., Ed., Saunders, Philadelphia, 125, 1996.
119. Garner, M.M. et al., Shell disease in river cooters (*Pseudemys concinna*) and yellow-bellied turtles (*Trachemys scripta*) in a Georgia (USA) lake, *J. Wildl. Dis.*, 33, 78, 1997.
120. Tangredi, B.P. and Evans, R.H., Organochlorine pesticides associated with ocular, nasal, or otic infection in the eastern box turtle (*Terrapene carolina carolina*), *J. Zoo Wildl. Med.*, 28, 97, 1997.
121. Ramsay, E.C. et al., A retrospective study of neoplasia in a collection of captive snakes, *J. Zoo Wildl. Med.*, 27, 28, 1996.
122. Oros, J. et al., Multicentric lymphoblastic lymphoma in a loggerhead sea turtle (*Caretta caretta*), *Vet. Pathol.*, 38, 464, 2001.
123. Roe, W.D., et al., Squamous cell carcinoma in a tuatara (*Sphenodon punctatus*), *N.Z. Vet. J.*, 50, 207, 2002.
124. Aguirre, A.A. et al., Organic contaminants and trace-metals in the tissues of green turtles (*Chelonia-mydas*) afflicted with fibropapillomas in the Hawaiian Islands, *Mar. Pollut. Bull.*, 28, 109, 1994.
125. Nagle, R.D., Rowe, C.L., and Congdon, J.D., Accumulation and selective maternal transfer of contaminants in the turtle *Trachemys scripta* associated with coal ash deposition, *Arch. Environ. Contam. Toxicol.*, 40, 531, 2001.
126. Roe, J.H. et al., Maternal transfer of selenium in Alligator mississippiensis nesting downstream from a coal burning power plant, *Environ. Toxicol. Chem.*, 23, 1969, 2004.
127. Rowe, C.L., Hopkins, W.A., and Congdon, J.D., Ecotoxicological implications of aquatic disposal of coal combustion residues in the United States: a review, *Environ. Mon. Assess.*, 80, 207, 2002.
128. Bishop, C.A. et al., The case for a cause-effect linkage between environmental contamination and development in eggs of the common snapping turtle (*Chelydra s. serpentina*) from Ontario, Canada, *J. Toxicol. Environ. Health*, 33, 521, 1991.
129. Crews, D., Bergeron, J.M., and McLachlan, J.A., The role of estrogen in turtle sex determination and the effect of PCBs, *Environ. Health Perspect.*, 103, 73, 1995.

130. Crain, D.A. and Guillette, L.J., Reptiles as models of contaminant-induced endocrine disruption. *Anim. Reprod. Sci.*, 53, 77, 1998.
131. Bergeron, J.M., Crews, D., and McLachlan, J.A., PCBs as environmental estrogens: turtle sex determination as a biomarker of environmental contamination, *Environ. Health Perspect.*, 102, 780, 1994.
132. Willingham, E. and Crews, D., Sex reversal effects of environmentally relevant xenobiotic concentrations on the red-eared slider turtle, a species with temperature-dependent sex determination, *Gen. Comp. Endocrinol.*, 113, 429, 1999.
133. Podreka, S. et al., The environmental contaminant DDE fails to influence the outcome of sexual differentiation in the marine turtle *Chelonia mydas*, *Environ. Health Perspect.*, 106, 185, 1998.
134. Portelli, M.J. et al., Effect of dichlorodiphenyl-trichloroethane on sex determination of the common snapping turtle (*Chelydra serpentina serpentina*), *Ecotox. Environ. Saf.*, 43, 284, 1999.
135. Milnes, M.R., Woodward, A.R., and Guillette, L.J., Morphological variation in hatchling American alligators (*Alligator mississippiensis*) from three Florida lakes, *J. Herpetol.*, 35, 264, 2001.
136. Guillette, L.J., Jr. et al., Reduction in penis size and plasma testosterone concentrations in juvenile alligators living in a contaminated environment, *Gen. Comp. Endocrinol.*, 101, 32, 1996.
137. Guillette, L.J., Jr. et al., Plasma steroid concentrations and male phallus size in juvenile alligators from seven Florida (USA) lakes, *Gen. Comp. Endocrinol.*, 116, 356, 1999.
138. Pickford, D.B. et al., Plasma dihydrotestosterone concentrations and phallus size in juvenile American alligators (*Alligator mississippiensis*) from contaminated and reference populations, *J. Herpetol.*, 34, 233, 2000.
139. Stoker, C. et al., Sex reversal effects on *Caiman latirostris* exposed to environmentally relevant doses of the xenoestrogen bisphenol A, *Gen. Comp. Endocrinol.*, 133, 287, 2003.
140. Congdon, J.D., Dunham, A.E., and Tinkle, D.W., Energy budgets and life histories of reptiles, in *Biology of the Reptilia*, Vol. 13, Gans, C., Ed., Academic Press, New York, 1982, 233.
141. Calow, P. and Sibly, R.M., A physiological basis of population processes: ecotoxicological implications, *Func. Ecol.*, 4, 283, 1990.
142. Calow, P., Physiological costs of combating chemical toxicants: ecological implications, *Comp. Biochem. Physiol.*, 100C, 3, 1991.
143. Congdon, J.D. et al., Resource allocation based life history trait values: a conceptual basis for studies of environmental toxicology, *Environ. Toxicol. Chem.*, 20, 1698, 2001.
144. Rowe, C.L., Hopkins, W.A., and Congdon, J.D., Integrating individual-based indices of contaminant effects: how multiple sublethal effects may ultimately reduce amphibian recruitment from a contaminated breeding site, *Sci. World*, 1, 703, 2001a.
145. Hopkins, W.A., Rowe, C.L., and Congdon, J.D., Elevated trace element concentrations and standard metabolic rate in banded water snakes, *Nerodia fasciata*, exposed to coal combustion wastes, *Environ. Toxicol. Chem.*, 18, 1258, 1999.
146. Rowe, C.L. et al., Elevated maintenance costs in an anuran (*Rana catesbeiana*) exposed to a mixture of trace elements during the embryonic and early larval periods, *Physiol. Zool.*, 71, 27, 1998a.

147. Rowe, C.L. et al., Metabolic costs incurred by crayfish (*Procambarus acutus*) in a trace element-polluted habitat: further evidence of a common response among diverse taxonomic groups, *Comp. Biochem. Physiol.*, 129C, 275, 2001b.
148. Calow, P. and Forbes, V.E., How do physiological responses to stress translate into ecological and evolutionary processes? *Comp. Bioch. Physiol.*, 120A, 11, 1998.
149. Steyermark, A.C., A high standard metabolic rate constrains juvenile growth, *Zoology*, 105, 147, 2002.
150. Bridges, C., Rowe, C.L., and Hopkins, W.A., Conservation genetics of amphibians, in *Amphibian Decline: An Integrated Analysis of Multiple Stressor Effects*, Linder, G., Crest, S., and Sparling, D.W., Eds., SETAC Press, Boca Raton, FL, 2003.
151. Rowe, C.L., Hopkins, W.A., and Bridges, C., Physiological ecology of amphibians in relation to susceptibility to natural and anthropogenic factors, in *Amphibian Decline: An Integrated Analysis of Multiple Stressor Effects*, Linder, G., Crest, S., and Sparling, D.W., Eds., SETAC Press, Boca Raton, FL, 2003.
152. Rowe, C.L. et al., Oral deformities in tadpoles (*Rana catesbeiana*) associated with coal ash deposition: effects on grazing ability and growth, *Freshwater Biol.*, 37, 723, 1996.
153. Rowe, C.L., Kinney, O.M., and Congdon, J.D., Deformities in tadpoles of the bullfrog (*Rana catesbeiana*) caused by conditions in a polluted habitat, *Copeia*, 1998, 244, 1998b.
154. Crain, D.A. et al., Alteration in steroidogensis in alligators (*Alligator mississippiensis*) exposed naturally and experimentally to environmental contaminants, *Environ. Health. Perspect.*, 105, 528, 1997.
155. Guillette, L.J., Jr. and Gunderson, M.P., Alterations in development of reproductive and endocrine systems of wildlife populations exposed to endocrine-disrupting contaminants, *Reproduction*, 122, 857, 2001.
156. Fernie, K.J. et al., In ovo exposure to polychlorinated biphenyls: reproductive effects on second-generation American kestrels. *Arch. Environ. Contam. Toxicol.*, 40, 544, 2001.
157. Custer, C.M. et al., Exposure and effects of chemical contaminants on tree swallows nesting along the Housatonic River, Berkshire County, Massachusetts, USA, 1998–2000, *Environ. Toxicol. Chem.*, 22, 1605, 2003.
158. Congdon, J.D. and Gibbons, J.W., The evolution of turtle life histories, in *Life History and Ecology of the Slider Turtle*, Gibbons, J.W., Ed., Smithsonian Institution Press, Washington, DC, 1990, 45.
159. Bain, D. et al., Effects of sublethal Fenitrothion ingestion on cholinesterase inhibition, standard metabolism, thermal preference, and prey-capture ability in the Australian central bearded dragon (*Pogona vitticips*, Agamidae), *Environ. Toxicol. Chem.*, 23, 109, 2004.
160. Peveling, R. and Demba, S.A., Toxicity and pathogenicity of *Metarhizium anisopliae* var. *aridum* (Deuteromycotina, Hyphomycetes) and Fiprinil to the fringe-toed lizard *Acanthodactylus dumerili* (Squamat: Lacertidae), *Environ. Toxicol. Chem.*, 22, 1434, 2003.
161. Steyermark, A.C., and Spotila, J.R., Body temperature and maternal identity affect snapping turtle (*Chelydra serpentina*) righting response, *Copeia*, 2001, 1050, 2001.
162. Burger, J. et al., Effects of lead on behavior, growth, and survival of hatchling slider turtles, *J. Toxicol. Environ. Health*, Part A 55, 495, 1998.

163. Hopkins, W.A., Winne, C.T., and DuRant, S.E., Differential swimming performance of two Natricine snakes exposed to a cholinesterase-inhibiting pesticide. *Environ. Pollut.*, 133, 531, 2005.
164. Goldman, D. and Yawetz, A., Cytochrome P-450 mediated metabolism of progesterone by adrenal microsomes of PCB-treated and untreated barn owl (*Tyto alba*) and marsh turtle (*Mauremys caspica*) in comparison with the guinea pig. *Comp. Biochem. Physiol.*, 99C, 251, 1991
165. Kirchin, M. and Winston, G.W., Microsomal activation of benzo[a]pyrene to mutagens by *Alligator mississippiensis* in vitro: induction by 3-methylcholanthrene. *Mar. Environ. Res.*, 34, 273, 1992.
166. Yawetz, A. et al., The effect of PCB on the induction of cytochrome P-450 enzyme, detected by monoclonal antibody against the P-4501A1 gene products, in barn owl and marsh turtle hepatic microsomes, *Water Sci. Technol.*, 27, 457, 1993b.
167. Ertl, R.P., Alworth, W.L., and Winston, G.W., Liver microsomal cytochromes P450-dependent alkoxyphenoxazone O-dealkylation *in vitro* by alligator and rat: activities, inhibition, substrate preference, and discrimination factors, *J. Biochem. Molec. Toxicol.*, 13, 17, 1999.
168. Mayeaux, M.M. and Winston, G.W., Antibiotic effects on cytochromes P450 content and mixed-function oxygenase (MFO) activities in the American alligator, *Alligator mississippiensis*, *J. Vet. Pharmacol. Ther.*, 21, 274, 1998.
169. Yawetz, A., Woodin, B.R., and Stegeman, J.J., Cytochromes P450 in liver of the turtle *Chrysemys picta picta* and the induction and partial purification of CYP1A-like proteins. *Biochim. Biophysica Acta*, 1381, 2, 1998.
170. Ertl, R.P., Stegeman, J.J., and Winston, G.W., Induction time course of cytochromes P450 by Phenobarbital and 3-methylcholanthrene pretreatment in liver microsomes of *Alligator mississippiensis*, *Biochem. Pharmcol.*, 55, 1513, 1998.
171. Schlezinger, J.J. et al., 3,3′,4,4′-tetrachlorobiphenyl oxidation in fish, bird and reptile species: relationship to cytochrome P450 1A inactivation and reactive oxygen production. *Comp. Biochem. Physiol.*, 125C, 273, 2000.
172. Stone, W.B., Kiviat, E., and Butkas, S.A., Toxicants in snapping turtles, *N.Y. Fish Game J.*, 27, 39, 1980.

chapter 5

Hepatic, Renal, and Adrenal Toxicology

Patricia McClellan-Green, Malin Celander, and Eva Oberdörster

Table of Contents

I. Introduction

This chapter will cover the basics of liver anatomy, its role in maintaining homeostasis, and its role in metabolizing and eliminating xenobiotics and reactive oxygen species (ROS). The second part of the chapter will cover kidney anatomy, its role in maintaining homoeostasis and interactions with the adrenal gland, and renal toxicity of xenobiotics.

II. Liver Anatomy

In vertebrates, the liver is usually the largest internal organ. Reptilian livers are quite different from mammalian livers, both in gross anatomy and in ultrastructure. In reptiles, a large bilobed liver is seen in all species, except in snakes where the liver has a more distinct fusiform or spindle shape. The shape and form of the reptile liver can vary greatly from species to species, and it usually grows to fill all the available space in the abdominal region between the heart and the stomach. For example, in Chelonians the liver is a large, ventral, saddle-shaped organ that spreads from side to side under the lungs,[1] and it generally accounts for approximately 2% to 5% of the total body mass.

In mammalian systems, the classical liver architecture has been historically depicted as a hexagonal lobule. In the middle or center of this lobule lies the hepatic venule (central vein), and at the edge the portal space occurs. This portal space contains the portal vein, hepatic arteriole, and the bile duct. The ultrastructure of reptile liver is somewhat different from that observed in other vertebrate species. Hering[2] first described the reptile liver as a structural network of anastomosing tubules with narrow vascular spaces containing five to six cells with basal nuclei. These comprised a tubular cross-section where the tubules connect with ducts near to the portal areas that may run parallel to the portal branches.

As discovered since that time, the lobule pattern for liver structure is not discernible in many species of reptiles or may be totally absent. Portal areas of the liver contain branches of the hepatic artery, the portal vein, and bile ducts. These structures are supported by connective tissue. The central vein, when it is present, is located in the center of the tubular structure where cords of polyhedral-to-cuboidal hepatocytes radiate outward from the vein. These cords are normally two cells thick and are often lined by endothelium sinusoids containing scattered Küpffer cells. Normally, endothelial cells found in these sections are minimal with few fenestrae, and pigment cells are rare in some species. However, Küpffer cells are numerous. Ito cells, or fat-storing cells, may also be numerous depending on the species of reptile (larger in crocodilians and turtles than some lizards).[3] The Küpffer cells observed in most reptile species tend to be more flattened or spindle-shaped than those occurring in other vertebrates.[4] In reptile species that have a less organized liver morphology, hepatocytes are often arranged in a series of branching cords that are interspersed with vascular spaces lined by endothelium-containing Küpffer cells. In addition, extensive occurrence of fibrous trabeculae has been observed in alligator liver. This structural matrix purportedly strengthens and adds stability to this critical organ during rapid bodily movements.[5]

The bile canaliculi, which are small channels functioning as terminals of the biliary system, extend throughout the connective tissue of the liver. Multiple branches of the bile duct have been observed near the bile canaliculi in many reptile species. In crocodilians, three hepatic ducts join together to form the bile duct.[6] The gallbladder may or may not be contiguous with the liver. In many lizards and chelonians, the gallbladder is located within the confines of the hepatic tissues. However, in snakes and some lizards, the gallbladder is located posterior to the main hepatic mass.[7] It is normally lined by simple or pseudostratified columnar epithelial cells. In alligators, bile drains from the liver through a shortened right hepatic duct and through an elongated left hepatic duct. The right hepatic duct drains directly into the gallbladder, whereas the left duct is split into two branches. One of these branches drains directly into the duodenum, and the other branch is connected to the right hepatic duct.[8]

An important feature, and one that is immediately apparent on physical examination of many reptile livers (e.g., sea turtles), is the large amount of melanin pigment distributed throughout the hepatocellular parenchyma. This pigment, which is usually contained in dense clusters within the tissue, has no known anatomic significance but may play a role in some physiologic processes, including potentially serving as a superoxide radical trap (see Section IV, B). The aggregations of specialized Küpffer cells containing this melanin increase with the age of the reptile and in some turtles constitute up to 20% of the liver cell volume.[9]

III. Liver Function in Homeostasis

The role of the liver surrounds four basic functions that are essential for homeostasis and metabolism:

- Storage and filtration of blood
- Synthesis of a variety of blood plasma constituents that are secreted into the blood, including endocrine functions
- Secretory and excretory activities related to the formation of bile
- Metabolism and storage of endogenous and exogenous compounds such as vitamins, steroids, fatty acids, and xenobiotics.

The liver of any species functions as the major center for metabolic activity, and metabolism of carbohydrates, lipids, and proteins will be discussed in the following sections. Metabolism of xenobiotics will be covered in Section IV.

A. Carbohydrate Metabolism

A major function of the liver is to serve as energy storage. Glucose is stored within the liver and muscles as glycogen, and glucose is released into the circulation after glycogenolysis when the plasma level of glucagon increases. Glycogen storage ensures a stable blood glucose level, which is important for adenosine triphosphate (ATP) production and as a cryoprotectant in poikilotherms, as discussed in more detail later in this section. In mammals, glycogenolysis is controlled by a complex metabolic pathway regulated by a hormonal system using second messengers such as cyclic adenosine monophosphate (cAMP), Ca^{2+}, and diacylglycerol. In Teleosts, this process is controlled by glucagons and catecholamines with cAMP as the sole intercellular messenger. The process in amphibians is similar to that observed in Teleosts with the exception that the neuropeptide hormone Arginine vasotocin plays a key role in the regulation of glycogenolysis.[10] In alligators (*Alligator mississipiensis*), lizards (*Tupinambis teguixin*), and some snakes (*Xenodon merremi*), laboratory studies have shown that the administration of glucagon results in the development of hyperglycemia, whereas the administration of insulin results in hypoglycemia.[11] Janssens and Giuliano[12] reported that both epinephrine (adrenalin) and glucagon stimulate glycogenolysis in the western netted dragon (*Amphibolurus nuchalis*). However, whether the neurohypophysial peptides arginine vasotocin or mesotocin are involved in glycogenolysis is still uncertain.

In mammals, the effects of adrenal hormones on glucose metabolism are mediated through α- and β-adrenergic receptors. These receptors also occur in birds, fish, amphibians, and reptiles,[13] but the role of epinephrine is less clear in reptiles. The glycogenolytic response to epinephrine exposure observed in the netted dragon is mediated through the β-adrenergic receptor with some evidence of α-adrenergic receptor involvement. This latter conclusion was based on the findings by Janssens and Giuliano[12] that -adrenergic antagonists such as phentolamine elicited a reduction in glycogen phosphorylase activation and in cAMP levels in liver tissue. The α-adrenergic receptor has been identified in the liver tissues of the lizard, *Podarcis sicula campestris*, and the turtle, *Pseudemys picta elegans*, through receptor binding studies.[14] Other investigators[15] have also identified the presence of low levels of the α-adrenergic receptor(s) in reptile liver tissue but were not able to confirm their involvement in hepatic glycogen breakdown. These studies focused both on receptor levels and hepatic glycogen levels in the freeze-tolerant hatchling painted turtle (*Chrysemys picta marginata*). Examination of hepatic glycogen levels clearly demonstrated a significant decrease (50%) in turtles exposed to freezing temperatures compared to controls not exposed to cold. On thawing, the hepatic glycogen levels returned to concentrations that more closely resembled that observed in the control animals. Parallel to the change in hepatic glycogen, β_2-adrenergic receptor binding to plasma membranes decreased by 40% after 48 h following exposure to freezing temperatures, whereas γ-glutamyl transpeptidase activity exhibited a pulse increase at 12 h and then decreased to control levels by 24 h. No involvement of the α-adrenergic receptor was detected. Hemmings and Storey[15] showed that, although the percent change in glycogen content was equivalent in both lobes of the turtle liver, the small lobe contained up to two-fold higher glucose levels per milligram protein or milligram liver tissue compared with the larger lobe. The physiologic basis for this difference in glycogen content still remains unknown.

B. Lipid Metabolism

Metabolic activity in the liver is responsible for the synthesis, oxidation, storage, and distribution of lipids. The liver aids the absorption of fats through the production and action of the bile salts. It also synthesizes and oxidizes fatty acids (FA), cholesterol, triacylglycerols (TA), and phospholipids (PL), which are the major components of cell membranes.[16] The liver is also responsible for the synthesis of most of the plasma lipoproteins and converts excess carbohydrates and proteins into lipids for energy storage.

The composition of lipids in hepatic tissues and membranes of reptiles is largely dependent on the diet, as Simandle and colleagues[17] demonstrated through a series of experiments using male desert iguanas (*Dipsosaurus dorsalis*). Animals fed diets high in saturated FA or polyunsaturated FA reflected these levels in hepatic lipid composition.

However, dietary TAs were not reflected in hepatic lipid composition, and no dietary effect was observed in the FA content of PL found in the liver.

Because both TA and PL are key components of cell membranes, the effect of altering FA composition may result in variations in fluidity and permeability of cell membranes.[18] Simandle and colleagues[17] showed that more saturated FA correlated to selection of warmer nighttime body temperatures to possibly take into account a change in membrane fluidity. Similar results in terms of dietary impacts on liver FA composition were observed in the American alligator (*A. mississipiensis*)[19]and the loggerhead sea turtle (*Caretta caretta*).[20] Analysis of the FA content of loggerhead sea turtle liver tissue identified an FA ratio of polyene n-3/n-6 and arachidonic acid similar to that observed in dolphins and marine birds, and it corresponded to the FA ratio of their diet. In birds, about 80% of the total liver lipid content consists of cholesterol. The remaining 20% is composed of TA and PL. In snakes, such as the hatchling water python (*Liasis fuscus*), approximately 70% of the total liver lipid pool consists of TA, with PL and cholesterol composing 8% and 14%. This reflects the dietary lipid composition (e.g., yolk sac) of newly hatched individuals.[21] Thus, FA content and chemical composition of total liver lipids varies with diet.

It has also been shown that the concentration of liver lipids can be influenced by both seasonal and reproductive status in reptiles. Lacy and colleagues[22] examined the lipid concentration in the liver of feral male and female tree lizards, *Urosaurus ornatus*, and found their body lipid concentrations varied with seasonal and reproductive status. Lipid storages, total liver lipids, the level of free FA, and the activity of enzymes involved in lipid metabolism, such as diacylglycerol acyltransferase, also fluctuate in a seasonal pattern. Examination of the data revealed that lipid storage in male and female lizards correlated with vitellogenesis, gravidity, mating, and territorial defensive behaviors.

C. *Protein Metabolism and Vitellogenesis*

The liver is a primary organ in protein metabolism, including many plasma proteins and peptide hormones. Key enzymes involved in various physiologic processes also are synthesized in the liver, and the liver is important for the breakdown (catabolism) of proteins.

The major yolk proteins in egg-laying vertebrates, including reptiles, are derived from the vitellogenin (Vtg) precursor protein that is synthesized in the liver. 17β-Estradiol (E_2) binds to the estrogen receptor (ER) to initiate vitellogenesis, and the production of E_2 is controlled by levels of aromatase (alternatively known as CYP 19), which can be located in the brain or the adrenal–kidney–gonad (AKG) complex. Aromatase levels vary between the sexes and among species during development and is reviewed in more detail in Chapter 6. Endocrine-disrupting chemicals can alter steroidogenesis by a number of mechanisms, including binding to hormone receptors, altering steroidogenesis, or altering cell signaling. A study in alligator has shown that even nitrates can interfere with steroidogenesis on several levels,[23] which could in turn affect vitellogenesis in the liver. Not only is vitellogenesis controlled by E_2, but stress hormones such as corticosterone can severely hinder reproductive capabilities of reptiles, in part resulting directly from reduction in vitellogenesis.[24] This complication by stress is further discussed later in this chapter.

After synthesis of the Vtg precursor, Vtg is secreted into the bloodstream as a lipo-glyco-phospho-protein and incorporated in the maturing oocyte where it undergoes cleavage. Ho and colleagues[25] reported the presence of a single Vtg protein (molecular weight ~200,000) produced in the liver in the painted turtle (*Chrysemys picta*). This protein was later cleaved into lipovitellin and phosvitin. In the garter snake (*Thamnophis sirtalis*), the European lizard (*Lacerta vivipora*), and the tropical lizard (*Anolis pulchellus*), Vtg is produced in the liver and secreted into the circulation as two phosphorylated polypeptides.[26,27] The size of these proteins differ among species, with the garter snake producing two proteins of 149 and 124 kDa, the European lizard producing two proteins of 220 and 110 kDa, and the tropical lizard producing two phosphoproteins of 226 and 116 kDa.

The liver is the major organ involved in detoxification, and therefore proteins involved in metabolic clearance and detoxification are produced in the liver. These include metalloproteins, enzymes involved in phase I and phase II metabolism, and enzymes involved in redox status. The cytochrome P450 (CYP) monooxygenases catalyze 95% of the phase I reactions. Other phase I enzymes are, for example, flavin-containing monooxygenases, hydrolases, and reductases. Phase II enzymes comprise glutathione S-transferases (GST), UDP-glucuronosyltransferase (UGT), and sulphotransferases.[28,29] Key enzymes in oxidative stress such as glutathione reductase, glutathione synthase, catalase (CAT), and superoxide dismutase (SOD) also are synthesized in the liver. Use of these enzymes as biomarkers is extensively reviewed in Chapter 4.

IV. Liver Function in Metabolism of Xenobiotics

In most vertebrate species, the liver is the site of the greatest level of metabolic clearance of endobiotic and xenobiotic compounds.

Xenobiotics usually enter the organism via the respiratory and gastrointestinal tracts where they can be absorbed and transported directly to the liver for "first-pass" metabolism. Lipophilic xenobiotics and endogenous waste products may accumulate to toxic levels if they are not efficiently excreted. Biotransformation through phase I and phase II metabolism usually generates more hydrophilic derivates to facilitate excretion. However, biotransformation may result in generation of activated (more toxic) intermediates. Reactive metabolites generated in this fashion may cause serious damage to the liver or other tissues, which can lead to chemical carcinogenesis. Understanding how reptiles metabolize xenobiotics is necessary in determining potential toxicity of xenobiotics.

A. *Phase I and II Enzymes*

Members of the CYP gene superfamily are the major enzymes involved in the oxidative metabolism of lipophilic compound including many endogenous compounds, such as steroids and fatty acids, and foreign compounds, such as natural products, food additives, pharmaceutical drugs, pesticides, and environmental contaminants in reptiles. The CYP enzymes act in concert with a reductase (either cytochrome b_5- or CYP NAD(P)H reductase). In reptiles, CYP activities were first identified in the mid1980s as elevated total CYP content or increased benzo[a]pyrene hydroxylase activities (presumably CYP1A) in liver microsomes from 3-methylcholanthrene (MC)–treated animals—for example, garter snakes (*Thamnophis sirtalis*), the spectacle caiman (*Caiman crocodylus*), and the American alligator (*A. mississipiensis*).[30–32]

The use of P450 isozymes as biomarkers is extensively reviewed in Chapter 4 Here, we review some examples of the variety of P450 isozymes. Ertl and colleagues[33] later showed the presence of six to seven different CYP isoforms in *A. mississipiensis* from Louisiana treated with prototypical CYP inducers by using antibodies raised against various mammalian CYP forms. Cross-reactivity among mammalian CYP1A1, CYP1A2, CYP2B, CYP2C, CYP2E1, and CYP2K antibodies and proteins in livers from phenobarbital (PB) or 3-MC–treated alligators were reported. No evidence of CYP4A immunoreactive proteins were detected in alligators treated with clofibrate. This was supported by lack of lauric acid hydroxylase activity, a biomarker of activity of CYP4A in mammals. Additional population-level differences in CYP1A-like proteins in alligators has been reported for *A. mississipiensis* from Florida. Gunderson and colleagues[34] showed that some individual alligators had CYP1A-like protein immunoreactivity in doublet bands, whereas others had only singlet bands. In the Louisiana alligators studied by Ertl and colleagues,[33] only singlet bands for CYP1A-like proteins were reported. This may indicate the presence of additional multiple CYP1A isoforms in different regional populations.

Xenobiotics may also be able to inhibit enzyme activity. In the study by Gunderson and colleagues,[34] CYP activities were also measured in wild alligators by using ethoxyresorufin-*O*-deethylase (EROD) and methoxyresorufin-*O*-deethylase (MROD). They found that in areas with lowest contamination, there was a significant negative correlation between body size and EROD activity, showing that EROD activity decreases in larger animals. This type of correlation was not seen in higher-contaminant areas (EROD was always low), indicating that xenobiotics may be able to inhibit liver enzymes in young animals. Mayeaux and Winston[35] showed a similar xenobiotic-induced decrease in CYP content and benzyloxyresorufin-*O*-dealkylase (BROD) activity (presumably CYP3A) in *A. mississipiensis* treated with oxytetracycline, ceftazidime, and enrofloxacin.

The presence of CYP-like proteins or activities has been identified in other reptile species by using diagnostic substrates or antibodies against mammalian and piscine CYP forms. For instance, CYP1A-like proteins or activities have been identified in several turtle species, including *Mauremys caspica rivulata*, *Chrysemys picta picta*, *Chrysemys scripta elegans*, and *Lepidochelys kempii*; the spectacle caiman (*C. crocodylus*); and the garter snake (*T. sirtalis*).[36–40] Herman and Oliw[41] examined the ability of liver microsomes from the yellow rat snake to oxygenate polyunsaturated FAs via CYP enzymes. Their data indicated that eicosapentaenoic acid was metabolized to a much greater extent than arachidonic acid or linoleic acid. Haasch and colleagues[42] reported elevated CYP1A messenger ribonucleic acid (mRNA) levels in polyaromatic hydrocarbon (PAH)–treated garter snakes and painted turtle by using rainbow trout liver CYP1A complementary deoxyribonucleic acid (cDNA). Moreover, CYP3A-immunoreactive proteins have been detected in painted turtle, Kemp's Ridley sea turtle (*Lepidochelys kempii*), and ball python (*Python regius*).[39,43,44] A partial CYP3A cDNA sequence denoted CYP3A42 (GenBank accession number AAG33693) was isolated from ball python liver. Phylogenetic analysis of the vertebrate CYP3 gene family revealed a high degree of sequence identity among vertebrate CYP3A genes. The diapsid sequences (chicken CYP3A37 and ball python CYP3A42) formed a cluster separate from the piscine and mammalian clusters.[44]

Further metabolism and detoxification of constitutive compounds and xenobiotics occur through mechanisms that function to increase metabolite excretion or counteract oxidative stress. Phase II enzymes conjugate activated xenobiotics to products that are more easily excreted from the body. The major phase II reactions include sulfate, glucuronide, amino acid, and glutathione conjugation. Of these, only activities involved in glutathione conjugation such as GST have been studied in reptiles. Hermes-Lima and Storey[45] reported the presence of moderate levels of GST activity in liver, lung, and muscle of freeze-tolerant *T. sirtalis*. Gunderson and colleagues[34] also showed moderate levels of GST in wild alligators collected from low to moderate contaminant-level sites in Florida, and several steroidogenic oxido-reductases and phase II enzymes have been used as biomarkers in alligators.[29]

B. *Antioxidant Defense System*

The liver is a target for oxidative stress, and the use of antioxidant enzymes and factors as biomarkers is extensively reviewed in Chapter 4. Here we focus on the interaction of the various enzymes and factors in the antioxidant defense system. Uncoupling of the CYP redox cycle may result in formation of ROS,[46] although the antioxidant defense system protects tissues from these ROS, as well as ROS produced through other metabolic pathways (cytochrome b_5 reductase, flavoprotein reductase) or the leakage of ROS from the electron transport chain.[47] The major types of ROS include the hydroxyl radical (OH•), the superoxide radical ($O_2\bullet^-$), and hydrogen peroxide (H_2O_2). Components of the antioxidant defense system include enzymes such as CAT, SOD, glutathione peroxidase, and glutathione reductase. Additional components include nonenzymatic antioxidant compounds such as vitamins A, C, and E; reduced glutathione (GSH); and uric acid (Figure 5.1). At high concentrations of H_2O_2, CAT is the primary enzyme responsible for reduction of H_2O_2 to H_2O, but at low concentrations of H_2O_2, glutathione peroxidase is the primary catalyst. SOD catalyzes the dismutation of $O_2\bullet^-$ to H_2O_2, and glutathione reductase catalyzes the reduction of small disulfides.

The nature of the antioxidant defense systems in reptiles appears to be dependent on species as well as environmental and nutritional factors. For example, liver CAT and SOD levels were three- and six-fold higher, respectively, in anoxia-tolerant freshwater turtles (*Trachemys scripta elegans*) compared with anoxia-tolerant garter snakes (*T. s. parietalis*).[45,48,49] de Brito-Gitirana and Storch[50] also reported a critical role for CAT during temperature challenge of the wall lizard, *Hemidactylus frenatus*. Thus, increasing the incubation temperature of the animals from 20°C to 30°C resulted in a two-fold

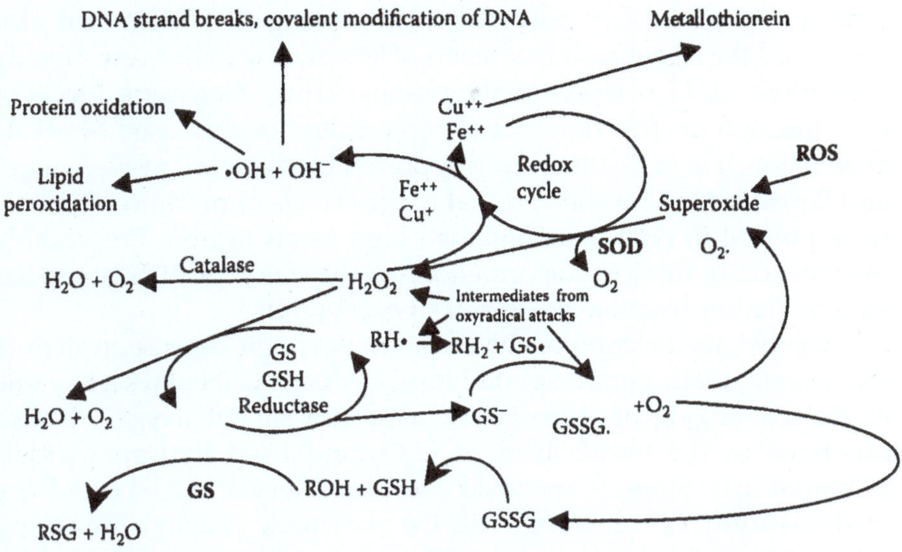

Figure 5.1 Summary of some of the antioxidant defenses found in reptiles. A unique pathway for this class of organisms is the scavenging of superoxide radicals by melanin. GSH, glutathione, reduced form; GSSG, glutathione, oxidized form; GPx, glutathione peroxidase; GST, glutathione S-transferase; SOD, superoxide dismutase; ROS, reactive oxygen species.

increase in hepatic CAT activities from 73.9 mU/mg protein to 143 mU/mg protein. No increase in total protein or in the uricase activity was observed. However, the higher CAT activity correlated with an increase in peroxisome size in the liver tissue.

In addition to CAT, the involvement of other enzymes in the antioxidant defense system of reptiles has been studied. In mammalian and reptile systems, the intracellular $O_2\bullet^-$ concentrations are maintained at low levels in the cells by SOD, and two major SOD isoforms have been identified. These include copper–zinc SOD (Cu-Zn-SOD), which is localized in the cytosol, and manganese SOD (Mn-SOD), which is found in the mitochondria.

In reptiles an additional mechanism is available for the scavenging of $O_2\bullet^-$. Melanins, which are pigmented biopolymers found in tissues, especially the liver of reptiles, provide a protective function against the toxicity of $O_2\bullet^-$. Sichel and colleagues[51] compared the melanin content in the liver of various vertebrate species, including the turtle *Testudo graeca* L. and the lizard *Lacerta sicula* Rafinesque, with their levels of Mn-SOD. In each instance, animals possessing higher levels of melanin exhibited lower levels of Mn-SOD. No differences were observed in the levels of Cu-Zn-SOD. They concluded that the melanin pigment present in liver tissue functioned as a free radical trap where $O_2\bullet^-$ was efficiently scavenged. Their results also confirmed the hypothesis that levels of Mn-SOD activity were directly dependent on $O_2\bullet^-$ levels in the tissues. Thus, decreasing levels of $O_2\bullet^-$ through the function of melanin resulted in decreased Mn-SOD production. These results were supported by the work of Henninger and Beresford,[52] who showed that higher levels of melanin pigments were present in cells that contained high levels of iron. Presumably the melanin in these tissues ameliorated the action of ROS generated via the Fenton reaction as a result of excess iron.

Another major component of the antioxidant defense system is the nonenzymatic peptide glutathione. Reduced GSH plays a key role in the scavenging of hydroxyl radicals and singlet oxygen. It also functions in the metabolism of H_2O_2 and lipid hydroperoxides. Willmore and Storey[53] analyzed the tissue levels of GSH and the oxidized form, GSSG, along with the enzymatic reactions involving this compound in anoxia-stressed *T. s. elegans*. Hypoxia (5°C for 20 h in deoxygenated tanks) resulted in a 25% decrease in GST activity, a 52% increase in glutathione reductase activity, and no change in glutathione synthetase activity. Liver γ-glutamyl transpeptidase activity decreased 71% under anoxic conditions, whereas glutaredoxin activity was unaffected. These results indicated changes in hepatic glutathione status during anoxia suggesting oxidative stress. Total hepatic glutathione content (GSH + GSSG) decreased 49% compared with normoxic controls. The GSSG content increased during anoxia, which together with the GSH levels led to a dramatic decrease in the GSH/GSSG ratio. This indicates that turtles that are tolerant to environmental stressors may decrease enzyme systems involved in production or oxidation of GSH.

V. Disease of the Liver

Attributing the presence of disease in reptiles to exposure to specific chemicals has not been done to date. The organ structure, tissue pathology, and biochemical mechanisms of reptiles are similar to higher vertebrate, and therefore the capacity to develop tumors or liver disease is present. The occurrence and development of neoplastic disease in reptiles is not well documented. For many years it was assumed that the development of tumors in reptiles was rare; however, the documentation of various diseases in reptiles has increased in recent years.[54,55] Catão-Dias and Nichols[56] published a report on the postmortem examination of snakes from the National Zoological Park, Washington, DC, spanning a 20-year period. Of the 291 snakes examined during this period, 12.4%, or 36 snakes, contained tumors. Liver and biliary tract tumors accounted for 18% (7 of 39) of the tumors. The type of tumors observed included hepatocellular carcinomas, cholangiocarcinomas, and multicentric lymphosarcoma. Neoplasms have also been reported in the monitor lizard (*Varanus exanthematicus*). Schultze and colleagues[57] reported the presence of hepatic adenocarcinoma in the monitor lizard, and Martorell and colleagues[58] reported the presence of hepatic spindle-cell sarcoma in this species as well.

VI. Kidney Anatomy

The reptilian kidney is also quite different from other vertebrates and can be highly adapted for water retention in desert species. In reptilian kidney development, an AKG complex is initially formed, which has metabolic and renal functions.[59] Therefore the kidney and adrenals will be considered together in this chapter. The kidneys develop initially as two pairs of kidneys, termed the mesonephros and metanephros, with the metanephros retained as the adult kidney.[60] The function of the mesonephros is unclear; however, it appears that in embryonic and neonatal iguanid lizards the mesonephros has important functions in water and ion balance regulation because the combined mesonephros and metanephros weight follows the predicted allometric relationship between kidney and body mass.[60] Normal vertebrate kidney development is guided by WT1 tumor suppressor gene, as has been demonstrated in turtle and alligator species.[61] Expression of WT1 in the red-eared slider turtle (*Trachemys scripta*) begins in the embryonic kidney–gonad complexes after the mesonephros pair has formed, and it continues until the gonad begins to differentiate.[61] In embryos incubated at 26°C, a male-producing temperature, WT1 expression is elevated as compared with embryos incubated at female-producing temperatures.[61]

In adult reptiles, kidney anatomy has been well studied in turtles and has been reviewed by Bradshaw and Bradshaw.[62] In turtles, the kidneys are lobed and contain nephrons that are composed of a glomerulus, short neck segment, proximal tubule, intermediate segment, distal convoluted tubule, and collecting duct.[62,63] The intermediate segment contains both proximal nonsecretory functions and distal mucus-secreting functions and may be homologous to the mammalian thin-ascending limb of the Loop of Henle.[62] The complex foldings of the proximal and distal tubules are quite distinct from that of mammalian tubules.[63] In addition, some lizards and snakes have aglomerular tubules that have secretory roles, and the lizard kidney also has a renal-portal system that allows for waste excretion during the absence of glomerular filtration (for review, see reference 62).

New techniques are being refined to study reptilian kidneys. These include the use of epifluorescence video microscopy to study glomerular arteriolar diameters, glomerular blood flow, and glomerular capillary pressure in real time.[64] From these methods, blood flow and filtration rates were measured in 12 garter snakes (*Thamnophis sirtalis*) and in a total of 100 glomeruli.[65] The glomeruli had both constant blood flow rates and times of highly variable rates, with 21% of nephrons with intermittent glomerular perfusion.[65] Whether nephrons showed intermittent flow depended heavily on the plasma osmolality and on low mean single-nephron blood flow rate and could be one mechanism for lowering glomerular filtration rate.[65]

Male snakes and lizards have a sex segment of the kidney.[66] A distal portion of the nephron is hypertrophied and is thought to be involved in production of seminal fluid.[66,67] This sexual segment shows seasonal variation in synthesis and secretion that is correlated with mating activity in the snake *Seminatrix pygaea*.[67] In turtles, although a true sexual segment does not exist, kidney products activate sperm motility during the mating season.[68] Therefore chelonians may have a more primitive functional renal sexual segment.[68]

VII. Kidney Function in Homeostasis

A. pH and Ion Balance

Blood–acid balance is also controlled in part by the reptilian kidney. In turtle urinary epithelia, an ATPase actively transports H+, and the presence of this transporter pump is regulated by elevated CO_2 levels.[69] The carnivorous reptiles such as crocodilians and alligators excrete large amounts of ammonium ions in an alkaline urine, although their diet is largely acidic.[70,71] In alligator, bicarbonate is secreted in the tubular lumen, whereas hydrogen is secreted at the peritubular site, creating an overall acid–base balance.[70,72] There is no evidence for a Na+/H+ antiporter system in alligators, although this is the primary bicarbonate transporter in mammals.[70] Carbonic anhydrase is found in the connecting segment and collecting duct of alligator kidney, indicating that the main area of bicarbonate addition to urine is in the distal tubule.[70] Normally, alligators have low plasma bicarbonate and low blood pH as a result of lactate, but the excreted urine is still alkaline.[71]

Overall, alligator kidney is both ammoniagenic and gluconeogenic and contains metabolic enzymes similar to what is found in mammals, including glutamate dehydrogenase, alanine aminotransferase, glutaminase I, phosphoenolpyruvate carboxykinase (PEPCK), malate dehydrogenase, and lactate dehydrogenase.[71] Studies in snake (*Thamnophis* spp.) kidney have confirmed an active glucose transporter system.[73]

In terms of calcium homeostasis, the calcium binding protein, calbindin-D28K, has been identified in the distal tubule cells of the red-eared slider turtle by immunoblotting and immunocytochemistry.[74] In addition, a vitamin D–dependent calcium binding protein (CaBP) of 28 kDa has been identified in kidneys of two species of lizards.[75]

B. Adrenals

1. The Renin–Angiotensin and Vasopressin Systems

In the mammalian system, the enzyme renin is released by the kidney's juxtaglomerular body in response to low blood pressure and low filtrate K+ levels. Renin acts on the plasma protein angiotensinogen (produced by the liver) to produce angiotensin. In the lungs, angiotensin is converted by angiotensin-converting enzyme (ACE) to angiotensin II, which has four primary modes of action. Angiotensin II acts on the following:

1. The brain to release arginine-vasopressin (also known as antidiuretic hormone) to cause water retention and increased blood pressure
2. The brain to induce thirst
3. The blood vessels to increase peripheral vasoconstriction and total peripheral resistance to increase blood pressure
4. The adrenals to release aldosterone, which causes tubular reabsorption of Na+, which in turns causes reabsorption of water and increased blood volume and blood pressure

During times of high blood volume, atrial natriuretic factor (ANF) released by the heart will cause excretion of excess water via the kidney by increasing glomerular filtration rate and decreasing Na^+-reabsorption. Therefore, ANF and the renin angiotensin system work together to maintain proper blood pressure.

Renin, angiotensin, angiotensin II, and ACE have all been identified in several freshwater turtle species and have mammalian-like functions,[76,77] and angiotensin I has been identified in alligator as well.[78] In turtle, angiotensin II has been shown to be a potent constrictor of preglomerular blood vessels in addition to its other function in increasing arterial blood pressure.[79] Reptilian kidneys excrete either isoosmotic or hypoosmotic urine,[62] and water can be secreted directly into the proximal tubule (bypassing glomerular filtration) under hydrated conditions.[80] It appears that arginine vasotocin acts as an antidiuretic hormone in reptiles (for review, see reference 62). Arginine vasotocin–like receptors have been located in the intermediate segment and proximal and distal convoluted tubules of reptiles,[62] and Bradshaw and Bradshaw[62] suggest that arginine vasotocin (AVT) has a dual function in reptiles: (1) dilution of urinary fluid in the thin-intermediate segment (prior to the collecting duct), and (2) facilitating water reabsorption in the final segments of the nephron. Atrial natriuretic peptide receptor has also been identified in the kidney and adrenal gland of freshwater turtle *Amyda japonica*.[77]

2. *Adrenal Functions and Stress*

In mammals, adrenals release numerous stress hormones for regulation of short-term stress (epinephrine, norepinephrine) and long-term stress (corticosteroids) and are highly conserved among vertebrate taxa. The reptilian stress response is reviewed in detail by Guillette and colleagues[81] and will be reviewed here in context of toxicant interactions.

Both epinephrine and norepinephrine are quickly metabolized and eliminated, but corticosteroids have a much more complex regulatory framework, including steroidal synthesis, control by the hypothalamus–pituitary axis (CRH→ACTH→corticosterone), and metabolism in peripheral tissues or the liver.[82] As defined by Pottinger,[82] long-term stressors affecting wildlife can be (1) physical (e.g., temperature, ultraviolet light exposure, parasitic infections); (2) chemical (e.g., low DO [dissolved oxygen], change in pH, exposure to toxicants); (3) physiologic (e.g., starvation, disease); or (4) psychological (e.g., threat of predation, territoriality). The behavioral and neuroendocrine processes involved in dealing with these types of stressors are discussed in Chapter 7.

Few studies have been done on stress-hormone levels in reptiles in relation to toxicant exposures. This is partly the result of the high variability in capture-stress responses, as was demonstrated by Gregory and colleagues[83] in loggerhead turtle (*Caretta caretta*). In the loggerhead, there was both size- and season-dependent changes in corticosterone levels, as well as differences among capture methods. As a result of this inherent difficulty in measuring corticosterone-toxicant alterations, few studies have been done in reptiles. Most work has been done in alligator (*Alligator mississipiensis*), which are surprisingly resistant to toxicant-induced changes in corticosteroids, although there are toxicant-induced changes in sex-steroid hormones in the same individuals.[84] In a later study, it was observed that there was a much greater variability in stress response, especially in female alligators, in higher-contaminant load lakes, indicating that homeostatic and stress-response physiology could be altered by contaminants.[85] Although some research has been done on the interactions of toxicants on the hypothalamus–pituitary–adrenal axis in fish, amphibians, birds, and mammals, no studies have focused on this axis in reptiles.[82]

Several researchers have noted variations in stress responses depending on sex, season, and reproductive status within the sexes, including in turtle, tuatara, snake, and alligator.[83,85–89] As with other vertebrates, high corticosterone levels can affect reproduction and development,[24,82,88,89] especially in viviparous species. Cree and colleagues[90] showed that embryonic growth and development could be altered by exogenous maternal administration of corticosteroid in gecko (*Hoplodactylus maculatus*). Capture stress has been well documented to inhibit vitellogenesis, and it can be notoriously difficult to obtain breeding in captivity.[24] Because toxicants can be adreno-toxic themselves or can lead to chemical stress (as defined earlier), there is the possibility that changes in corticosterone could affect not only the adult, but also the production of offspring.

As reviewed by Pottinger,[82] there are numerous adreno-toxicants, including cadmium, DDT and its metabolites, and dioxins. Because kidneys are a target organ for Cd, it is possible that this toxicant could also affect adrenal glands in reptiles; however, this type of research has been done only in mammalian, fish, and amphibian systems.

C. *Hormone Metabolism and Hormone Receptors*

The reptilian adrenal–kidney complex is involved in steroidogenesis for both sex steroids and glucocorticoids,[91] and ACTH can induce steroid secretion in turtle and crocodilian adrenal–kidney complex.[92] The steroidogenesis includes aromatase activity for converting androgens to estrogens during embryonic development, although it appears that brain aromatase activity is more important in temperature-dependent sex determination.[93] (See Chapter 6 for additional details.)

In turtle, thyroxine (T4) is transported by the vitamin D–binding protein (along with vitamin D), which is highly expressed in many tissues, including the kidney.[94] In mammalian systems the liver is the primary metabolizing organ of the thyroid hormones, but it appears that in reptiles the kidney performs this function. Metabolism of thyroid hormones has been reported in kidney of several reptile species, including reverse T3 and T4 deiodination in saltwater crocodile, *Crocodylus porosus*,[95,96] and two forms of monodeiodinases in the red-eared slider.[97] According to Hugenberger and Licht,[97] "the kidney may play a critical role in the metabolism of thyroid hormones in the turtle," indicating that renal toxicants could potentially alter thyroid hormone metabolism and therefore act as endocrine-disrupting chemicals.

In addition, hormone receptors have been found in the kidney of various reptilians, including a progesterone receptor in the red-eared slider[98] and a prolactin receptor in the common rat snake, *Ptyas mucosa*.[99,100]

VIII. Kidney Function in Metabolism of Xenobiotics and in Antioxidant Defense

It has been a long-standing premise that drugs should not be injected into the caudal body of reptiles because they will be carried by the renal-portal system to the kidneys before entering the central circulation. In the kidney they could be either rapidly metabolized or cause nephrotoxicity.[101] Holz and colleagues[101] studied whether there was a difference in clearance of drugs based on forelimb or hind limb injections of the red-eared slider (comparing two different drugs cleared by the kidney in mammals). One of these drugs (carbenicillin) had significantly delayed bioavailability if injected in the hind limb, confirming this renal-portal system in the turtle.[101]

The kidney in snakes seems to be an important xenobiotic metabolizing organ. Hunter and colleagues[102] studied the metabolism of azithromycin in ball pythons (*Python regius*) and found that kidney tissue contained a greater number of metabolites than liver tissue. Gopher snakes (*Pitophis melanoleucus* catenifer) that were administered high doses of gentamycin (50 mg/kg/day) showed nephrotoxicity, specifically proximal tubule necrosis, which led to visceral gout.[103]

Freshwater turtles have been studied for their ability to survive anoxic submergence using antioxidant defenses such as GSH. In the red-eared slider, anoxia exposure resulted in depression of gamma-glutamyl transpeptidase (gamma-GTPase) activity in the kidney to only 2% of control. During aerobic recovery, GST activity decreased in the kidney by 56%.[53] Turtles may be able to regulate GSH levels in tissues by regulating the activities of GSH-using enzymes.[53]

It is possible that anoxia may alter the ability of turtles to handle other stressors, which may affect the antioxidant defense system.

IX. Kidney as a Target Organ for Metals

The accumulation of heavy metals specifically into the kidney of various reptilian groups has been extensively studied, especially in the crocodilians and marine turtles. High levels of mercury have been found in the kidney of juvenile alligators from Florida, South Carolina, and Georgia.[104–106] In marine turtles, trace elements and metals have been shown to accumulate in the kidney as well.[107,108] In a controlled feeding study on banded water snakes, *Nerodia fasciata*, accumulation of heavy metals was found to be dose dependent (with the exception of Cu) and highly variable among sizes and between sexes.[109] Over the 2-year study, although Se levels were exceedingly high, the snakes showed no physiologic changes as compared with controls.[109] In a Cd-injection study, the painted turtle was able to eliminate all but 9% of the dose after 1 week, and Cd was preferentially accumulated in a number of organs, including the kidney, which also had a modest induction of cadmium-binding protein.[110]

X. Summary and Connections to Other Chapters

Because of the importance of liver, kidney, and adrenals in maintaining homeostasis, toxicant-induced changes to enzymes, proteins, and the antioxidant defense system could significantly alter the ability of the organism to thrive. Of special note should be the interaction of the liver, kidneys, and adrenals with other parts of the endocrine system, such as glucocorticoids inhibiting the production of Vtg, ultimately leading to unsuccessful or limited reproduction. Up-regulation or down-regulation of certain enzymes and proteins may be useful in biomonitoring and is covered more closely in Chapter 4. Because the liver and kidneys are also important in reproductive functions, some additional detail on vitellogenesis and steroidogenesis is covered in Chapter 6, and behavioral and neuroendocrine changes associated with stress are covered in Chapter 7.

Literature Cited

1. Boyer, T.H. and Boyer, D.M., Turtles, tortoises and terrapins, in *Reptile Medicine and Surgery*, Mader, D.R., Ed., W.B. Saunders Company, Philadelphia, 1998, pp. 61–78.

2. Hering, E., Ueber den Bau der Wirbeltierleber, *Arch. Mikrosk. Anat.*, 3, 88–114, 1867.
3. Storch, V., Braunbeck, T., and Waitkuwait, W., The liver of the West African crocodile *Osteolamaemus tetrapis*. An ultrastructural study, *J. Submicr. Cytol. Pathol.*, 21, 317–327, 1989.
4. Frye, F.L., *Biomedical and Surgical Aspects of Captive Reptile Husbandry*, Krieger Publishing Company, Malabar, FL, 1991.
5. Beresford, W.A., Fibrous trabeculae in the liver of alligator (*Alligator mississippiensis*), *Anat. Anz.*, 175 (4), 357–359, 1993.
6. Lane, T.J., Crocdilians, in *Reptile Medicine and Surgery*, Mader, D.R., Ed., W.B. Saunders Company, Philadelphia, 1998, pp. 78–94.
7. Mader, D.R., *Reptile Medicine and Surgery*, W.B. Saunders Company, Philadelphia, 1998.
8. Xu, G., Elsey, R.M., Lance, V.A., Javors, F., Chen, T.S., Salen, G., and Tint, G.S., A study on biliary ductal system and bile fistula in the american alligator, *Alligator mississippiensis*, *J. Exp. Zool.*, 277, 554–561, 1997.
9. Christiansen, J., Grzybowski, J., and Kodama, R., Melanomacrophage aggregations and their age relationships in the yellow mud turtle, *Kinosternon flavescens* (Kinosternidae), *Pigment Cell. Res.*, 9 (4), 185–190, 1996.
10. Janssens, P.A. and Grigg, J.A., Adrenergic regulation of glycogenolysis in liver of *Xenopus laevis, in vitro, Comp. Biochem. Physiol.*, 77C, 403–408, 1984.
11. Penhos, J.C. and Ramey, E., Studies on the endocrine pancreas of amphibians and reptiles, *Am. Zool.*, 13, 667–698, 1973.
12. Janssens, P.A. and Giuliano, M., Hormonal regulation of hepatic glycogenolysis in *Amphibolurus nuchalis*, the western netted dragon: an in vitro study, *J. Comp. Physiol.*, 159, 323–331, 1986.
13. Nilsson, S., in *Autonomic Nerve Function in the Vertebrates*, Farner, D., Heinrich, B., Johansen, K., Langer, H., Neuweiler, G., and Randall, D., Eds., Springer-Verlag, Berlin Heidelberg, 1983.
14. Fabbri, E., Barbin, L., and Capuzzo, A., Coexistence of alpha 1 and beta adrenergic receptors in the liver of the frog *Rana esculenta*, the toad *Bufo bufo*, the lizard *Podarcis sicula campestris*, and the turtle *Pseudemys picta elegans*, *Gen. Comp. Endocrinol.*, 107 (3), 351–358, 1997.
15. Hemmings, S.J. and Storey, K.B., Hepatic changes in the freeze-tolerant turtle, *Chrysemys picta marginata*, in response to freezing and thawing, *Cell Biochem. Func.*, 18, 175–186, 2000.
16. Kester, J., Liver, in *Encyclopedia of Toxicology*, Wexler, P., Ed., Academic Press, San Diego, 1998, pp. 253–261.
17. Simandle, E.T., Espinoza, R.E., Nussear, K.E., and Tracy, C.R., Lizards, lipids and dietary links to animal function, *Physiol. Biochem. Zool.*, 74, 625–640, 2001.
18. Tosheva, R.T., Jankowski, W.A., Stoykova, L.O., and Chojnacki, T., Dolichols in the liver of some poikilothermic vertebrates and birds, *Comp. Biochem. Physiol.*, 98B, 397–402, 1991.
19. Staton, M.A., Edwards, H.M.J., Brisbin, I.L.J., Joanen, T., and McNease, L., Essential fatty acid nutrition of the American alligator (*Alligator mississippiensis*), *J. Nutr.*, 120, 674–685, 1990.
20. Guitart, R., Silvestre, A.M., Guerrero, X., and Mateo, R., Comparative study on the fatty acid composition of two marine vertebrates: striped dolphins and loggerhead turtles, *Comp. Biochem. Physiol.*, 124, 439–443, 1999.

21. Speake, B.K., Thompson, M.B., Thacker, F.E., and Bedford, G.S., Distribution of lipids from the yolk to the tissues during development of the water python (*Liasis fuscus*), *J. Comp. Physiol.*, 173, 541–547, 2003.
22. Lacy, E., Sheridan, M., and Moore, M., Sex differences in lipid metabolism during reproduction in free-living tree lizards (*Urosaurus ornatus*), *Gen. Comp. Endocrinol.*, 128 (3), 180–192, 2002.
23. Guillette, L.J. and Edwards, T., Is nitrate an ecologically relevant endocrine disruptor in vertebrates? *Integr. Comp. Biol.* 45, in press, 2005.
24. Morales, M. and Sanchez, E., Changes in vitellogenin expression during captivity-induced stress in a tropical anole, *Gen. Comp. Endocrinol.*, 103 (2), 209–219, 1996.
25. Ho, S.M., L'Italien, J., and Callard, I.P., Studies on reptilian yolk: chrysemis vitellogenin and phosvitin, *Comp. Biochem. Physiol.*, 65B, 139–144, 1980.
26. Bast, R.R. and Gibson, A.R., Characterization of reptilian vitellogenin: subunit composition and molecular weights of vitellogenin from the colubrid snake, *Thamnophis sirtalis* (L), *Comp. Biochem. Physiol.*, 80B, 409–418, 1985.
27. Morales, M.H., Baerga-Santini, C., and Cordero-Lopez, N., Synthesis of vitellogenin polypeptides and deposit of yolk proteins in *Anolis pulchellus*, *Comp. Biochem. Physiol.*, 114B, 225–231, 1996.
28. Nebert, D., Drug-metabolizing enzymes in ligand-modulated transcription, *Biochem. Pharmacol.*, 47 (1), 25–37, 1994.
29. Guillette, L.J., Jr. and Iguchi, T., Contaminant-induced endocrine disruption and reproductive alterations in reptiles, *Pure Appl. Chem.*, 75 (11–12), 2275–2286, 2003.
30. Schwen, R.J. and Mannering, G.J., Hepatic cytochrome P-450-dependent monooxygenase systems of the trout, frog and snake-III. Induction, *Comp. Biochem. Physiol.*, 71B, 445–453, 1982.
31. Andersson, T. and Nilsson, E., Characterization of cytochrome P-450-dependent activities in hagfish, dogfish, perch and spectacle caiman, *Comp. Biochem. Physiol.*, 94B, 99–105, 1989.
32. Jewell, C., Cummings, L., Ronis, M., and Winston, G., Induction of the hepatic microsomal mixed-function oxygenase (MFO) system of *Alligator mississippiensis* by 3-methylcholanthrene (3-MC), *Mar. Environ. Res.*, 28, 73–79, 1989.
33. Ertl, R., Bandiera, S., Buhler, D., Stegeman, J., and Winston, G., Immunochemical analysis of liver microsomal cytochromes P450 of the American alligator, *Alligator mississippiensis*, *Toxicol. Appl. Pharmacol.*, 157 (3), 157–165, 1999.
34. Gunderson, M., Oberdörster, E., and Guillette, L.J., Jr., Phase I and II liver enzyme activities in juvenile alligators (*Alligator mississippiensis*) collected from three sites in the Kissimmee–Everglades drainage, Florida (USA), *Comp. Biochem. Physiol.*, 139C, 39–46, 2004.
35. Mayeaux, M.H. and Winston, G.W., Antibiotic effects on cytochromes P450 content and mixed-function oxygenase (MFO) activities in the American alligator, *Alligator mississippiensis*, *J. Vet. Pharmacol. Ther.*, 21, 274–281, 1998.
36. Yawetz, A., Goldman, D., Stegeman, J.J., Woodin, B., Adin, A., and Gasith, A., The effect of PCB on the induction of cytochrome P450 enzyme, detected by monoclonal antibody against the P4501A1 gene products in barn owl and marsh turtle hepatic microsomes, *Water Sci. Technol.*, 27, 457–464, 1993.
37. Yawetz, A., Woodin, B.R., and Stegeman, J.J., Cytochrome P450 in liver of the turtle *Chyrsemys picta picta* and the induction and partial purification of CYP1A-like proteins, *Biochim. Biophys. Acta*, 1381, 12–26, 1998.

38. Stegeman, J.J. and Hahn, M.E., Biochemistry and molecular biology of monooxygenases: current perspectives on forms, functions, and regulation of cytochrome P450 in aquatic species, in *Aquatic Toxicology*, Malins, D.C. and Ostrander, G.K., Eds., CRC Press, Boca Raton, FL, 1994, pp. 87–206.
39. Goldman, R. and McClellan-Green, P.D., Isolation and characterization of P450 enzymes from Kemps Ridley (*Lepidochelys kempii*) liver, in *22nd Annual Meeting Society of Environmental Toxicology and Chemistry, Baltimore, MD, November 11–15, 2001*, 2001.
40. Ronis, M.J.J., Andersson, T., Hansson, T., and Walker, C.H., Differential expression of multiple forms of cytochrome P450s as molecular probes for the evolution of P-450 gene families I and II, *Mar. Environ. Res.*, 28, 131–135, 1989.
41. Herman, C. and Oliw, E., Liver microsomes from the yellow rat snake (*Elaphe obsoleta*) and American bullfrog (*Rana catesbeiana*) oxidize polyunsaturated fatty acids by NADPH-dependent hydroxylation and epoxidation, *J. Exp. Zool.*, 280 (1), 1–7, 1998.
42. Haasch, M.L., Wejksnora, P.J.S., and Lech, J.J., Cloned rainbow trout liver P(1)450 complementary DNA as a potential environmental monitor, *Toxicol. Appl. Pharmacol.*, 98, 362–368, 1989.
43. Yawetz, A., Woodin, B.R., Smolowitz, R.M., and Stegeman, J.J., Induction, fractionation and localization of cytochrome P450 isozymes in the liver of the freshwater turtle, *Chrysemys picta picta*, *Mar. Environ. Res.*, 35, 205–206, 1993.
44. McArthur, A.G., Hegelund, T., Cox, R.L., Stegeman, J.J., Liljenberg, U.O., Sundberg, P., and Celander, M.C., Phylogenetic analysis of the cytochrome P450 3 (CYP3) gene family, *J. Molec. Evol.*, 57, 200–211, 2003.
45. Hermes-Lima, M. and Storey, K., *In vitro* oxidative inactivation of glutathione S-transferase from a freeze tolerant reptile, *Molec. Cell. Biochem.*, 124, 149–158, 1993.
46. Poulos, T. and Raag, R., Cytochrome P450cam: crystallography, oxygen activation, and electron transfer, *FASEB J.*, 6 (2), 674–679, 1992.
47. Yu, B., Cellular defenses against damage from reactive oxygen species, *Physiol. Rev.*, 74 (1), 139–162, 1994.
48. Willmore, W.G. and Storey, K.B., Antioxidant systems and anoxia tolerance in a freshwater turtle *Trachemys scripta elegans*, *Molec. Cell. Biochem.*, 170, 177–185, 1997.
49. Hermes-Lima, M. and Storey, K.B., Role of Antioxidant defenses in the tolerance of freezing and anoxia by garter snakes, *Am. J. Physiol.*, 265, R646–R652, 1993.
50. de Brito-Gitirana, L. and Storch, V., Temperature induced alterations in the liver of wall lizard (*Hemidactylus frenatus*): morphological and biochemical parameters, *Micron* 33, 667–672, 2002.
51. Sichel, G., Corsaro, C., Scalia, M., Sciuto, S., and Geremia, E., Relationship between melanin content and superoxide dismutase (SOD) activity in the liver of various species of animals, *Cell Biochem. Func.*, 5, 123–128, 1987.
52. Henninger, J.M. and Beresford, W.A., Is it coincidence that iron and melanin coexist in hepatic and other melanomacrophages? *Histol. Histopathol.*, 5, 457–594, 1990.
53. Willmore, W. and Storey, K., Glutathione systems and anoxia tolerance in turtles, *Am. J. Physiol.*, 273 (1 Pt 2), 219–225, 1997.

54. Lucké, B. and Schlumberger, H.G., Neoplasia in cold-blooded vertebrates, *Physiol., Rev.*, 29, 91–126, 1949.
55. Frye, F.L., Common pathologic lesions and disease processes: neoplasia., in *Reptile Care. An Atlas of Diseases and Treatments*, Frye, F. L., Ed., T. H. F. Publication Inc., Neptune City, NJ, 1991, pp. 576–619.
56. Catão-Dias, J.L. and Nichols, D.K., Neoplasia in snakes at the National Zoological Park, Washington, DC (1978–1997), *J. Comp. Pathol.*, 120, 89–95, 1999.
57. Schultze, A.E., Mason, G.L., and Clyde, V.L., Lymphosarcoma with a leukemic blood profile in a savannah monitor (*Varanus exanthematicus*), *J. Zòo Wildl. Med.*, 30, 158–164, 1999.
58. Martorell, J., Ramis, A., and Espada, Y., Use of ultrasonography in the diagnosis of hepatic spindle-cell sarcoma in a savannah monitor (*Varanus exanthematicus*), *Vet. Rec.*, 150, 282–284, 2002.
59. Willingham, E. and Crews, D., The red-eared slider turtle: an animal model for the study of low doses and mixtures, *Am. Zool.*, 40 (3), 421–428, 2000.
60. Beuchat, C. and Braun, E., Allometry of the kidney: implications for the ontogeny of osmoregulation., *Am. J. Physiol.*, 255 (5 Pt 2), 760–767, 1988.
61. Spotila, L. and Hall, S., Expression of a new RNA-splice isoform of WT1 in developing kidney-gonadal complexes of the turtle, *Trachemys scripta*, *Comp. Biochem. Physiol. B. Biochem. Mol. Biol.*, 119 (4), 761–767, 1998.
62. Bradshaw, S. and Bradshaw, F., Arginine vasotocin: site and mode of action in the reptilian kidney, *Gen. Comp. Endocrinol.*, 126 (1), 7–13, 2002.
63. Solomon, S., The morphology of the kidney of the green turtle (*Chelonia mydas* L.). *J. Anat.*, 140 (Pt 3), 355–369, 1985.
64. Yokota, S., Ophidian kidney preparation for the measurement of glomerular dynamics in real-time, *Pflugers Arch.*, 415 (4), 501–503, 1990.
65. Yokota, S. and Dantzler, W., Measurements of blood flow to individual glomeruli in the ophidian kidney, *Am. J. Physiol.*, 258 (6 Pt 2), 1313–1319, 1990.
66. Fox, H., The urogenital system of reptiles, in *Biology of the Reptilia*, Gans, C. and Parsons, T., Eds., Academic Press, London, 1977, pp. 157.
67. Sever, D., Stevens, R., Ryan, T., and Hamlett, W., Ultrastructure of the reproductive system of the black swamp snake (*Seminatrix pygaea*). III. Sexual segment of the male kidney, *J Morphol.*, 252 (3), 238–254, 2002.
68. Kuchling, G., *The Reproductive Biology of the Chelonia*, Springer, New York, 1999.
69. Al-Awqati, Q., Gluck, S., Reeves, W., and Cannon, C., Regulation of proton transport in urinary epithelia, *J. Exp. Biol.*, 106, 135–141, 1983.
70. Ventura, S., Northup, T., Schneider, G., Cohen, J., and Garella, S., Transport and histochemical studies of bicarbonate handling by the alligator kidney, *Am. J. Physiol.*, 256 (2 Pt 2), 239–245, 1989.
71. Lemieux, G., Craan, A., Quenneville, A., Lemieux, C., Berkofsky, J., and Lewis, V., Metabolic machinery of the alligator kidney, *Am. J. Physiol.*, 247 (4 Pt 2), 686–693, 1984.
72. Lemieux, G., Berkofsky, J., Quenneville, A., and Lemieux, C., Net tubular secretion of bicarbonate by the alligator kidney. Antimammalian response to acetazolamide, *Kidney Int.*, 28 (5), 760–766, 1985.
73. Barfuss, D. and Dantzler, W., Glucose transport in isolated perfused proximal tubules of snake kidney, *Am. J. Physiol.*, 231 (6), 1716–1728, 1976.
74. Mutema, G. and Rhoten, W., Occurrence and localization of calbindin-D28K in kidney and cerebellum of the slider turtle, *Trachemys scripta*, *Anat. Rec.*, 239 (2), 185–190, 1994.

75. Rhoten, W., Bruns, M., and Christakos, S., Presence and localization of two vitamin D-dependent calcium binding proteins in kidneys of higher vertebrates, *Endocrinology,* 117 (2), 674–683, 1985.
76. Cipolle, M. and Zehr, J., Characterization of the renin-angiotensin system in the turtle *Pseudemys scripta, Am. J. Physiol.*, 247 (1 Pt 2), 15–23, 1984.
77. Kim, S., Kim, S., and Cho, K., Overlapping distribution of receptors for atrial natriuretic peptide and angiotensin II in the kidney and the adrenal gland of the freshwater turtle, *Amyda japonica, Gen. Comp. Endocrinol.*, 108 (1), 119–131, 1997.
78. Takei, Y., Silldorff, E., Hasegawa, Y., Watanabe, T., Nakajima, K., Stephens, G., and Sakakibara, S., New angiotensin I isolated from a reptile, *Alligator mississippiensis, Gen. Comp. Endocrinol.*, 90 (2), 214–219, 1993.
79. Brown, S., Stephens, G., and Todt, M., Systemic and renal effects of angiotensin II in the freshwater turtle *Pseudemys scripta elegans, Am. J. Physiol.*, 245 (6), 837–842, 1983.
80. Fleishman, D., Nikiforov, V., Saulus, A., Vasilieva, V., and Borkin, L., Lithium secretion in kidneys of amphibians and reptiles under hydrated conditions, *Comp. Biochem. Physiol. A. Physiol.*, 118 (4), 1259–1265, 1997.
81. Guillette, L., Cree, A., and Rooney, A., Biology of stress: interactions with reproduction, immunology and intermediary metabolism, in *Health and Welfare of Captive Reptiles*, Warwick, C., Grye, F., and Murphy, J., Eds., Chapman and Hall, London, 1995, pp. 32–81.
82. Pottinger, T.G., Interaction of endocrine-disrupting chemicals with stress responses in wildlife, *Pure Appl. Chem.*, 75 (11–12), 2321–2333, 2003.
83. Gregory, L., Gross, T., Bolten, A.B., Bjorndal, K.A., and Guillette, L.J., Plasma corticosterone concentrations associated with acute captivity stress: effects of capture method, season, size class, and sex in wild Loggerhead sea turtles (*Caretta caretta*), *Gen. Comp. Endocrinol.*, 104, 312–320, 1996.
84. Guillette, L., Crain, D., Rooney, A., and Woodward, A., Effect of acute stress on plasma testosterone, estradiol-17b and corticosterone concentrations in juvenile alligators living in control and contaminated lakes, *J. Herpetol.*, 31, 347–353, 1997.
85. Gunderson, M., Kools, S., Milnes, M., and Guillette, L.J., Effect of acute stress on plasma b-corticosterone, estradiol-17b and testosterone concentrations in juvenile American alligators collected from three sites within the Kissimmee–Everglades drainage basin in Florida, *Comp. Biochem. Physiol.*, 135C (3), 365–374, 2003.
86. Cree, A., Guillette, L.J., Cockrem, J.F., Brown, M.A., and Chambers, G.K., Absence of daily cycles in plasma sex steroids in male and female tuatara (*Sphenodon punctatus*), and the effects of acute capture stress on females, *Gen. Comp. Endocrinol.*, 79 (1), 103–113, 1990.
87. Mahmoud, I. and Licht, P., Seasonal changes in gonadal activity and the effects of stress on reproductive hormones in the common snapping turtle, *Chelydra serpentina, Gen. Comp. Endocrinol.*, 107, 359–372, 1997.
88. Moore, I., Greene, M., and Mason, R., Environmental and seasonal adaptations of the adrenocortical and gonadal responses to capture stress in two populations of the male garter snake, *Thamnophis sirtalis., J. Exp. Zool.*, 289, 99–108, 2001.
89. Moore, I. and Mason, R., Behavioral and hormonal responses to corticosterone in the male red-sided garter snake, *Thamnophis sirtalis sirtalis, Physiol. Behav.*, 72, 669–674, 2001.

90. Cree, A., Tyrrell, C., Preest, M., Thorburn, D., and Guillette, L.J., Protecting embryos from stress: corticosterone effects and the corticosterone response to capture and confinement during pregnancy in a live-bearing lizard (*Hoplodactylus maculatus*), *Gen. Comp. Endocrinol.*, 134 (3), 316–329, 2003.
91. White, R. and Thomas, P., Adrenal-kidney and gonadal steroidogenesis during sexual differentiation of a reptile with temperature-dependent sex determination, *Gen. Comp. Endocrinol.*, 88 (1), 10–19, 1992.
92. Nothstine, S., Davis, J., and DeRoos, R., Kidney extracts and ACTH on adrenal steroid secretion in a turtle and a crocodilian, *Am. J. Physiol.*, 221 (3), 726–732, 1971.
93. Willingham, E., Baldwin, R., Skipper, J., and Crews, D., Aromatase activity during embryogenesis in the brain and adrenal-kidney-gonad of the red-eared slider turtle, a species with temperature-dependent sex determination, *Gen. Comp. Endocrinol.*, 119 (2), 202–207, 2000.
94. Whitworth, D., Hunt, L., and Licht, P., Widespread expression of the mRNA encoding a novel vitamin D/thyroxine dual binding protein in the turtle *Trachemys scripta*, *Gen. Comp. Endocrinol.*, 118 (2), 354–358, 2000.
95. Shepherdley, C., Richardson, S., Evans, B., Kuhn, E., and Darras, V., Characterization of outer ring iodothyronine deiodinases in tissues of the saltwater crocodile (*Crocodylus porosus*), *Gen. Comp. Endocrinol.*, 125 (3), 387–398, 2002.
96. Shepherdley, C., Richardson, S., Evans, B., Kuhn, E., and Darras, V., Thyroid hormone deiodinases during embryonic development of the saltwater crocodile (*Crocodylus porosus*), *Gen. Comp. Endocrinol.*, 126 (2), 153–164, 2002.
97. Hugenberger, J. and Licht, P., Characterization of thyroid hormone 5′-monodeiodinase activity in the turtle (*Trachemys scripta*), *Gen. Comp. Endocrinol.*, 113 (3), 343–359, 1999.
98. Custodia-Lora, N. and Callard, I.P., Progesterone and progesterone receptors in reptiles, *Gen. Comp. Endocrinol.*, 127 (1), 1–7, 2002.
99. Cheng, C., Lee, H., Ng, T., and Wong, C., Presence of prolactin receptors in kidney and large intestine of the snake *Ptyas mucosa*, *Gen. Comp. Endocrinol.*, 79 (3), 351–360, 1990.
100. Ng, T., Lee, H., Cheng, C., and Wong, C., Partial purification of prolactin-like substance from snake (*Ptyas mucosa*) pituitaries, *Endocrinol. Jpn.*, 37 (6), 777–786, 1990.
101. Holz, P., Barker, I., Burger, J., Crawshaw, G., and Conlon, P., The effect of the renal portal system on pharmacokinetic parameters in the red-eared slider (*Trachemys scripta elegans*), *J. Zoo. Wildl. Med.*, 28 (4), 286–293, 1997.
102. Hunter, R., Koch, D., Coke, R., Goatley, M., and Isaza, R., Azithromycin metabolite identification in plasma, bile, and tissues of the ball python (*Python regius*), *J. Vet. Pharmacol. Ther.*, 26 (2), 117–121, 2003.
103. Montali, R., Bush, M., and Smeller, J., The pathology of nephrotoxicity of gentamicin in snakes. A model for reptilian gout, *Vet. Pathol.*, 16 (1), 108–115, 1979.
104. Jagoe, C., Arnold-Hill, B., Yanochko, G., Winger, P., and Brisbin, I.J., Mercury in alligators (*Alligator mississippiensis*) in the southeastern United States, *Sci. Total Environ.*, 213 (1–3), 255–262, 1998.
105. Khan, B. and Tansel, B., Mercury bioconcentration factors in American alligators (*Alligator mississippiensis*) in the Florida everglades, *Ecotoxic. Environ. Saf.*, 47 (1), 54–58, 2000.

106. Yanochko, G., Jagoe, C., and Brisbin, I.J., Tissue mercury concentrations in alligators (*Alligator mississippiensis*) from the Florida Everglades and the Savannah River site, South Carolina, *Arch. Environ. Contam. Toxicol.*, 32 (3), 323–328, 1997.
107. Anan, Y., Kunito, T., Watanabe, I., Sakai, H., and Tanabe, S., Trace element accumulation in hawksbill turtles (*Eretmochelys imbricata*) and green turtles (*Chelonia mydas*) from Yaeyama Islands, Japan, *Environ. Toxicol. Chem.*, 20 (12), 2802–2814, 2001.
108. Sakai, H., Saeki, K., Ichihashi, H., Kamezaki, N., Tanabe, S., and Tatsukawa, R., Growth-related changes in heavy metal accumulation in green turtle (*Chelonia mydas*) from Yaeyama Islands, Okinawa, Japan, *Arch. Environ. Contam. Toxicol.*, 39 (3), 378–385, 2000.
109. Hopkins, W., Roe, J., Snodgrass, J., Staub, B., Jackson, B., and Congdon, J., Effects of chronic dietary exposure to trace elements on banded water snakes (*Nerodia fasciata*), *Environ. Toxicol. Chem.*, 21 (5), 906–913, 2002.
110. Rie, M., Lendas, K., and Callard, I., Cadmium: tissue distribution and binding protein induction in the painted turtle, *Chrysemys picta*, *Comp. Biochem. Physiol. C. Toxicol. Pharmacol.*, 130 (1), 41–51, 2001.

chapter 6

Developmental and Reproductive Effects

Emily Willingham

Table of Contents

I. An Overview

A. Focusing on Endocrine Disruptors

Why development and reproduction? The events of development lay the groundwork for everything that follows for an organism. Toxicants can completely disrupt development, but in some cases their effects are more subtle. Rather than manifesting acutely, the effects of some compounds do not manifest until later in life, as circuits laid down during development under endogenous and exogenous influences are activated. Often, these mechanisms relate in some way to reproduction — fertility, mating behavior, sex-based behaviors, parental behaviors — and their proper coordination can influence the success of the individual and population. Thus, developmental events and reproduction go hand in hand, and toxicants that disrupt one — i.e., developmental processes — ultimately can disrupt the other.

Reptiles have unique characteristics when it comes to reproduction, development, and toxicants, especially toxicants that affect endocrine communication. Reptilians encompass a diverse group of organisms — snakes, lizards, crocodilians, and the tuatara. They also encompass a broad diversity of reproductive characteristics ranging from oviparity to viviparity to ovoviviparity and a range of sex-determining mechanisms and levels of parental care. They are a widely distributed taxon with varied roles in ecosystems that expose them in different ways to toxicants. Reptiles can be aquatic, arboreal, or terrestrial, or a combination of these, and thus can experience routes of exposure more generally than some other taxa.

When considering toxicants in the context of reptiles, it is important to keep this variation in mind and to avoid the pitfalls of extrapolation that is too broad or conclusions that are too narrow. Endocrine systems in the taxon can vary widely, and research indicates an equally varied response to toxicants at the developmental level, which will be discussed in detail later in this chapter.

The method of bearing offspring can make a difference in terms of exposure; offspring that emerge from an egg laid early in development experience both the egg environment and the environment outside the egg. Offspring retained in the mother for a length of time are exposed to the maternal environment, influenced by what she eats, drinks, and breathes.

B. Temperature-Dependent Sex Determination

The life history strategy for determining sex can also influence susceptibility to certain toxicants, especially those that disrupt endocrine pathways. In many species, the sex-determining pathway exists under the influence of temperature, and an organismal decision point about sex is not reached until the middle-third of development, rather than occurring at conception. This delay of the sex-determining decision confers some presumed benefits on the organism, and an entire cottage industry of investigations into its evolutionary advantages exists.[1–5] Most researchers subscribe to some variation on the Charnov-Bull hypothesis[6] that the delaying of sex determination allows for the development of the sex that will be better fit for the environment at birth.

This temperature-dependent pathway relies on triggers that are part of the endocrine system, rather than the solely genetic trigger that exists in chromosomally determined sex. The lability that temperature allows also leaves a pitfall: the pathway that determines the organism's sex can be shifted — in some cases completely to the opposite sex — by perturbations in the environment that do not reverse sex in chromosomally determined species. For example, in the red-eared slider turtle (*Trachemys scripta elegans*), placing estrogenic compounds on the eggshells of incubating embryos can cause them to develop as females, even when the temperature signals "male."[7] In a chromosomally determined species, this outcome is less likely; thus, species that rely on temperature may experience greater susceptibility to adverse effects at the individual and population levels compared with other species. These organisms also provide an excellent model for assessing the endocrine-disrupting capabilities of contaminants and have been used successfully for these purposes.[8–11]

Endocrine disruptors are by no means the only compounds that act as toxicants during reptilian development or that affect reproduction, but for two reasons I will focus on them here. First, the field of endocrine disruption is rapidly growing as we begin to recognize the huge implications of the ability of these contaminants — even at very low concentrations — to exert adverse effects at the population level via sometimes-subtle developmental action. The growth of this field is best described by its growth and representation in journals; a search on "endocrine disruption" in a biological abstracts database reveals 0 hits before 1995 and 264 hits from 1995 to 2003. Of those 264 results, 252 were from 1998 onward. The field has a broad representation in peer-reviewed publications from *Nature* to toxicology publications. Second, the subtlety itself requires focus and investigation. Traditional toxicology studies address satisfactorily the grossly teratologic effects of acute-acting toxicants, but they leave unaddressed developmental effects that can manifest as adverse in the second generation because of endocrine disruption. Exploration and discussion of endocrine disruptors, therefore, take center stage in any review of reptilian toxicology at the beginning of the 21st century.

II. Signal Interference: Disrupting a Delicate Developmental Balance

In the end, developmental processes — including the events of reproductive maturity — are all about proper signaling. The right time, the right place, the right signaler, the right recipient, the right level of signal — all these combine in the delicate dance that is vertebrate development. Given the complexities, it is astonishing that it ever works correctly, but one key characteristic is that, often, the next step cannot occur until the previous step has been properly completed. Toxicants that disrupt this balance, this order, often do so by adjusting the organism's sense of signal.

As an example in this chapter, I will be using estrogen signaling. The literature on estrogenic compounds is abundant, but there is a growing literature — although not reptile specific — on other types of endocrine disruption, including disruption of thyroid, androgen, and corticosteroid pathways. Two good places to start a review of the broader picture in endocrine disruption are the Crisp and colleagues[12] review and the Global Assessment of the State of the Science of Endocrine Disruptors produced by the International Programme on Chemical Safety[13] in cooperation with the World Health Organization, among others.

My comment about the lack of reptile-specific studies is telling. Amphibians are well represented in toxicologic literature because of their now-famous declines and manifestations of developmental abnormalities and because they are used in standard toxicity tests. Birds and fish also have received a great deal of attention (e.g., see reference 12). But reptiles? The number of studies on their reproduction and development is minimal, and only a few species are represented, as the examples in this chapter will show.

Chemicals that disrupt the endocrine system are called, as a group, "endocrine disruptors." Before there was widespread agreement on this nomenclature and before the discovery of other kinds of disruption, researchers often referred to these compounds as environmental estrogens. Many of them — including pesticides, polychlorinated biphenyls (PCBs), plasticizers, and pharmaceuticals — do exert estrogenic effects or result in endpoints related to estrogen exposure.[12] Some examples include the ability of some combinations of PCBs to reverse the sex of red-eared slider turtles from male to female under a male temperature signal[14] or the feminization of juvenile male American alligators (*Alligator mississippiensis*) in Lake Apopka, FL, in response to a cocktail of such compounds.[9,10,15,16] But how does it happen? What are the mechanisms by which toxicants operate during vertebrate development to alter endocrine outcomes?

A. Developmental Mechanisms

Several different mechanisms of action — all related to developmental signaling — have been proposed for endocrine-disrupting compounds. For example, exposure to a compound could change circulating sex steroid levels disrupting steroidogenic enzyme activity or neuroendocrine signaling systems. PCBs — acknowledged endocrine disruptors — can interfere with gonadotropin secretion in the anterior pituitary by affecting hypothalamic activity[17]; a change in gonadotropin secretion will manifest at the level of the gonad, potentially affecting reproduction. Other possible mechanisms of action include steroid receptor activation or inhibition. Some studies demonstrate steroid receptor activation in the absence of ligand binding. Thus far, steroid receptor activation remains the focus of study for elucidating the mechanism of action for these compounds, although I describe two studies later in this chapter with a different focus.

Several *in vitro* assays have established the ability of a number of endocrine-disrupting compounds to bind estrogen or androgen receptors,[18–20] and these findings can correlate with *in vivo* investigations. For example, nonylphenol and p,p′-DDE [1,1′-(dichloroethenylidiene) bis 4-chlorobenzene (DDE)] are two estradiol-17ß competitors,[18] and they both reverse sex in the red-eared slider[8,21]; however, some compounds seem to exhibit estrogenic or antiandrogenic effects yet do not bind well to estrogen receptors (ER) and androgen receptors (AR).

The discovery that ER can be activated in the absence of ligand binding[22,23] offers another possible mechanism. Some endocrine-disrupting compounds may operate independently from steroid-receptor binding and work via cell membrane receptors. Research shows that the binding of epidermal growth factor (EGF) to its cell membrane receptor can lead to ER activation.[22,23] The binding triggers a phosphorylation cascade that ultimately triggers ER activity. Compounds that trigger this pathway may also exhibit disruptive properties during development, completely in the absence of ligand binding to steroid receptors. Regardless of how the interference occurs, these disruptions occur because during embryogenesis, endocrine pathways involve accurate signals occurring at the right time and place in development.

In terms of reptilian development, only a few reptile-based studies that focus on a relatively small group of species exist and much work remains left to do in the area of reptilian endocrinology. Some studies are informative and illustrative, however. One characteristic that the species studied seem to have in common with many other vertebrates is that regardless of trigger, the sex-determining and developmental pathway for females involves estrogen. It comes as no surprise, then, that the reptilian models of endocrine disruption have focused on these female-based pathways.

Often, the assumption is that disruption of these steroid-involved pathways is via the receptor for that steroid; in other words, the signal that is disrupted is the communication between the steroid hormone and its protein mediator.[24] Other mechanisms of disruption, however, involve less direct interference, including effects on the proteins, such as aromatase, that metabolize steroids from one form to another. The role of aromatase is to biometabolize testosterone into estradiol-17ß, and this enzyme appears to play a strong role in the sex-determining signal in temperature-dependent reptiles. As the following example illustrates, it is possible for disruption to occur via interference with aromatase activity, rather than direct action at the receptor.

1. A Mechanistic Example Using the Reptilian Developmental Model
As has been mentioned, in the red-eared slider turtle, the incubation temperature the embryo experiences during the midtrimester of development determines sex.[25] High temperatures produce females and low temperatures produce males. The physiologic equivalent of temperature appears to be the steroidal milieu of the egg — including the steroid estradiol-17ß and its precursor, testosterone — and concurrently, the activity of steroidogenic enzymes such as aromatase, which converts testosterone to estradiol-17ß.[7] Natural estrogens bind the estrogen receptor in a dose-dependent fashion.[7,26] Female development can be blocked by administration of aromatase inhibitors, resulting in male development at female-producing temperatures.[27] Additionally, in the European pond turtle, another species with temperature-dependent sex determination, aromatase inhibition results in both the development of male gonadal structure and changes in aromatase activity in the gonad.[28]

Aromatase appears to be closely involved in the female developmental pathway in the red-eared slider turtle.[29] Research with other reptiles with temperature-dependent sex determination points to the brain as the organ that transduces the temperature signal into an aromatase response.[30] Salame-Mendez and colleagues[30] measured estradiol-17ß in the diencephalon/mesencephalon and telencephalon regions of the brain of the olive ridley sea turtle. They demonstrated differences at different incubation temperatures — and thus between sexes — in the temperature-sensitive developmental period, with a much higher estradiol-17ß concentration in the brains of putative females. They also found that the gonads showed no differences in estradiol-17ß levels between temperatures. The different levels of steroid and enzyme activity at a critical period in gonadal development suggest that enzymes in the brain may play a role in the process.

In a study assessing the response of aromatase to temperature, we compared the pattern of aromatase activity at male- and female-producing temperatures in the brain and adrenal–kidney–gonad (AKG) of the red-eared slider turtle through development.[29] The pattern in the AKG was one of increased activity through development in both sexes but with no significant differences. In the brain, however, putative females exhibited a significantly higher level of aromatase activity at the beginning of the temperature-sensitive period.

The work established that differences in aromatase levels in the brain appeared to be associated with the female-developmental pathway in this reptile. In a related study, we examined whether this pathway was altered as a result of exposure to endocrine disruptors known to produce female turtles at a male-producing temperature.[31] Our previous work[8] had shown that the ubiquitous PCB Aroclor 1242 and the persistent pesticide chlordane can shift the sex-determining pathway from male to female in the red-eared slider turtle and that these compounds can also alter levels of circulating steroid hormones in hatchlings exposed during embryogenesis. We examined the ability of these persistent contaminants to disrupt the environmental developmental signal of temperature and alter aromatase activity in the brain or AKG of developing red-eared slider turtles. Aroclor 1242 significantly altered aromatase activity in the embryonic brain 24 h after exposure and significantly altered activity in the and AKG just before hatching. It also ultimately resulted in significant reversal of sex to female in hatchlings. Chlordane, a suspected antiandrogen in this species, did not affect aromatase activity in the brain or AKG at any time, supporting our hypothesis that it operates via a different pathway by blocking the androgen receptor.

This work with the red-eared slider turtle identified an endocrine pathway and potential endpoints along the pathway where toxicants might knock development off course. Aromatase activity — possibly in response to environmental signaling from temperature or contaminants — modulates steroid levels, thus modulating the endocrine signal and shifting development from normal pathways. It is an example of endocrine disruption in reptiles that has been observed in many other species (see Table 6.1 for examples of contaminants, affected species, and outcomes).

B. Species-Specific Considerations

To illustrate here the care that must be taken in extrapolating across taxa, I want to describe the results Lou Guillette and his group found when they examined aromatase levels in the brain and AKG of the developing American alligator. They assessed levels under control conditions and in the presence of estradiol-17ß treatment.[32] In contrast to the findings with the red-eared slider turtle, they found no sex-based differences in aromatase levels in the alligator brain, but they saw an increase in aromatase in female gonads late in the temperature-sensitive window. These findings underscore the species differences and similarities; both the red-eared slider turtle and the American alligator use temperature-dependent sex determination, rely on aromatase and estrogen as major players in the process of determination and differentiation, and are sensitive to contaminant influence; however, the pathways differ between the species.

Other groups have examined steroid levels during embryonic development to address whether levels of these powerful developmental signals change in response to environmental contaminants[10,15,16] or have investigated whether contaminants bind to the signal receptor (e.g., see references 18 and 20). There are many possible steps at which disruption could occur, but because steroids are the signaling molecules that these compounds often mimic, steroid levels and receptor effects are the logical first place to look for changes.

III. A Toxicologic Shift

To paraphrase Paracelsus, "The dose makes the poison." We now know that in the endocrine-disruptor context, it would be more accurate to say, "The dose and the developmental stage make the poison." Development is an ongoing process, and all of its processes are major. A failure of gastrulation would, of course, be catastrophic, and compounds that exert this level of influence are also probably toxic at periods beyond embryogenesis. Again, my focus in this chapter is on those compounds that exert their toxic effects specifically in the embryonic milieu, with later manifestations, and those compounds are primarily the endocrine-disrupting compounds. Developmental processes specifically susceptible to these are those that rely on

Table 6.1 Some Examples of Recent Investigations into Developmental Exposures and Outcomes in Reptiles

Species	Outcome	Contaminant(s)	Citation
American alligator (*Alligator mississippiensis*)	Feminized juvenile males, altered steroid concentrations	Various (field study)	9,10, 15,16
American alligator	Altered bone composition	Various (field study)	59
American alligator	Aromatase activity	Estradiol	32
American alligator	Thymus and spleen histology	Various (field study)	57
American alligator	Thyroid hormone disruption	Various (field study)	28
Caiman (*Caiman latirostris*)	Sex reversal/female development	bisphenol A	11
Fence lizards (sp. *Sceloporus*)	Disrupted sex development	Ethinyl estradiol	70
Fence lizards	Heavy metal uptake in eggs	Cadmium	71
Green sea turtle (*Chelonia mydas*)	No effect of p,p′-DDE on sex	p,p′-DDE	43
Painted turtle (*Chrysemys picta*)	Vitellogenin induction	Estrogenic feedlot effluent	34a
Red-eared slider turtle (*Trachemys scripta elegans*)	Altered circulating steroid levels	Chlordane, trans-Nonachlor, Aroclor 1254	40
Red-eared slider turtle	Altered steroidogenic pathway	Aroclor 1254	31
Red-eared slider turtle	Female development/sex reversal, synergy	PCB mixtures	14
Red-eared slider turtle	Growth rates/mass at hatching	trans-Nonachlor, chlordane, DDE, atrazine	53,73
Red-eared slider turtle	Sex reversal/female development	PCBs, DDT metabolites, atrazine, chlordane, others	8,73
Red-eared slider turtle	Shell and yolk contaminants	Heavy metals	54
Snapping turtle (*Chelydra serpentina*)	No effect of p,p′-DDE on sex	p,p′-DDE	44
Snapping turtle	Sexual dimorphism, demasculinization	Organochlorines	55,56
Water snakes (sp. *Nerodia*)	Accumulation	Chlorinated hydrocarbons	64
Water snakes	Accumulation in eggs/greater comparative accumulation vs. amphibian	PCBs	65
Water snakes	Lab study, organochlorine uptake in eggs	Lindane, aldrin, dieldrin, DDT	66

endocrine-signaling, including gonadogenesis, sex determination and differentiation, neural development, and growth, among others.

Because embryonic development is a highly sensitive period, the signaling milieu is carefully calibrated throughout. When discussing toxicology, it has become important to distinguish the toxicology of adult animals not experiencing sensitive developmental periods to that of embryonic organisms, whose experiences can result in profound and permanent changes, as is well known to any developmental biologist. To use a human example, an adult might find that after a night of overindulgence, the worst thing that happens is a headache in the morning; the effects are acute and transient for the most part. For an embryonic organism, however, exposure to ethanol results in alterations in the signaling milieu that can translate into permanent effects, including — in the most severe cases — the manifestations of fetal alcohol syndrome. Because of these differences in the result of exposure based on the state of the organism, toxicology has experienced a bifurcation of model, and reptiles have been part of the vanguard of the new approach.

The traditional toxicologic paradigm emphasizes a carcinogenic–survival model in which an adult animal is assessed for acute effects, cancer, or mortality as a result of exposure to a toxicant. In this traditional model, single compounds are tested with an assumption that increasing dose will result in an increased response until death or the response plateaus. Additionally, the model assumes that a threshold for the compound in question exists below which no adverse effects will be observed in the organism.[33] When embryonic development is considered, it is usually in the context of mortality or gross teratogenic effects. Even in assays that include a life-cycle component, the outcomes in question are lethality or gross effects.

The newer, developmental model used in endocrine-disruption studies takes a completely different approach. Chronic exposure — rather than acute — is modeled in the lab because that type of exposure is environmentally relevant (see references 9, 10, 15, 16, and 34b). Additionally, acute effects are not necessarily the endpoints of interest; instead, the focus is often on latent effects, such as those that were mapped during development but do not manifest until reproductive maturity — for example, mating behavior, infertility, or other reproductive disruptions. Unlike the traditional model, the developmental model does not assume a threshold. When the pathways under discussion involve endocrine signaling, a correct assumption is that threshold of appropriate effect has already been achieved endogenously, and the addition of compounds that mimic this endogenous signal will increase the signal beyond threshold (see reference 34b for an example using reptiles). Also, under this model, the effects have the potential to be transgenerational, as has been observed in humans and diethylstilbestrol exposure,[35–38] and rather than a focus on single compounds, there is an acknowledgment that no contaminant occurs alone and that mixtures must be assessed to better reflect real-life conditions.

A. Reptilian Models

For two reasons reptiles have proved to be excellent choices as study organisms for endocrine disruption and to apply the developmental model. First, they exhibit life history characteristics such as temperature-dependent sex determination and an accessible egg that make them able to be manipulated in the lab. Second, they are, in fact, exposed to these compounds in the environment, often via contaminants in the yolk or soil (see Chapter 3), thus making studies of them applicable in real-life scenarios and for careful extrapolation among vertebrates.

Species with temperature-dependent sex determination are among the most studied reptiles in toxicology and endocrine-disruption studies because the endpoint of disruption is obvious, key events in the pathways at the genetic and molecular levels have been elucidated, the eggs are accessible, and a growing body of literature provides a base. Although some crocodilians and other turtle species have been used as models, in this chapter I discuss the red-eared slider turtle as a developmental model for endocrine disruption involving reproductive endpoints that can be used to address mechanisms and mixtures.

The red-eared slider turtle is found throughout the southeastern United States. Living at the sensitive interface of land and water, it inhabits a number of areas known to be contaminated with suspected endocrine-disrupting compounds. As in many turtles, gonadal sex in the slider turtle is determined by the temperature of the incubating egg rather than by sex chromosomes.[7] Natural estrogens have the ability to direct sex development to the female pathway at a temperature that would normally produce males. Because of these features, the red-eared slider turtle is an excellent animal model for the study of the effects of endocrine-disrupting toxicants.

1. No Threshold/Low Doses/Ecologic Relevance

The standard approach for toxicologic studies involves assumption of a threshold dose, or the dose below which no adverse effects are seen. Studies with the red-eared slider turtle have shown that exogenous estradiol-17ß — even when applied in doses as low as 0.4 ng/10 g egg — will affect sex development during embryogenesis.[34b] In this study by Sheehan and colleagues,[34b] extrapolation of the dose-response curve using a modified Michaelis-Menten equation indicated that estradiol-17ß would not have a threshold in this species because of the endogenous presence of estradiol-17ß.

Research indicates that only 2% of estradiol-17ß applied to a turtle egg gets into the yolk, and of that, only 0.1% (or 0.4 pg in the example cited earlier) ends up in the embryo,[31] implying that very low doses can have an effect. We tested the low-dose idea using the eight compounds Heinz and colleagues[39] identified in the yolk of alligator eggs from Lake Apopka, FL. The compounds were administered to red-eared slider turtle eggs in the ecologically relevant concentrations identified in the alligator yolk. Five of the compounds — the PCB mixture Aroclor 1242, trans-Nonachlor, cis-Nonachlor, p,p'-DDE, and chlordane — altered sex ratio outcomes when applied to eggshells during development.[8]

2. *Endpoints: Hidden Disruption*

In addition to sex reversal as an endpoint, we sought to uncover whether morphologically normal turtles exhibited unseen disruption. To do so, we examined basal steroid levels and steroid levels in response to follicle-stimulating hormone (FSH) administration in 6-week-old hatchling males and females treated during embryogenesis with Aroclor 1242, chlordane, or trans-Nonachlor. Aroclor-exposed males had significantly lower testosterone levels than did controls, and chlordane-exposed females had significantly lower progesterone, testosterone, and 5-alpha-dihydrotestosterone levels relative to controls.[40] Thus, these normal-looking males and females were altered in subtle — but important — ways by this developmental exposure.

Males treated with Aroclor 1242 and trans-Nonachlor also displayed an elevated estradiol-17ß response to FSH administration vs. control males, demonstrating that the normal-appearing animals had altered physiologic responses as a result of a single dose applied to the eggshell during a sensitive period. These results point to the possibility that endocrine-disrupting compounds can cause alterations that may affect reproductive success later in life.

3. *Considering Mixture Effects During Development*

The models, to be accurate and relevant, must also assess mixture effects, and here again the reptiles lead the way. Lake Apopka in Florida has become a classic example of how environmental contamination can affect reproduction of animals in nature. The resemblance of gonadal and penile abnormalities of the American alligator in this lake to those described in mice treated with the potent, synthetic estrogen diethylstilbestrol (DES)[35,41] led to detailed studies documenting that chronic pollution by agricultural runoff — exacerbated by a chemical spill of dicofol — was the most likely cause of the observed reproductive abnormalities.[9,10] Dicofol and its components have been shown to bind the ER from the alligator[20] and therefore may mimic estrogens in the alligator. In addition to o,p′-DDE/o,p′-DDT contamination, PCB mixtures resembling Aroclor 1242 and a variety of pesticides have been detected in alligator eggs, including dieldrin, toxaphene, cis/trans Nonachlor, chlordane, and pp′DDD.[39] Exposure of alligator[9,10,15,16] embryos to combinations and concentrations of these compounds results in anomalous reproductive development. We found the same thing when we applied this cocktail of chemicals to red-eared slider turtle eggs.

In a study that encompassed both ecologically relevant concentrations and a chemical mixture, we applied all eight of the compounds identified in the Lake Apopka alligator eggs to red-eared slider turtle eggs. When all eight compounds were applied in a single-dose mixture, they significantly increased the ratio of females to males, although the percentage of females was less than that produced by some of the compounds singly.[8] Bergeron and colleagues[14] assessed the effects of mixtures of PCBs and found that one pair of PCBs exerted a more profound sex-reversing effect than did the hydroxylated congeners singly. Mixture studies like these are rare, however, and more are needed.

I point out here that these animals are being used as models of effects (i.e., we are not assuming that turtles or alligators are necessarily exposed via transfer across the eggshell, although that has been demonstrated in some species). The studies simply address mechanisms and endpoints that result from exposure. In the following review of reptile toxicology studies, I mention some of the pathways of exposure in reptiles.

4. *Developmental Synergy*

Reptiles also make good *in vivo* models for synergy, and developmental periods may be more sensitive to greater-than-expected results from molecular signals. Synergism occurs when the effect of two factors together is greater than the sum of their separate effects. This phenomenon has been demonstrated both with natural and environmental estrogens in the red-eared slider turtle. Bergeron and colleagues[42] showed that estradiol-17ß synergizes with estriol, a natural estrogen with powerful effects in the red-eared slider turtle. At some doses, the two compounds together resulted in twice the sex reversal shown by estradiol-17ß alone.

Synergism between environmental estrogens was first shown with PCBs in the red-eared slider. As metabolites of other PCBs, hydroxylated PCBs can persist in aquatic environments. Bergeron and colleagues[14] applied pairs of PCBs to eggs and found that two of the compounds tested, 2′,4′,6′-trichloro-4-biphenylol (3-PCBOH) and 2′,3′,4′,5′-tetrachloro-4-biphenylol (4-PCBOH), both hydroxylated PCBs, resulted in female or intersex hatchlings from eggs incubated at a male-producing temperature. Various mixtures of PCBs except 3- and 4-PCBOH had no effect on sex ratio, but the 3/4 combination behaved synergistically, resulting in a significant increase in female and intersex hatchlings at a dose of less than 1 ppm. When administered alone, 3-PCBOH and 4-PCBOH required at least a 10-fold higher dose to show sex reversal.

B. *Species-Specific Considerations*

Despite some of the success of reptiles as developmental models, as usual, the caveat applies. Although there is a paucity of developmental and reproductive endpoint studies using reptiles, those that have been done point to a wide variation in species response. Podreka and colleagues[43] found no effect of p,p′-DDE on turtle sex in the developing green sea turtle (*Chelonia mydas*), although this metabolite did affect sex in the red-eared slider turtle.[8] Additionally, Portelli and colleagues[44] found no effect of p,p′-DDE on sex development in the snapping turtle (*Chelydra serpentina*), although the concentrations used were several orders of magnitude higher than those used in Willingham and Crews.[8]

At the molecular level, two studies examining species-specific response of estrogen receptor to various environmental estrogens, including PCBs and DDT metabolites, found that the ERs of different species show different affinities for these compounds. For example, the green anole (*Anolis carolinensis*) estrogen receptor binds less well to bisphenol A or o,p′-DDT than does the rainbow trout (*Onchorynchus mykiss*) ER.[45] Also, in a comparison of PCB-estrogen receptor binding affinity, a few PCB congeners bound with greater affinity to the rainbow trout estrogen receptor than to either the human or green anole receptor,[46] underscoring how different responses among species can be.

These findings highlight two points about developmental toxicology and endocrine disruptors. First, responses can be species specific, and we should exercise caution in extrapolating findings and always do so with a caveat. Second, the dose makes the poison, and in endocrine situations, it is not uncommon to find a hormetic response in which a very low dose or very high dose elicits similar effects and effects that differ from those at a mid-range dose.[47] These issues complicate developmental toxicology and endocrine-disruptor studies and require careful identification of the dose-response curve for a compound or mixtures of compounds on a species-by-species basis. Such an effort would be a huge undertaking, and because of various limitations we must rely on reptilian models as our guides.

IV. The State-of-the-Science for Developmental Investigations in Reptiles

Other chapters in this book cover issues of risk and exposure for reptiles. Here, I want briefly to mention that reptiles occupy various corners of the food web and are exposed via a variety of mechanisms to contaminants. One characteristic of endocrine-disrupting contaminants is their lipophilic nature,[12] and as would be predicted, these compounds accumulate most readily in fatty tissues and yolk and also in embryos of snakes, turtles, crocodilians, and lizards.[39,48–52] The exposure to the embryo via maternal contaminants and the embryo's own contaminant load are of most direct interest here and have also been covered in Chapter 3.

As a result of these exposures, there are some studies of developmental effects of toxicants in reptiles, both from the field and from the lab. In the following I discuss a few examples of these studies. I also present studies of developmental and reproductive disruption that go beyond estrogenic effects.

A. Turtles

The red-eared slider turtle has been used extensively as an example in this chapter (it is one of the few reptiles extensively studied as a developmental model), but there are further examples of developmental disruption that is not necessarily estrogenically mediated. Red-eared slider turtle hatchlings exposed to known endocrine disruptors experienced changes in growth rate that were not sex based.[53] The implication of this study is the involvement of other endocrine axes, including the thyroid axis and the growth-hormone axis. Although studies with the red-eared slider turtle are generally intended as model studies, fieldwork indicates that eggs from this species do take up contaminants from the environment; researchers found higher concentrations of lead, mercury, and selenium in the egg yolk than in the shell,[54] providing another starting point for developmental toxicologic examinations. Heavy metals have the potential for acute, chronic, or latent toxicity and deserve much more investigation under the developmental model, with a focus on mixtures and interactions and low doses.

Snapping turtles from contaminated Great Lakes sites show demasculinized sexual morphology compared with animals from reference sites[55] and exhibited higher serum organochlorine values. Additionally, de Solla and colleagues[56] found differences in sexually dimorphic morphology of adult snapping turtles from pesticide-contaminated sites compared with control sites. Any of these manifestations could result from developmental disruptions or cause reproductive problems, and the demasculinization is likely to have arisen during the embryonic period.

Irwin and colleagues[34a] identified *in situ* effects of estrogenic feedlot effluent on painted turtles (*Chrysemys picta*); they found that male turtles were producing vitellogenin, a yolk protein usually produced only by females and used as a marker of inappropriate estrogenic induction in males. Their findings and global trends in reptilian toxicology are discussed in Chapter 4. Although this vitellogenin production may not necessarily be the result of embryonic influences, they serve as a twofold warning: the population may be imperiled because feminized males may experience reduced fertility, and any embryos that are produced may also be experiencing an endocrine-altering environment via yolks from their mothers. This possibility deserves investigation.

B. *Crocodilians*

Crocodilians are known to experience sex reversal when exposed to certain compounds, but they also experience thymic changes[57] and changes in thyroid hormone status,[58] both of which may be traceable to developmental influences. Stoker and colleagues[11] found that environmentally relevant levels of bisphenol A affect sex determination in the caiman (*Caiman crocodilus*), a crocodilian with temperature-dependent sex determination and the latest entry of that group in developmental toxicology. Bisphenol A is a compound found in everything from baby bottles to dental sealants, and this work uses the caiman as a developmental vertebrate model, rather than strictly as an ecologic species model. Lind and colleagues[59] found altered bone composition in female juvenile American alligators from contaminated Lake Apopka, a finding that indicates an altered hormonal milieu for these animals from a very early stage. These animals are juvenile, and development is an ongoing process that includes, for example, crucial stages such as puberty and reproductive maturity; thus, examinations of developmental stages beyond embryonic development may also be informative, as will be assessment of reproductive endpoints such as fertility, mating success, and fecundity.

C. *Squamates*

As the previous list illustrates, turtles and crocodilians have some representation in developmental toxicologic studies, although much more work on the developmental effects of these compounds on reptile ecology remains to be done. Where are the snakes and lizards? Snake species — often obligate carnivores exposed via their diets[60,61] and therefore potentially bioaccumulating contaminants — require further investigation. Only a few studies exist describing the effects of contaminants on snakes at any biological level, and much work remains in this area. Squamates deserve our attention both as developmental vertebrate models and as ecologically relevant organisms (see reference 62); because they reach reproductive maturity far faster than the turtles or crocodilians, they can be better model choices for reproductive studies. Most contaminant studies involving squamates are older (>15–20 years) and generally involve acute toxic and lethal effects of pesticides,[63] although there are some more recent studies examining contaminant levels in organisms. For example, Bishop and Rouse[64] measured chlorinated hydrocarbon concentrations in Great Lakes water snakes (sp. *Nerodia*), including in gravid females, and found a pattern of contamination similar to that of herring gull eggs. Additionally — and here is another example of species-specific considerations — a study comparing PCB uptake of bullfrogs (*Rana catesbeiana*) and green frogs (*Rana clamitans*) with water snakes found that the snakes experienced significantly higher bioaccumulation of the compounds, and the compounds were found in the eggs of all species.[65] In one study modeling organochlorine uptake in snake eggs, the researchers found that the compounds lindane, heptachlor, aldrin, and dieldrin were taken up into the eggs, thus demonstrating a potential route of exposure to these endocrine-disrupting compounds in a snake.[66]

Lizards — which usually eat insects[63] — are in a similar position to that of snakes, although a couple of species have been identified as potential squamate models of disruption and contaminant effects.[67–70] One group has proposed the eastern and western fence lizards (*Sceloporus* spp.) as lab models for contaminant evaluation and, in conducting their own lab studies, found that exposure to the synthetic estrogen 17α-ethinylestradiol affected sex differentiation processes in males, even inhibiting development of embryonic secondary sex characteristics.[70] The fence lizards appear to do well in the lab environment and provide a number of parameters that relate to developmental influences.[70,71] In another lab-based study involving fence lizards, researchers found that the eggs were able to take up the heavy metal cadmium, which disrupted normal thyroid hormone levels in hatchlings, providing another example of the fence lizard as a developmental model.[71] Other work has demonstrated the ability of reptile eggs to take up contaminants from the terrestrial environment (e.g., arsenic[72]), making the necessity of investigating the developmental effects of contaminants in these taxa all the more urgent.

V. Future Considerations

Studies with other vertebrates have shown that developmental exposures can affect a spectrum of developmental endpoints, from cognition to gonadogenesis and sex differentiation to behavior, among others (see references 12 and 13 for review, and see Chapter 7 in this book for information about cognitive and behavioral effects). These outcomes not only affect the individual organism, but can also have effects at the population and community levels (e.g., see reference 73) that are latent in nature; for more discussion of population- and community-level effects of toxicants in general, see Chapter 10. As the previous discussions indicate, our focus for the future involves both an expansion of species investigated and a refinement of models, dose-response curves, and mixture considerations. In every context, the effects of exposure during development are paramount, whether we are elucidating endpoints or mechanisms and pathways. Also of greater consideration in the future for reptiles and other species are pharmaceutical and personal care products in the environment; the list of these is long,[74] and because many of them are intended to be physiologically active, investigations of their effects in wildlife are imperative. As a last consideration, I suggest taking interactions of external signals into account when assessing developmental effects in these ectothermic animals. Many of these species — regardless of sex-determining mechanism — are subject to developmental temperature modulations of many endocrine-governed parameters including mass at hatching,[75,76] growth rate,[77–79] swiftness,[1,80] temperature choice,[79,81] sex behavior,[82] size or length,[3] and predator avoidance.[1,84] Given that global warming is of concern and that predictions of temperature increases are within the range that can result in a complete shift of sex-determining pathways in some species, no ecologic assessment of the developmental effects of an endocrine-disrupting contaminant will be complete without consideration of temperature as a potential interacting factor.

Literature Cited

1. Janzen, F.J., Experimental evidence for the evolutionary significance of temperature-dependent sex determination, *Evolution*, 49, 864–873, 1995.
2. Janzen, F.J. and Paukstis, G.L., Environmental sex determination in reptiles: ecology, evolution, and experimental design. *Q. Rev. Biol.*, 66 (2), 149–179, 1991.
3. Harlow, P. and Shine, R., Temperature-dependent sex determination in the frill-neck lizard, *Chlamydosaurus kingii* (Agamidae). *Herpetologica*, 55, 205–212, 1999.
4. Freedberg, S., Ewert, M.A., and Nelson, C.E., Environmental effects on fitness and consequences for sex allocation in a reptile with environmental sex determination. *Evol. Ecol. Res.*, 3, 953–963, 2002.
5. Shine, R., Why is sex determined by nest temperature in may reptiles?, *Trends Ecol. Evol.*, 186–189, 1999.
6. Charnov, E.L. and Bull, J.J., When is sex environmentally determined? *Nature*, 266, 828–830, 1977.
7. Crews, D., Temperature-dependent sex determination: the interplay of steroid hormones and temperature, *Zool. Sci.*, 13, 1–13, 1996.
8. Willingham, E.J. and Crews, D., Sex reversal effects of environmentally relevant pesticide concentrations on the red-eared slider turtle, a species with temperature-dependent sex determination, *Gen. Comp. Endocrinol.*, 113, 429–435, 1999.
9. Guillette, L.J., Jr. et al., Developmental abnormalities of the reproductive system of the American alligator (*Alligator mississippiensis*) from contaminated and control lakes in Florida, *Environ. Health Perspect.*, 102, 680–688, 1994.
10. Guillette, L.J., Jr. et al., Reduction in penis size and plasma testosterone concentrations in juvenile alligators living in a contaminated environment, *Gen. Comp. Endocrinol.*, 101, 32–42, 1996.
11. Stoker, C. et al., Sex reversal effects on *Caiman latirostris* exposed to environmentally relevant doses of the xenoestrogen bisphenol A, *Gen. Comp. Endocrinol.*, 133(3), 287–296, 2003.
12. Crisp, T.M. et al., Environmental endocrine disruption: an effects assessment analysis, *Environ. Health Perspect.*, 106 (Suppl.), 11–56, 1998.
13. Damstra, T. et al., Eds., *Global Assessment of the State-of-the-Science of Endocrine Disruptors*, International Programme on Chemical Safety/World Health Organization, 2002.
14. Bergeron, J.M., Crews, D., and McLachlan, J.A., PCBs as environmental estrogens: turtle sex determination as a biomarker of environmental contamination, *Environ. Health Perspect.*, 102, 780–786, 1994.
15. Guillette, L.J., Jr. et al., Serum concentrations of various environmental contaminants and their relationship to sex steroid concentrations and phallus size in juvenile American alligators, *Arch. Environ. Contam. Toxicol.*, 36, 447–455, 1999.
16. Guillette, L.J., Jr. et al., Plasma steroid concentrations and male phallus size in juvenile alligators from seven Florida lakes, *Gen. Comp. Endocrinol.*, 116, 356–372, 1999.

17. Kahn, I.A. and Thomas, P. Aroclor 1254–induced alterations in hypothalamic monamine metabolism in the Atlantic croaker (*Micropogonias undulatus*), *Neurotoxicology*, 18, 553–560, 1997.
18. Danzo, B.J., Environmental xenobiotics may disrupt normal endocrine function by interfering with binding of physical ligand to steroid receptors and binding proteins, *Environ. Health Perspect.*, 105, 294–301, 1997.
19. Kelce, W.R. et al., Persistent DDT metabolite p,p′-DDE is a potent androgen receptor antagonist, *Nature*, 375, 581-585, 1995.
20. Vonier, P.M. et al., Interaction of environmental chemicals with the estrogen and progesterone receptors from the oviduct of the American alligator, *Environ. Health Perspect.*, 104, 1318–1322, 1996.
21. Willingham, unpublished data.
22. Bunone, G. et al., Activation of unliganded estrogen receptor by EGF involves the MAP kinase pathway and direct phosphorylation, *EMBO*, 15, 2174–2183, 1996.
23. Curtis, S.W. et al., Physiological coupling of growth factor and steroid receptor signaling pathways: estrogen receptor knockout mice lack estrogen-like response to epidermal growth factor, *Proc. Nat. Acad. Sci.*, 93, 12626–12630, 1996.
24. McLachlan, J.A., Environmental signaling: what embryos and evolution teach us about endocrine disrupting chemicals, *Endocrine Rev.*, 319–341, 2001.
25. Wibbels, T., Bull, J.J., and Crews, D., Chronology and morphology of temperature-dependent sex determination, *J. Exp. Zool.*, 260, 371–381, 1991.
26. Wibbels, T. and Crews, D., Specificity of steroid hormone-induced sex determination in a turtle, *J. Endocrinol.*, 133, 121–129, 1992.
27. Crews, D. and Bergeron, J.M., Role of reductase and aromatase in sex determination in the red-eared slider (*Trachemys scripta*), a turtle with temperature-dependent sex determination, *J. Endocrinol.*, 143, 279–289, 1994.
28. Richard-Mercier, N. et al., Endocrine sex reversal of gonads by the aromatase inhibitor letrozole (CGS 20267) in *Emys orbicularis*, a turtle with temperature-dependent sex determination, *Gen. Comp. Endocrinol.*, 100, 314–326, 1995.
29. Willingham, E.J. et al., Aromatase activity during embryogenesis in the brain and adrenal-kidney-gonad of the red-eared slider turtle, *Trachemys scripta elegans*, *Gen. Comp. Endocrinol.*, 119, 202–207, 2000.
30. Salame-Mendez, et al., Response of the diencephalon but not the gonad to female-promoting temperature with elevated estradiol-17ß levels in the sea turtle *Lepidochelys olivacea*, *J. Exp. Zool.*, 280, 304–313, 1998.
31. Willingham, E.J. and Crews, D., The red-eared slider turtle: an animal model for the study of low doses and mixtures, *Am. Zool.*, 40, 421–428, 2000.
32. Milnes, M.R., Roberts, R., and Guillette, L.J., Jr., Effects of incubation temperature and estrogen exposure on aromatase activity in the brain and gonads of embryonic alligators, *Environ. Health Perspect.*, 110 (Suppl. 3), 393–396, 2002.
33. Crews, D., Willingham, E.J., and Skipper, J.K., Endocrine disruptors: present issues, future problems. *Q. Rev. Biol.*, 75 (3), 243–260, 2000.

34a. Irwin, L.K., Gray, S.L., and Oberdorster, E., Vitellogenin induction in painted turtle, *Chrysemys picta*, as a biomarker of exposure to environmental levels of estradiol-17ß, *Aquat. Toxicol.*, 55, 49–60, 2001.

34b. Sheehan, D.M. et al., No threshold dose for estradiol-17ß-induced sex reversal of turtle embryos: how little is too much?, *Environ. Health Perspect.*, 107, 155–159, 1999.

35. McLachlan, J.A., Prenatal exposure to diethylstilbestrol in mice: toxicological studies, *J. Toxicol. Environ. Health*, 2, 527–237, 1977.
36. Klip, H. et al., Hypospadias in sons of women exposed to diethylstilbestrol in utero: a cohort study, *Lancet*, 359, 1102–1107, 2002.
37. Newbold, R.R. et al., Increased tumors but uncompromised fertility in the female descendants of mice exposed developmentally to diethylstilbestrol, *Carcinogenesis*, 19, 1655–1663, 1998.
38. Newbold, R.R. et al., Proliferative lesions and reproductive tract tumors in male descendants of mice exposed developmentally to diethylstilbestrol, *Carcinogenesis*, 21 (77), 1355–1663, 2000.
39. Heinz, G.H., Percival, H.F., and Jennings, M.L., Contaminants in American alligator eggs from Lake Apopka, Lake Griffin, and Lake Okeechobee, Florida, *Environ. Monitor. Assess.*, 16, 277–285, 1991.
40. Willingham, E.J. et al., Embryonic treatment with xenobiotics disrupts steroid hormone profiles in hatchling red-eared slider turtles (*Trachemys scripta elegans*), *Environ. Health Perspect.*, 108 (4), 329–332, 2000.
41. McLachlan, J.A. et al., Reduced fertility in female mice exposed transplacentally to diethylstilbestrol (DES), *Fertil. Steril.*, 38, 364–371, 1982.
42. Bergeron, J.M. et al., Developmental synergism of steroidal estrogens in sex determination, *Environ. Health Perspect.*, 107, 93–97, 1999.
43. Podreka, S. et al., The environmental contaminant p,p′-DDE fails to influence the outcome of sexual differentiation in the marine turtle *Chelonia mydas*, *Environ. Health Perspect.*, 106 (4), 1998.
44. Portelli, M.J. et al., The effect of p,p′-DDE on sex determination of the common snapping turtle (*Chelydra serpentina serpentina*), *Ecotoxicol. Environ. Saf.*, 43, 284–291, 1999.
45. Matthews, J. and Zacharewski, T., Differential binding affinities of PCBs, HO-PCBs, and aroclors with recombinant human, rainbow trout (*Onchorhynkiss mykiss*), and green anole (*Anolis carolinensis*) estrogen receptor, using a semi-high throughput competitive binding assay, *Toxicol. Sci.*, 53 (2), 326–329, 2000.
46. Matthews, J. et al., Differential estrogen receptor binding of estrogenic substances: a species comparison, *J. Steroid Biochem. Mol. Biol.*, 74 (4), 223–234, 2000.
47. Calabrese, E.J. and Baldwin, L.A., Toxicology rethinks its central belief, *Nature*, 421, 691–692, 2003.
48. Meeks, R.L., The accumulation of 36CL ring-labeled DDT in a freshwater marsh, *J. Wildl. Manage.*, 32, 376–398, 1968.
49. Stone, W.B., Kiviat, E., and Butkas, S.A., Toxicants in snapping turtles, *N.Y. Fish Game J.*, 1, 40–49, 1980.
50. Ryan, J.J. et al., 2,3,7,8-tetrachlorodibenzo-p-dioxin and related dioxins and furans in snapping turtle (*Chelydra serpentina*) tissues from the Upper St. Lawrence River, *Chemosphere*, 15, 537–548, 1986.
51. Bishop, C.A. et al., The case for a cause-effect linkage between environmental contamination and development in eggs of the common snapping turtle (*Chelydra s. serpentina*) from Ontario, Canada, *J. Toxicol. Environ. Health*, 33, 521–547, 1991.
52. Struger, J. et al., Environmental contaminants in eggs of the common snapping turtle (*Chelydra serpentina serpentina*) from the Great Lakes–St. Lawrence River Basin of Ontario, Canada (1981, 1984), *J. Great Lakes Res.*, 19, 681–694, 1993.

53. Willingham, E.J., Embryonic exposure to low-dose pesticides: effects on fitness parameters in the hatchling red-eared slider turtle, *J. Toxicol. Environ. Health*, 64, 257–272, 2001.
54. Burger, J. and Gibbons, J.W., Trace elements in egg contents and egg shells of slider turtles (*Trachemys scripta*) from the Savannah River Site. *Arch. Environ. Contam. Toxicol.*, 34 (4), 382–386, 1998.
55. de Solla, S.R. et al., Organochlorine contamination, sex hormones, and sexual morphology of common snapping turtles (*Chelydra serpentina serpentina*), *Environ. Health Perspect.*, 105, 1–7, 1998.
56. de Solla, S.R., Bishop, C.A., and Brooks, R.J., Sexually dimorphic morphology of hatchling snapping turtles (*Chelydra serpentina*) from contaminated and reference sites in the Great Lakes and St. Lawrence River Basin, North America, *Environ. Toxicol. Chem.*, 21 (5), 922–929, 2002.
57. Rooney, A.A., Bermudez, D.S., and Guillette, L.J., Jr., Altered histology of the thymus and spleen in contaminant-exposed juvenile American alligators, *J. Morphol.*, 256, 349–359, 2003.
58. Crain, D.A. et al., Sex steroid and thyroid hormone concentrations in juvenile alligators (*Alligator mississippiensis*) from contaminated and reference lakes in Florida, USA, *Environ. Toxicol. Chem.*, 17 (3), 446–452, 1998.
59. Lind, P.M. et al., Abnormal bone composition in female juvenile American alligators from a pesticide-polluted lake (Lake Apopka, Florida), *Environ. Health Perspect.*, 112 (3), 359–362, 2004.
60. Bauerle, B., Spencer, D.L., and Wheeler, W., The use of snakes as a pollution indicator species, *Copeia*, 366–368, 1975.
61. Lower, W.R. et al., Movement and fate of 2,3,7,8-tetrachloridbenzo-p-dioxin in fauna at Times Beach, Missouri, *Chemosphere*, 20, 1021–1025, 1990.
62. Campbell, K.R. and Campbell, T.S., The accumulation and effects of environmental contaminants on snakes: a review, *Environ. Monitor. Assess.*, 70 (3), 253–301, 2001.
63. Campbell, K.R. and Campbell, T.S., A logical starting point for developing priorities for lizard and snake ecotoxicology: a review of available data, *Environ. Toxicol. Chem.*, 21 (5), 894–898, 2002.
64. Bishop, C.A., and Rouse, J.D., Chlorinated hydrocarbon concentrations in plasma of the Lake Erie water snake (*Nerodia sipedon insularum*) and northern water snake (*Nerodia sipedon sipedon*) from the Great Lakes basin in 1998, *Arch. Environ. Contam. Toxicol.*, 39 (4), 500–505, 2000.
65. Fontenot, L.W. et al., Bioaccumulation of polychlorinated biphenyls in ranid frogs and northern water snakes from a hazardous waste site and a contaminated watershed, *Chemosphere*, 40 (8), 803–809, 2000.
66. Canas, J.E. and Anderson, T.A., Organochlorine contaminants in eggs: the influence of contaminated nest material, *Chemosphere*, 47 (6), 585–589, 2002.
67. Fossi, M.C. et al., The lizard *Gallotia galloti* as a bioindicator of organophosphorus contamination in the Canary Islands, *Environ. Pollut.*, 87, 289–294, 1995.
68. Hall, R.J. and Clark, D.R., Jr., Responses of the iguanid lizard *Anolis carolinensis* to four organophosphorus pesticides, *Environ. Pollut. Ser. A*, 28, 45–52, 1982.
69. Meenakshi, V. and Karpagaganapathi, P.R., Toxicity and behavioural responses of *Calotes versicolor* (Daud) administered with phosphamidon, *Indian J. Environ. Toxicol.*, 6, 50, 1996.

70. Talent, L.G. et al., Evaluation of western fence lizards (*Sceloporus occidentalis*) and eastern fence lizards (*Sceloporus undulates*) as laboratory reptile models for toxicological investigations, *Environ. Toxicol. Chem.*, 21 (5), 899–905, 2002.
71. Brasfield, S.M., Development of a terrestrial vertebrate model for assessing the bioavailability of cadmium in the fence lizard (*Sceloporus undulates*) and *in ovo* effects on hatchling size and thyroid function, *Chemosphere*, 54 (11), 1643–1651, 2004.
72. Marco, A., Lopez-Vicente, M., and Perez-Mellado, V., Arsenic uptake by reptile flexible-shelled eggs from contaminated nest substrates and toxic effect on embryos, *Bull. Environ. Contam. Toxicol.*, 72 (5), 983–990, 2004.
73. Willingham, E.J., Different incubation temperatures result in differences in mass in female red-eared slider turtle hatchlings, *J. Thermal Biol.*, 30 (1), 61–64, 2005.
74. Daughton, C.A. and Ternes, T.A. Pharmaceutical and personal care products in the environment: agents of subtle change?, *Environ. Health Perspect.*, 107 (6), 907–938, 1999.
75. Reece, S.E. et al., The effects of incubation environment, sex and pedigree on the hatchling phenotype in a natural population of loggerhead turtles, *Evol. Ecol. Res.*, 4, 737–748, 2002.
76. Rhen, T. and Lang, J., Incubation temperature and sex affect mass and energy reserves of hatchling snapping turtles, *Chelydra serpentine*, *Oikos*, 86 (2), 311–319, 1999.
77. Janzen, F.J. and Morjan, C.L., Egg size, incubation temperature, and post-hatching growth in painted turtles (*Chrysemys picta*). *J. Herpetol.*, 36, 308–311, 2002.
78. Andrews, R.M., Mathies, T., and Warner, D.A., Effect of incubation temperature on morphology, growth, and survival of juvenile *Sceloporus undulates*, *Herpetol. Monogr.*, 14, 420–431, 2000.
79. Rhen, T. and Lang, J., Temperature during embryonic and juvenile development influences growth in hatchling snapping turtles, *Chelydra serpentine*, *J. Therm. Biol.*, 24, 33–41, 1999.
80. Elphick, M.J. and Shine, R., Long-term effects of incubation temperatures on the morphology and locomotor performance of hatchling lizards (*Bassiana duperreyi*, Scincidae), *Biol. J. Linnean Soc.*, 63, 429–447, 1998.
81. O'Steen, S., Embryonic temperature influences juvenile temperature choice and growth rate in snapping turtles *Chelydra serpentine*, *J. Exp. Biol.*, 201, 439–449, 1998.
82. Flores, D., Tousignant, A., and Crews, D., Incubation temperature affects the behavior of adult leopard geckos (*Eublepharis macularius*), *Physiol. Behav.*, 55, 1067–1072, 1994.
83. Shine, R. and Downes, S.J., Can pregnant lizards adjust their offspring phenotypes to environmental conditions? *Oecologia*, 119, 1–8, 1999.

chapter 7

Neurotoxicology and Behavioral Effects in Reptiles

Joanna Burger

Table of Contents

I. Introduction

Only recently have scientists, managers, and public policy makers become aware of the importance of understanding the behavior of organisms (not just their numbers) in the maintenance of healthy populations. Behavior is a result of complex interaction between genetic and environmental components, acting separately or in combination on multiple organ systems. Stressors that affect morphology and physiology usually have secondary effects on the behavior of organisms, which in turn affect reproductive success and survival. Environmental stresses include physical disruptions (e.g., habitat loss, human disturbance), climatic factors (e.g., temperature, rainfall, and other climate changes), and chemical and radiologic exposures.

Remarkably, toxicology studies with reptiles have lagged behind those of other vertebrate groups. For example, from 1990 to 1999, reptiles made up only 1% of the vertebrate studies in the journal *Environmental Toxicology and Chemistry*; there were seven times as many studies with amphibians.[1] This is surprising given the concern for declining reptile populations globally.[2] Although more than 95% of reptiles are lizards and snakes, they are the least studied group of reptiles.[3] Even when there are toxicologic studies with reptiles, they primarily focus on tissue levels of contaminants, rather than neurobehavioral or other biological effects,[3–5] and focus mainly on turtles.[6] A review by Pauli and Money[7] noted that there has been little research on the effects of pesticides in reptiles since Hall's[4] review. Effects studies often measure only mortality,[3,7] not sublethal neurobehavioral effects. Because reptiles, such as alligators and some large snakes, are at the top of food chains, understanding how chemicals affect their behavior is important. Because they are high on the food chain, they accumulate relatively high levels and thus can be expected to show effects before other reptiles that accumulate contaminants at a slower rate or at lower levels. It is also important to study reptiles at a variety of trophic levels to understand the potential adverse effects on these species and on predators that eat them. Unlike many top trophic-level birds and mammals, reptiles are often quite sedentary and can represent local exposure and effects, as well as being amenable to long-term studies with marked individuals.

There are many reasons why reptiles are underrepresented in toxicologic studies in general and neurobehavioral studies in particular: (1) many reptiles do not have short generation times (they have a long latency to sexual maturity), (2) many have small clutch sizes, (3) many have a long interbrood period, and (4) larger species are difficult to keep and breed in laboratories.[1] Further, it is challenging to develop neurobehavioral endpoints for reptiles because they are generally less active than homeotherms. In a way, however, these disadvantages are exactly the reasons why studies of the effects of chemicals on behavior of reptiles are essential. Because reptiles reach sexual maturity after a long period, live a long time, and have small clutches, they are more vulnerable to the effects of chemicals than other, smaller, short-lived species. There is more time for bioaccumulation and biomagnification, a greater potential for transmittal of chemicals to their eggs,[8–12] and more time for sublethal deficits to affect reproductive success and survival. Further, their relatively low fecundity and late sexual maturity result in populations affected by chemicals having a lower resiliency and less ability to recover from catastrophic events.[1] Reptiles, then, are ideal indicators of the effects of toxic chemicals in the environment.[13]

In this chapter I review the major issues involved in examining neurobehavioral effects of chemicals in reptiles, explore the fate to effects continuum, discuss the major aspects of neurobehavioral studies (including endpoint selection), examine studies of neurobehavioral deficits caused by chemicals, and suggest areas for future research. Reptile behavioral ecotoxicology and neurotoxicology have largely been ignored to date.[1]

II. Major Issues with Neurotoxicology and Behavioral Effects Studies

There are a number of major issues surrounding the study of the effects of environmental and toxic chemical stressors on behavior (Table 7.1). These include duration of studies, behavior that affects growth and reproductive success rather than survival, individual vs. population effects, chronic vs. acute exposure (and effects), species, gender or age differences, and reproductive life history information. Each of these should be considered when planning studies of the effects of chemicals (or other stressors) on reptile behavior. They should also be considered whether the study is primarily laboratory based or field based.

A. Study Design

The duration of behavioral or neurotoxicologic studies depends not only on the overall objectives of the study but on expected environmental

Table 7.1 Factors to Consider When Designing Neurotoxicologic and Behavioral Studies of Reptiles

Study design
Duration of studies
Endpoints: sublethal effects vs. survival
Chronic vs. acute exposure and effects
Critical periods
Multigenerational effects
Host differences
Gender differences
Age differences
Size and weight differences
Nutritional status
Individual differences
Family or species vulnerabilities
Reproductive and life history strategies (oviparous vs. viviparous)
Habitat usage
Longevity
Individual vs. population

variations. For example, determination of the acute effects of a chemical spill on survival may be as simple as monitoring reptile populations for a few months after the spill. However, if the objective is to understand the long-term effect of a chemical spill on the behavior and reproductive success of a suite of reptile species, then several years may be necessary to compare reproductive success before and after the spill for organisms near the spill and at a reference site.

The first decision that must be made for neurotoxicologic studies relates to selection of measurement endpoints. Whereas in the past many toxicologic studies merely documented the lethal effects of chemicals and calculated an LD-50 (dose that is lethal to 50% of test animals), a suite of endpoints may more accurately measure chemical effects. The endpoints selected should be biologically relevant (not just easily measured), such as locating and capturing prey, feeding, finding shelter or appropriate habitats, recognizing mates, or avoiding predators (Table 7.2). If exposure results in behavioral deficits, then growth, reproductive success, or ultimately survival may be affected. Being able to measure these behavioral endpoints requires understanding the ecology and life history of reptiles in the wild and of their behavioral requirements for reproduction and survival. Ideally, behavioral endpoints can be selected that can be applied to both laboratory and field studies. For example, the response of snakes to a simulated predator (an approaching person) can be examined in

Table 7.2 Possible Neurobehavioral Endpoints for Determining the Effects of Chemicals on Reptiles[1]

Feeding behavior
Prey selection
Latency to notice prey or other food items
Latency to begin a response
Latency to capture prey
Percent successful captures (or failure rate)
Prey-capture attempt rate
Time required to handle (or subdue) prey
Time required to completely consume food item
Finding shelter
Latency to respond
Type of shelter selected
Dose response with temperature and chemical exposure
Suitability of shelter selected
Time to remain in the shelter
Habitat selection
Time to choose between two habitat types
Time to reach a specific habitat
Time in a given habitat
Appropriateness of choice
Basking
Time to choose places
Types of places
Total basking times
Ratio of basking to nonbasking
Relationship of basking to temperatures
Thermoregulation
Latency to find shade when heat stressed
Latency to find sun when cold stressed
Time to remain in shade when heat stressed
Type of shade selected
Recognition
Choice experiments of laboratory caretaker and noncaretaker
Learning to respond to the presence of a caretaker or sound that indicated the presence of food
Learning to respond to familiar habitat or shelter
Recognition of gender, neighbors, mates
Aggressive behavior
Types of behavior exhibited
Latency, duration, intensity, and frequency of aggression
Changes in body posture, color, stance, or structures (e.g., dewlap extension variables)

Table 7.2 (continued) Possible Neurobehavioral Endpoints for Determining the Effects of Chemicals on Reptiles[1]

Antipredator behavior
Choice, y-maze experiments of predator and nonpredator odors
Response to different prey models (other snakes, birds, people)
Latency to respond
Strike score or accuracy, strike height
Response to approaching predator (distance to respond, response, latency to reach shelter)
Balance and locomotion
Ability to move on a narrow surface (for the species)
Time to move up an incline
Time to move 1 m when approaching food
Time to right itself when placed on its back
Speed (including sprint speed)
Endurance (could be tested on treadmill)
Total distance moved
Swimming ability (speed, duration)
Activity patterns
Social behavior
Changes in interaction rates
Latency to respond to intruders

[1] These are not meant as an exhaustive list but as suggestions for behaviors that can be measured in a laboratory that are relevant to behavior in the wild. (After Burger,[14,50–53] Burger and colleagues,[54] Burger unpublished, and Clotfelter and colleagues.[72])

the laboratory setting and in the field.[14,15] In both cases, measurement endpoints included latency and distance for an initial response, latency to move, time to disappear, and place to hide. These are relatively complex behaviors that could be affected by chemicals).

There are critical windows of development for the effect of chemicals. Females sequester or transfer contaminants from themselves to their eggs (or offspring) during development.[8–12] The neurobehavioral effects of toxic chemicals are most severe during fetal and neonatal development, and studies should be concentrated during this period. It is during the developmental period that chemicals have the greatest effect on the developing brain and thus on subsequent neurobehavior.[3,5] That is not to say that chemicals do not affect behavior at any life stage, but the effect during development can influence further differentiation and development as well as subsequent behavior.

B. Host Differences

There are a series of individual host characteristics that can affect the neurobehavioral response of reptiles to chemicals, including gender, age, size, nutritional status, and genetic differences.[16] All these factors can affect uptake, bioaccumulation, and effects. For example, in American alligators (*Alligator mississippiensis*) there were (1) gender differences for lead levels but not for cadmium, selenium, manganese, tin, or mercury: (2) size differences for manganese and mercury but not for other metals; and (3) gender differences in cadmium levels in alligators in some lakes but not in others.[17] These differences in tissue concentrations, which relate to age, gender, or other host-related factors, are sometimes the result of individual differences in exposure, metabolism, or susceptibility. Finally, potential susceptibility differences of reptiles as a function of high trophic level need to be examined.[10]

Most laboratory studies examine the neurobehavioral effects of one toxic chemical, but in the wild reptiles are faced with a mixture of chemicals, most of which are below the "No observed adverse effect level" (NOAEL).[18] To determine the true effect of chemicals in the wild, exposure must include the amount accrued at every life stage, the burden inherited from its mother, and the burden built up during its lifespan. Determining the neurobehavioral effects from these different exposures is a daunting task, considering the individual's exposure to a mixture of chemicals.

C. *Family and Species Vulnerabilities*

There are a number of characteristics of families (or species) that make them particularly vulnerable to the effects of toxic chemicals. Life history strategies (oviparous vs. viviparous) affect the vulnerability of different life stages. Habitat usage also confers vulnerabilities. Reptiles exposed in aquatic environments have the potential for dermal exposure and inadvertent oral (or gill) exposure, whereas those moving through soil can be similarly exposed.

D. *Individual vs. Populations*

Although ecotoxicology studies, particularly in laboratory settings, involve individuals, the overall effect of neurobehavioral deficits in reptile populations is always paramount. Thus, studies should be designed to allow predictions about the possible effects of a given exposure to population stability. Multigenerational studies in the laboratory and field are particularly important for a complete understanding of the subtle neurobehavioral effects of toxic chemicals. Although this approach has not been used with reptiles, it has been effective with birds that were maintained in captivity.[19,20] Multigenerational studies could be conducted with reptiles that mature relatively rapidly and can be maintained in the laboratory, such as lizards.

III. Fate to Effects: The Problem with Reptile Studies

Understanding the effects of a given chemical on neurobehavior requires carefully controlled dose-response studies in the laboratory with clear-cut measurement endpoints, ideally relating to behavior in the wild. Whereas dose-response studies in have generally focused on lethal or large sublethal effects, there is a current need to examine the dose-response curve at the low end of exposures and at levels that might mimic those found in nature. That is, we should not be trying just to show that a given chemical has a specific effect, but rather we should try to show the levels that cause specific effects as well as the no observable effect levels. Studies that show dose-response curves, including the no effect levels, will be most useful. It is this information that will be helpful in mechanism studies and will provide managers and regulators with useful information on which to make decisions.

Relating laboratory studies to effects observed in the wild requires knowing (1) the dose, effects, and tissue levels in target tissues from laboratory studies and (2) effects and tissue levels in wild reptiles. This is a critical step to understanding the effects of chemicals on reptiles in the wild, and although both types of data (laboratory tissue levels to effects, field levels and effects) are available for neurobehavioral deficits in birds,[21–23] they are not generally available for reptile studies. Mainly the problem lies with the equipment needed for analysis; often the scientists conducting dose-response studies with neurobehavior are not equipped to measure the levels of toxic chemicals in tissues, and it is costly to have these analyses done in another laboratory. The lack of contaminant data from target organs of animals exposed in the laboratory (with known behavioral deficits) is currently the biggest data gap for neurotoxicology studies in reptiles. This problem is not limited to reptiles; it applies to other vertebrates as well.[24]

Finally, traditional neurobehavioral toxicity studies often examine the effect of a chemical at different dosages regardless of developmental state. Yet for neurobehavioral studies, where developing embryos and developing young are most vulnerable, it is critical to examine different life stages, particularly during the critical period of neurodevelopment — in reptiles, from hatching through the first months.

IV. The Nervous System

Compared with other vertebrates, little relatively recent research has concentrated on the neuroanatomy and neurophysiology of the reptilian brain. A review published in 2000 of aspects of reptilian anatomy and physiology did not even have a section on the brain or neurophysiology.[25] This perhaps corresponds to the relatively few studies of the effects of contaminants on reptiles.

A. Comparative Neuroanatomy

The nervous system is a complicated and efficient method of coordinating bodily activities by sending or receiving "messages" or impulses from or to specific areas of the body. It is an integrative system, using nerve impulses. During vertebrate development there are three major brain regions (prosencephalon, mesencephalon, and rhombencephalon) on which specialized outgrowths are added, depending on the vertebrate group. In general, the cerebral hemispheres of the prosencephalon are associated with smell, the tectum of the mesencephalon is associated with vision, and the cerebellum of the hindbrain outgrowth is associated with the ear and lateral line. Differences among vertebrates occur because of amplifications of one region over another, giving the animals different abilities. In reptiles, the tectum is of great importance, as is the development within the hemispheres, which is the center of sensory integration.

The hemispheres of reptiles are advanced in size and complexity over those of amphibians, and some of the gray matter tends toward a superficial position. The basal ganglia, which have moved inward to occupy a considerable area of the brain floor, are no longer purely olfactory. Strong projection fiber bundles run upward to them from the thalamus and back from them to the brainstem, and the basal nuclei (corpus striatum) are important correlation centers. The task for neurotoxicologists is to relate differences in brain anatomy or morphology caused by toxic chemicals to behavioral effects. Baxter[26] showed that the basal ganglia system was actively correlated with dominant display routines in *Anolis* lizards, suggesting that this system orchestrates the context-specific selection and rapid launching of complex behaviors while inhibiting inappropriate competing responses. The number of forelimb pushups was strongly correlated with the amount of radiolabeled deoxyglucose uptake in the dorsolateral basal ganglia.[26]

Attention has focused on the degree of brain lateralization (asymmetries) in lower vertebrates, including reptiles. For a long time, brain lateralization was assumed to be unique to humans and was associated with handedness and language. Evidence for brain lateralization is now widespread in birds and mammals.[27] However, asymmetries occur in snakes that coil themselves consistently in one direction or another,[28] and lizards consistently use their left eye more than their right eye during intense aggressive encounters.[29] In *Anolis* the small interhemispheric projections do not allow a functional integration of information stored in the two hemispheres, and aggressive responses seem to be mainly activated by the right hemisphere.[27] Work on lateralization and neural correlates of reproductive behavior in tree lizards (*Urosaurus ornatus*) shows (1) lateralization of amygdala volume among aggressive males, less aggressive males, and nonaggressive females with the right side being larger than the left and (2) that the amygdala volume varies between sexes and among morphs such that aggressive males, less aggressive males, and females have larger, intermediate, and small amygdalae, respectively.[30–32] Examination of neurobehavioral lateralizations in reptiles is a fertile ground for future research with a number of different species.

B. *Neuroendocrine Axis*

Although the brain sends and receives messages from various parts of the body, the endocrine system is also an integrative system that uses chemical messages (hormones) produced by the endocrine glands. Whereas nerve impulses are targeted and rapid, endocrine messages are slower and more diffuse. The two systems are connected; the endocrine hypophysis cerebri is strongly influenced by the adjacent hypothalamus of the brain (and some endocrine hormones are produced in the ganglia of that brain region). The neural correlates of reproductive behavior are active research areas for most vertebrates, although much less has been done in reptiles than other vertebrates (but see Hartman and Crews[33] and Rhen and colleagues[34]). Work has focused on the hypothalamic–pituitary–adrenal axis in lizards.[35] The hypothalamic–pituitary–adrenal axis plays a role in stress responses and flight-or-fight behavior (see also Chapter 5).

C. *From Brain to Behavior*

The field of examining brain anatomy and physiology in relationship to the effects of chemicals on behavior is wide open for reptiles. This is an approach that will be important in the future. Interdisciplinary studies that span neuroanatomy, neurophysiology, behavior, and behavioral development will help elucidate effects and mechanisms and also will aid in our understanding of the population and evolutionary consequences of chemical exposure.

V. Reptilian Models of Neurobehavioral Deficits

As indicated earlier, there are few neurobehavioral effects studies in reptiles using chemicals as the stressor,[1,3] except for the very detailed and thorough work on the effects of endocrine disruptors by Guillette and Iguchi.[36] The lack of neurobehavioral effects studies with reptiles is surprising because, for example, metals play a critical role in the functioning of the central nervous system; many are known neurotoxicants that cause a range of neurologic and behavioral dysfunctions.[3,5,10,37] Herein I review the limited literature on neurobehavioral effects, explore the example of temperature-dependent effects on behavior as a model for future studies, discuss studies with slider turtles (*Trachemys scripta*) that examined neurobehavioral deficits with lead exposure, examine other models for neurobehavioral studies, and briefly summarize endocrine disruption and neurotoxicology.

A. Neurobehavioral Effects

Most effects studies with reptiles in laboratories focus on lethal doses, on time to death, or on biochemical responses (reviewed in Campbell and Campbell[5,38]). There are also some studies in the field where the effects of applications of chemicals were examined; usually only lethality was examined, or population declines were attributed to contaminants.[39] Where lizard behavior (daily activity pattern, reproduction, growth rate) was examined in the field, there were no observed effects as a function of chemical exposure (DDE [1,1′-(dichloroethenylidiene) bis 4-chlorobenzene] methyl parathion, parathion[40]). Similar observations have not been made with snakes,[5] perhaps because snake behavior is more difficult to observe in the wild. An application of aminocarb resulted in snakes being half as active in the 2 months after application compared with preapplication,[41] but tissue levels were never obtained, and changes in activity could have been seasonal or the result of other changes. Polychlorinated biphenyls (PCBs) were thought to cause a neurologic disorder in several snakes in a zoo in Poland, associated with level in the liver of 0.05 mg/kg.[42] Radiation studies generally examine lethality, sterility, and sex ratios as a function of age rather than behavior.[38] Irradiation possibly caused altered perch heights among *Anolis* lizards in Puerto Rico.[43]

There is a hint that some reptiles may show a narrow effects range — that is, there may be a steep dose-response curve. Further, no observed adverse effects may be present at doses that would cause adverse effects in birds and mammals. For example, crocodiles (*Crocodylus porosus*) that sustained high blood levels of lead for several months (up to 363 μg/dL) showed no adverse effects[44]; these blood levels would have sublethal and lethal effects in birds and mammals.[37] Some sensitive birds have severe effects or die at 20 μg/dL blood lead levels.[37] The sensitivity of reptiles to contaminants, relative to birds and mammals, requires more extensive study because too few data have been gathered to reach any conclusion at present.

B. Temperature-Dependent Neurobehavioral Effects in Pine Snakes (Pituophis melanoleucus)

Whereas for most topics in this volume toxicity refers to toxic chemicals, in its broader sense neurotoxicity refers to any stressor that is toxic to the behavior and functioning of an organism. In the following I discuss briefly a series of experiments that examine the effect of incubation temperature on neurobehavioral development in pine snakes as an example of stressor effects on behavior and as a model for the kinds of neurobehavioral studies that could be conducted with toxic chemicals. It should be noted that in reptiles other than snakes, incubation temperature (among other abiotic factors) determines sex of the offspring[45–47] as well as egg size, hatching size, and growth.[48] Snakes are exceptions and have chromosomal sex determination.

Pine snakes in the New Jersey pine barrens are isolated from other conspecifics by several hundred miles. Females excavate their own nests because there are no excavating, nonpredatory mammals that can provide adequate nest sites.[49] Female pine snakes excavate burrows that can have tunnels a meter or so long, and then they deposit the eggs at the end of the tunnel. When the young hatch, they remain in the burrow for a few days to 2 weeks and then emerge directly above. Because New Jersey pine snakes are at the northern limit of their range, the placement of eggs is critical for adequate sun penetration and heat for embryonic development.[50] In a series of laboratory experiments, I examined the effect of incubation temperature on behavioral development by using a range of temperatures that can occur in nests in the New Jersey pine barrens. I incubated eggs at temperatures between 22°C and 32°C. These experiments showed that there is a range of behavioral tests that can be conducted on young snakes that relate to behaviors needed for survival in the wild. The experiments formed the basis for many of the endpoints given in Table 7.2.

There were temperature-dependent differences in nearly every behavioral test I could design, with hatchlings from the low-incubation–temperature regimen showing behavioral deficits that were not present in snakes incubated at medium or high temperatures.[14,50,52,53,56] Table 7.3 summarizes some of the results of these experiments and describes briefly the tests. As is clear, there is a remarkable diversity of controlled experiments that can be conducted with neurobehavioral endpoints in hatchling snakes that relate directly to survival in the field. Further, these experiments can be related directly to the ability of the young to reach a hibernaculum in the fall, having eaten a mouse, which increases their survival prospects (Figure 7.1). That is, at lower incubation temperature, eggs require a longer incubation period, hatching takes more time, emerging from the underground nest to the surface of the ground takes more time, time to first shed after hatching takes more time, and finding a mouse to eat takes more time. Thus, a hatchling from a nest incubated at low temperatures barely has enough time to hatch, shed its skin, and find food before it must hibernate. Most pine snakes are in a hibernaculum by early November.

These studies indicate that temperature affects a wide range of behaviors that directly influence survival in the wild. Chemicals may well adversely affect behavior in a similar manner as temperature — that is, both the timing and the quality or effectiveness of the behavior may be altered.

C. *Neurobehavioral Deficits from Lead Exposure in Slider Turtles (Trachemys scripta)*

Reptiles are exposed to contaminants during development because females can transfer them to their eggs.[12,57] Yet for many reptiles, the metal loads in hatchlings, derived from their mothers, are low (e.g., slider turtles[9]). In a series of experiments during the course of 2 years, slider turtle hatchlings were exposed (intramuscular injection) to 0.05 and 0.1 mg/g (1995) and 0.25, 1.0, and 2.5 mg/g lead (1996).[54] In both years, a control group received a saline injection of the same volume as the experimental turtles. The doses used initially were those that resulted in neurobehavioral deficits in birds and that occur in the wild in some individuals.[58,59] Using environmentally relevant doses for neurobehavioral studies is an important step in our understanding of the effects of contaminants in reptiles. Survival declined markedly as a function of dose, with survival at 4 months being 92% in controls and 0 at high doses. Behavioral tests

Table 7.3 Behavioral Deficits as a Function of Incubation Temperatures in Pine Snakes[1]

Behavior	Low	Medium	High	X^2 P
Temperature (°C)	21–23	26–28	31–32	
Righting response	1.8 ± 0.2	0.4 ± 0.1	0.5 ± 0.1	0.001
Drinking (swallows/min)	1.0 ± 0.1	1.4 ± 0.3	1.6 ± 0.3	0.0001
Activity score	3.2 ± 0.1	3.9 ± 0.1	3.1 ± 0.1	0.0001
Incline (% sliding off)	26	12	17	0.001
Antipredator in Lab[2]				
% initially respond	79	96	97	0.001
% move rapidly away	45	94	95	0.001
Distance moved (m)	2.2 ± 0.3	4.1 ± 0.3	4.4 ± 0.2	0.001
Antipredator in Field				
Response latency(s)	135 ± 20	17 ± 3	5 ± 1	0.001
Strike height (cm)	1.3 ± 0.1	5.6 ± 0.3	6.0 ± 0.3	0.001
Y-Maze Choices				
% Avoid predatory snake[3]	25	100	100	0.0001
% Choosing another pine snake[4]	80	100	100	0.05
Time to choose between mouse/no mouse (s)	23.1 ± 5.6	12.4 ± 3.9	7.1 ± 2.2	0.05
Behavior in Nests				
No. of exit holes[5]	1.2 ± 0.3	2.6 ± 0.5	1.9 ± 0.3	0.0002
No. extra chambers[6]	0.9 ± 0.2	5.6 ± 0.5	3.1 ± 0.3	0.0002
No. of snakes remaining in the nest[7]	2.4 ± 0.2	0.8 ± 0.1	1.0 ± 0.2	0.0001
Thermoregulation[8]				
Latency to move (s)	12.1 ± 3.7	4.5 ± 1.3	13.1 ± 2.9	0.0001
Latency to shade (s)	73.0 ± 11	34.1 ± 0.3	45.2 ± 5.8	0.002

1 Given are means (and standard errors) for selected tests from Burger,[14,51] Burger and Zappalorti,[49] and Burger and Gochfeld[55] (and Burger unpublished). The usual sample size was 50± for each group, but often it was more than 100 per group.

2 Tested with approaching person in the laboratory. In the field experiment, snakes emerged naturally from their underground nest, and a person approached and moved a pencil toward the snake.

3 Choosing between a king snake (*Lampropeltis gelulus*) and no odor.

4 Choosing between a pine snake and a corn snake (*Elaphe guttata*), neither is a predator on pine snakes.

5 Eggs are located in a chamber below ground. Once the eggs hatch, the hatchlings often remain in the nest but make exit holes to the ground surface. They also make side tunnels and chambers away from the nest, presumably as an antipredator behavior. If a predator digs up a nest, it eats the contents but does not necessarily dig in every direction where a hatchling may be resting in an extra chamber.

6 Number of extra chambers for resting, away from the nest.

7 In the experimental design, there were three hatchlings from the same female and of the same incubation temperature in each artificial nest. Snakes that remained underground without emerging for 7–10 days could either remain in the nest (vulnerable to predators) or could go into side burrows they constructed.

8 Snakes were placed in the middle of an apparatus that provided a structure and another place where there was shade (but no structure).

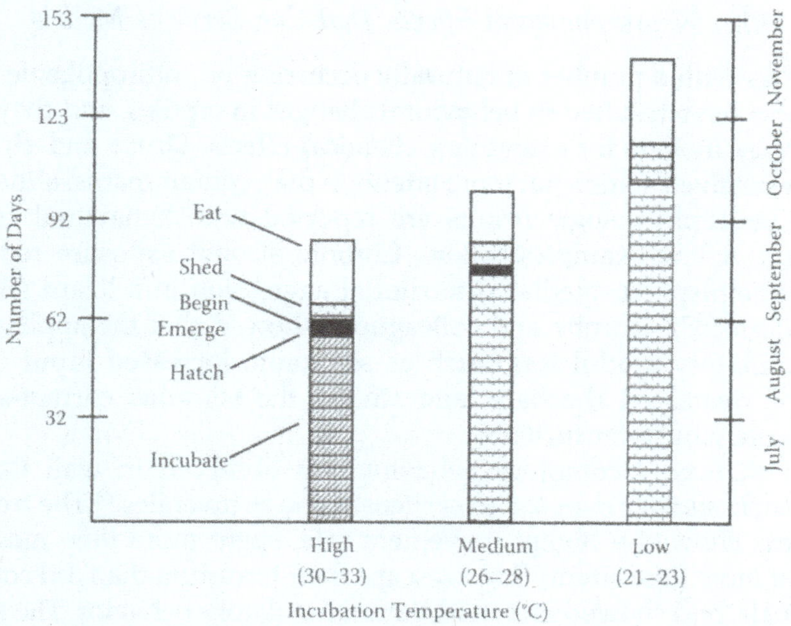

Figure 7.1 **Effect of incubation temperature (°C) on behaviors necessary before young pine snake hatchlings can enter a hibernaculum. Most enter hibernacula in mid-October. As this figure indicates, snakes hatched from low incubation temperatures would not have enough time to reach hibernacula in time for winter. (After Burger 1991.)**

were performed at 1, 2, 3, and 4 weeks postinjection. Time to right themselves when placed on their backs, as might occur if they fell down an embankment, was directly related to dose (at 0.25 mg/g and higher). Hatchlings in the high-dose group averaged 91 sec (standard error = 23) to right themselves, whereas controls took only 55 sec (standard error = 7). Time to reach cover when exposed to thermal stress also varied. Overall these experiments showed a rather tight response to exposure to chemicals in that they showed no effect up to doses of 0.1 mg/g but showed marked effects thereafter, including eventual lethality by 2 months in the high-dose group.[54] This also indicates the importance of continuing experiments for several months. In hatchling birds, mortality occurs rather quickly after lead exposure (if it is going to), and if they survive for 2–3 weeks, they do not die from lead exposure at 2 or 3 months. The difference may be that birds grow rapidly, allowing for dilution of the effect, whereas young reptiles do not grow as rapidly. Further, young birds excrete metals into their growing feathers,[60] allowing them an additional route of elimination.

D. Other Neurobehavioral Effects That Can Serve as Models

Studies with a number of naturally occurring or anthropogenic substances have resulted in behavioral changes in reptiles, and they can serve as models for examining chemical effects. Drugs and alcohol have received somewhat more attention than contaminants, although few neurophysiologic results are reported with behavioral correlates.[61] A few examples follow. Chronic alcohol exposure reduces right-hemisphere–mediated territorial aggression in a lizard (*Anolis carolinensis*[62]). Hornby and colleagues[63] showed that the application of excitatory modulators, such as serotonin, increased input resistance, decreased rheobase, and shifted the stimulus current-spike frequency in slider turtles.

Enhanced locomotory behavior was observed in wall lizards (*Podarcis muralis*) exposed to corticosterone as juveniles.[64] The treated lizards showed a higher movement rate, spent more time moving, spent more time attempting to escape their terrarium than did control animals, and showed impaired thermoregulatory behavior. The same patterns were obtained in nature and in the laboratory,[64] again demonstrating the potential for selecting endpoints that can be used in the laboratory and in nature. Studies with corticosterone and aggression have proved useful in examining subsequent social interactions in lizards,[65] suggesting that examining the relationship between hormones and aggression could be a useful model for examining neurobehavioral effects of toxic chemicals. Tree lizards are being developed as models for the study of brain, behavior, and physiologic interactions.[30–32] Knapp and Moore[66] found that different male morphs have different corticosterone–testosterone interactions, which directly affects territorial behavior. It should be noted, however, that in garter snakes (*Thamnophis sirtalis parietalis*), exogenous corticosterone suppressed mating behavior without depressing testosterone, which may indicate that corticosterone does not act directly on the hypothalamic–pituitary–gonadal axis.[35]

In sand lizards (*Lacerta agilis*), short-term exposure to organic and inorganic lead (lead acetate, intraperitoneal application) induced changes in neurosecretory and ependymosecretory cells of the suboptic nucleus, paraventricular nucleus, and ependymal cells.[67,68] Prolonged lead intoxication resulted in degeneration of the cell nuclei. Unfortunately, no tissue concentrations were measured.

Neuroanatomic changes as a function of chemical exposure can also be examined. O'Bryant and Wade[69] demonstrated gender differences in nerve cross-sectional areas as a function of season and androgen treatment, suggesting the importance of taking into account host-related factors in examining the effects of naturally occurring and anthropogenic chemicals. The rich literature on endocrine active substances (both natural and anthropogenic) can also serve as a model for studies on the effects of chemicals on neurotoxicity.

E. Endocrine Disruption and Neurotoxicology

Although this volume, and many others, describes contaminant-induced endocrine disruption leading to behavioral and reproductive effects, we should bear in mind that the endocrine and neurobehavioral systems are interrelated. Successful reproduction requires temporal coordination of physiology, vitellogenesis, and mating behavior (receptivity). Xenobiotic interference can result in disintegration and abnormal behavior.[34] Likewise, sexual behavior feeds back to the endocrine system,[33] so that behavioral toxicity could directly cause reproductive failure.

Endocrine-disrupting chemicals clearly affect behavior in reptiles (see Guillette and Iguchi[36] and this volume), and the relationship among endocrine disruption (largely studied on a mechanistic and morphologic basis), behavior of reptiles in the wild, and neurotoxicity will be a fertile research area in the future. Although the elegant studies of Guillette and others form the basis for much of our ecologic knowledge about the effects of endocrine disruptors in wild alligator populations,[36] there is a vast literature on chemically induced endocrine disruption, including many papers with reptiles.[13,70]

VI. Summary and Research Directions

A. Summary

Overall, reptiles are underrepresented in ecotoxicologic studies, and where there are studies, they primarily comprise tissue levels of contaminants. Effects studies, particularly neurobehavioral effects, are notably absent. Yet, reptiles are excellent subjects for neurotoxicologic studies because they are generally sedentary (except for sea turtles), are often high on the food chain, and have a long lifespan that results in bioaccumulation and biomagnification and allows for repeated studies of individuals. As with all laboratory studies of the effects of toxic chemicals, it is critical to measure both dose and final tissue concentrations so that the behavioral deficits found in the laboratory can be related to levels found in reptiles in the wild.

There is a range of measurement endpoints that can be used to assess the toxic effects of chemicals on neurobehavioral development

Table 7.4 Types of Studies Needed for Neurotoxicology of Reptiles

1. Laboratory dose and effects studies of a range of chemicals suspected of causing adverse neurobehavioral deficits in the wild
2. Dose and effects studies that also measure tissue levels in the kidney, liver, muscle, and brain for comparison with levels found in wild reptiles
3. Tissue assays of toxic chemicals in wild reptiles to provide baseline information
4. Tissue assays of chemicals in wild reptiles (kidney, liver, muscle, brain) where neurobehavioral effects are observed
5. Brain levels of chemicals and regional distribution of chemicals within the brain (cerebrum, cerebellum, basal ganglia, brainstem)
6. Multigenerational studies
7. Laboratory studies with mixtures of chemicals that correspond to the exposures reptiles face in the wild
8. Laboratory studies with media from the wild that represent the exposure of wild reptiles
9. Dose-, effects-, and tissue-level studies with reptiles in the wild where individuals can be followed for several weeks, months, or years
10. Developmental studies that examine dose-response curves with different developmental stages
11. Collaborative studies with neurotoxicologists interested in mechanisms of disruption

in reptiles, largely derived from temperature-dependent behavioral studies (Table 7.2). However, toxic chemicals, rather than temperatures, can be used as the stressor. Measurement endpoints include behaviors that relate to survival and reproductive success in the wild and can be used both in the laboratory and in the field. Useful assessment endpoints relate to searching for food (or prey), capturing and handling food, finding shelter or appropriate habitats, recognizing mates or neighbors, and avoiding predators. For each of these behaviors there are measurement endpoints that include latency to respond, time for a particular behavior, and accuracy of the behavior.

B. *Future Research Needs*

Examining the neurotoxicologic effects of chemicals in reptiles provides a research opportunity to make major contributions (Table 7.4). Because many reptiles are long lived and sedentary, they will be useful as bioindicators of ecosystem health; sentinels of potential human exposure; and tools for understanding the ecologic, behavioral, and neurologic basis for sublethal deficits and effects on survival, reproduction, and longevity. Major research needs are as follows (Table 7.4):

1. It is imperative to have a range of laboratory dose and effects studies for a number of chemicals suspected of causing adverse neurobehavioral deficits in the wild. When a given contaminant is suspected of causing adverse effects in reptiles in the wild, laboratory studies are needed to establish whether the chemical causes a similar effect in a range of different species.

2. Dose and effects studies often do not examine the levels of contaminants in target tissues. Tissue levels of contaminants should be measured in at least the kidney, liver, muscle, and brain for comparison with levels found in wild reptiles and for comparison with similar effects caused by other chemicals.
3. There is a need for an extensive database on the levels of chemicals in different tissues for a wide range of species in the wild. The database should include levels of PCBs; lead; mercury; DDT; and other contaminants in liver, kidney, blood, and muscle of a wide range of species. This would allow managers to evaluate levels found in species of interest and would provide baseline information on contaminants for managers, regulators, and the public.
4. Although establishing a baseline of contaminant levels in reptiles that do not show adverse effects is important (see earlier), it is equally important to assay the tissue levels of chemicals in wild reptiles (kidney, liver, muscle, brain) where neurobehavioral effects are observed — that is, what levels of contaminants in the wild are associated with adverse effects?
5. A review of the literature on contaminant levels in tissues of reptiles in the wild[5] indicates that the tissues usually examined for toxic chemical levels include the tail, skin, liver, kidney, muscle, and gonads, and only rarely does it include the brain (see Storelli and Marcorigiano[11]). However, to understand the mechanisms of neurobehavioral deficits, we need many studies of brain levels of chemicals. We also need to localize where toxic chemicals are residing in the brain and to examine temporal sequencing of nerve-cell migration and differentiation during development. For example, herring gull chicks (*Larus argentatus*) behaviorally impaired with lead also show disruption of the temporal expression of synaptic neural cell-adhesion molecules in their brains.[71]
6. It is relatively difficult to perform multigenerational studies for reptiles that are long lived and have a long period to first breeding. However, it is critical to determine if there are multigenerational effects and the nature of these effects as a function of type of chemical. Such studies could be performed in lizards, for example, where generation time is short and colonies can be easily maintained in laboratories.
7. There is a need for studies in the laboratory that use mixtures of chemicals that correspond to the exposures reptiles face in the wild. Although it is easier in the laboratory to isolate the

effect of one chemical on neurobehavioral development, animals are exposed to mixtures in the wild. There is a need to conduct neurobehavioral studies with mixtures of chemicals known to cause deficits (such as lead, mercury, cadmium) and with media reptiles are exposed to in the wild (such as water and sediment from contaminated lakes). Such mixtures can be artificially made.

8. In addition to concocting mixtures in the laboratory of chemicals that are known to cause neurobehavioral deficits, studies can be conducted using media (soil, sediment, water) from the wild that naturally contain a mixture of chemicals and that are thought to cause neurobehavioral deficits. Using this method, effects could be attributed to particular mixtures, and the chemical composition could then be assessed.
9. More studies are needed that examine dose, effects, tissue levels with reptiles in the wild where individuals can be followed for several weeks, months, or years. Nearly all toxicologic studies are performed in the laboratory, with a very few conducted in semiwild conditions (fenced enclosures). There is a need to conduct such studies in the wild, however. Although it is difficult, reptiles might lend themselves to such studies because some species are sedentary, are long lived, and can be recaptured. Species such as lizards, snakes, or turtles could be exposed to toxic chemicals (either in provisioned food or by injection) and then followed and observed, using some of the endpoints in Table 7.2. Animals could be relocated either by sight or with the use of radiotransmitters.

Studies on the effects of chemicals on neurobehaviors are generally performed in laboratory settings where environmental variables can be controlled, animals can be regularly observed, behavioral measures can be quantified, and (later) tissues can be collected for contaminant analysis. For some species, such studies can be done in the semiwild in enclosures of varying sizes that allow for more movement and interactions between individuals. I suggest that it is important to conduct studies with truly wild reptiles where the only intervention is exposure. This paradigm has worked well with birds, where hatchling gulls were exposed to lead and then left entirely to their parents to raise within their nesting colonies.[58] In this experiment, the same behavioral observations conducted in the laboratory were made in the field, and before fledging, the control and experimental birds were collected for tissue assaying of lead.

10. There is a need for studies that examine dose-response curves at different developmental stages for a wide range of reptiles. Such studies could also prove useful for comparison with those of birds and mammals, where critical periods of development indicate increased vulnerability in neonates.
11. Increasingly, research questions require collaborations among different disciplines and approaches. Such collaborative studies with neurotoxicologists interested in mechanisms of disruption would be particularly fruitful. Some of these are self-evident or were discussed earlier, but others require further elaboration, such as collaborations among behavioral ecologists, toxicologists, and neuroanatomists. Collaborative studies should be developed among biologists interested in reptiles, behavior, development, neurophysiology, and neuroanatomy.

In conclusion, behavior deficits in laboratory and wild reptiles should be correlated with tissue levels of contaminants (including brain levels), with specific regions of the brain, and with morphologic and mechanistic changes. In a sense, this is putting the whole picture together, from dose-response studies; to dose, response, tissue levels; to dose, response, brain tissue levels and brain function alterations. Although each type of study stands alone, the links must be forged before we will understand the neurobehavioral effects of toxic chemicals on reptiles.

Acknowledgments

My research on neurobehavioral deficits, toxicology, reptiles, and risk has benefited from discussions with L. J. Guillette Jr., J. W. Gibbons, R. Zappalorti, M. Gochfeld, K. Campbell, J. Snodgrass, M. Gallo, B. Goldstein, K. Reuhl, E. Orlando, and K. Cooper. I also thank C. Dixon, T. Shukla, S. Shukla, C. Jeitner, S. Burke, and R. Ramos for technical assistance throughout my work. Over the years my research has been supported by grants from NIEHS Center grant (ESO 5022); Environmental Protection Agency; Wildlife Trust; Consortium for Risk Evaluation with Stakeholder Participation (CRESP) through the Department of Energy (AI # DE-FC01-95EW55084, DE-FG 26-00NT 40938); the New Jersey Office of Science, Research, and Technology; New Jersey Department of Environmental Protection; and the Environmental and Occupational Health Sciences Institute. All research was conducted under a Rutgers University protocol. The views and opinions expressed in this paper are solely those of the authors and not of any of the funding agencies.

Literature Cited

1. Hopkins, W.A. Reptile toxicology: challenges and opportunities on the last frontier in vertebrate ecotoxicology, *Environ. Toxicol. Chem.*, 19, 2391, 2000.
2. Gibbons, J.W., Scott, D.E., Ryan, T.J., Buhlmann, K.A., Tuberville, T., Metts, B.S., Greene, J.L., Mills, T., Leiden, Y., Poppy, S., and Winne, C.T. The global decline of reptiles, déjà vu amphibians, *BioSci.*, 50, 653, 2000.
3. Campbell, K.R. and Campbell, T.S. A logical starting point for developing priorities for lizard and snake ecotoxicology: a review of available data, *Environ. Toxicol. Chem.*, 21, 894, 2002.
4. Hall, R.J. Effects of environmental contaminants on reptiles: a review, *U.S. Fish Wildl. Serv. Rep.*, 228, 1, 1980.
5. Campbell, K.R. and Campbell, T.S. The accumulation and effects of environmental contaminants on snakes: a review, *Environ. Monitor. Assess.*, 70, 253, 2001.
6. Sparling, D.W., Bishop, C.A., and Linder, G. The current status of amphibian and reptile ecotoxicological research, in *Ecotoxicology of Amphibians and Reptiles*, Sparling, D.W., Linder, G., Bishop, C.A., Eds., SETAC Press, Pensacola, FL, 2000, pp. 1–14.
7. Pauli, B.D. and Money, S. Ecotoxicology of pesticides in reptiles, in *Ecotoxicology of Amphibians and Reptiles*, Sparling, D.W., Linder, G., Bishop, C.A., Eds., SETAC Press, Pensacola, FL, 2000, pp. 269–290.
8. Culley, D.D. and Applegate, H.G. 1966. Pesticides at Presidio. IV. Reptiles, birds, and mammals, *Tex. J. Sci.*, 16, 301, 1966.
9. Burger, J. and Gibbons, W. Trace elements in eggs and eggshells of slider turtles (*Trachemys scripta*) from the Savannah River Site, *Arch. Environ. Contam. Toxicol.*, 34, 382, 1998.
10. Linder, G. and Grillitsch, B. Ecotoxicology of metals, in *Ecotoxicology of Amphibians and Reptiles*, Sparling, D.W., Linder, G., Bishop, C.A., Eds., SETAC Press, Pensacola, FL, 2000, pp. 325–408.
11. Storelli, M.M. and Marcotrigiano, G.O. Heavy metal residues in tissues of marine turtles, *Mar. Pollut. Bull.*, 46, 397, 2003.
12. Brasfield, S.M., Bradham, K., Wells, J.B., Talent, L.G., Lanno, R.P., and Janz, D.M. Development of a terrestrial vertebrate model for assessing bioavailability of cadmium in the fence lizard (*Sceloporus undulata*) and in ovo effects on hatching size and thyroid function, *Chemosphere*, 54, 1643, 2004.
13. Crain, D.A. and Guillette, L.J., Jr. Reptiles as models of contaminant-induced endocrine disruption, *Anim. Reprod. Sci.*, 53, 77, 1998.
14. Burger, J. Incubation temperature has long-term effects on behavior of young pine snakes (*Pituophis melanoleucus*), *Behav. Ecol. Sociobiol.*, 24, 201, 1989.
15. Burger, J. The behavioral response of basking Northern water (*Nerodia sipedon*) and Eastern garter (*Thamnophis sirtalis*) snakes to pedestrians in a New Jersey park, *Urban Ecosys.*, 5, 119, 2001.
16. Peakall, D. and Burger, J. Methodologies for assessing exposure to metals: speciation, bioavailability of metals and ecological host factors, *Ecotoxicol. Environ. Saf.*, 56, 110, 2003.
17. Burger, J., Gochfeld, M., Rooney, A.A., Orlando, E.F., Woodward, A.R., and Guillette, L. J., Jr. Metals and metalloids in tissues of American alligators in three Florida lakes, *Arch. Environ. Contam. Toxicol.*, 38, 501, 2000.

18. Crews, D., Putz, O., Thomas, P., Hayes, T., and Hosdeshell, K. Wildlife as models for the study of how mixtures, low doses, and the embryonic environment modulate the action of endocrine-disrupting chemicals, in *Implications of Endocrine Active Substances for Humans and Wildlife*, Miyamoto, J., and Burger, J., Eds., *Pure Appl. Chem.*, 75, 2305, 2003.
19. Heinz, G. Effects of low dietary levels of methylmercury on mallard reproduction, *Bull. Environ. Contam. Toxicol.*, 11, 386, 1974.
20. Heinz, G. Mercury poisoning in wildlife, in *Noninfectious Diseases of Wildlife*, Fairbrothers, A., Locke, L.N., and Hoff, G.L., Eds., Iowa State University Press, Ames, IA, 1996, p. 118.
21. Furness, R.W. Birds as monitors of pollutants, in *Birds as Monitors of Environmental Change*, Furness, R.W., Greenwood, J.J.D., Eds., Chapman & Hall, London. 1993, pp. 86–143.
22. Franson, J.C. Interpretation of tissue lead residues in birds other than waterfowl, in *Environmental Contaminants in Wildlife: Interpreting Tissue Concentrations*, Beyer, W.N., Heinz, G.H., Redmon-Norwood, A.W., Eds., Lewis Publishers, Boca Raton, FL, 1996, pp. 265–280.
23. Burger, J., Kannan, K., Giesy, J.P., Grue, C.E., and Gochfeld, M. Effects of environmental pollutants on avian behavior, in *Behavioral Ecotoxicology*, Dell'Omo, G., Ed., Wiley & Sons, Chichester, UK, 2002.
24. Burger, J., Diaz-Barriga, F., Marafante, E., Pounds, J., and Robson, M. Methodologies to examine the importance of host factors in bioavailability of metals, *Ecotoxicol. Environ. Saf.*, 56, 20, 2003.
25. Palmer, B.D. Aspects of reptilian anatomy and physiology, in *Ecotoxicology of Amphibians and Reptiles*, Sparling, D.W., Linder, G., Bishop, C.A., Eds., SETAC Press, Pensacola, FL, 2000, pp. 111–140.
26. Baxter, L.R. Jr. Basal ganglia systems in ritualistic social displays: reptiles and humans; function and illness, *Physiol. Behav.*, 79, 451, 2003.
27. Bisazza, A., Rodgers, L.J., and Vallortigara, G. The origins of cerebral asymmetry: a review of evidence of behavioural and brain lateralization in fishes, reptiles, and amphibians, *Neurosci. Behav. Rev.*, 22, 411, 1998.
28. Davis, T.A. Reversible and irreversible lateralities in some animals, *Behav. Brain Sci.*, 2, 291, 1978.
29. Deckel, A.W. Laterality of aggressive responses in *Anolis*, *J. Exp. Zool.*, 272, 194, 1995.
30. Kabelik, D., Weiss, S.L., and Moore, M.C. How plastic is the brain? Hormonal control of brain morphology and function in the tree lizard, *Urosaurus ornatus*, *Horm. Behav.*, 41, 473, 2002.
31. Kabelik, D., Weiss, S.L., and Moore, M.C. Seasonal plasticity in hormone levels, brain morphology, and aggressive behavior of the tree lizard *Urosaurus ornatus*, *Horm. Behav.*, 44, 57, 2003.
32. Kabelik, D., Weiss, S.L., and Moore, M.C. Neural dimorphisms within and between sexes in the tree lizard, *Urosaurus ornatus*, *Horm. Behav.*, 46, 93, 2004.
33. Hartman, V. and Crews, D. Sociosexual stimuli affect ER- and PR-mRNA abundance in the hypothalamus of all-female whiptail lizards, *Brain Res.*, 741, 344, 1996.
34. Rhen, T., Sakata, J.T., Woolley, S., Porter, R., and Crews, D. Changes in androgen receptor mRNA expression in the forebrain and oviduct during the reproductive cycle of female leopard geckos, *Eublepharis macularius*, *Gen. Comp. Endocrinol.*, 132, 133, 2003.

35. Moore, I.T., and Mason, R.T. Behavioral and hormonal responses to corticosterone in the male red-sided garter snake, *Thamnopohis sirtalis parietalis, Physiol. Behav.*, 72, 669, 2001.
36. Guillette, L.J. and Iguchi, T. Contaminant-induces endocrine and reproductive alterations in reptiles. in *Implications of Endocrine Active Substances for Humans and Wildlife*, Miyamoto, J., and Burger, J., Eds. *Pure Appl. Chem.*, 75, 2275, 2003.
37. Eisler, R. Lead hazards to fish, wildlife, and invertebrates: a synoptic review, *Biol. Rep.*, 14, 1, 1988.
38. Campbell, K.R. and Campbell, T.S. Lizard contaminant data for ecological risk assessment, *Rev. Environ. Contam. Toxicol.*, 165, 39, 2000.
39. Portelli, M.J. and Bishop, C.A. Ecotoxicology of organic contaminants in reptiles: a review of the concentrations and effects of organic contaminants in reptiles, in *Ecotoxicology of Amphibians and Reptiles*, Sparling, D.W., Linder, G., Bishop, C.A., Eds., SETAC Press, Pensacola, FL, 2000, pp. 495–544.
40. Saxon, J.G. *The biology of the lizard, Cnemidophorus tessalatus, and effects of pesticides upon the population in the Presidio Basin, Texas*, Ph.D. dissertation, Texas A & M University, College Station, TX, 1970.
41. Bracher, G.A. and Bider, J.R. Changes in terrestrial animal activity of a forest community after application of aminocarb (Matracil), *Can. J. Zool.*, 60, 1981, 1982.
42. Wojcik, J., Czekaj, H., Kozaczynski, W., and Niewiadowska, A. Polychlorinated biphenyls: a possible agent of neurologic disorders observed in young snakes, *Bull. Vet. Inst. Pulawy*, 39, 53, 1995.
43. Turner, F.B. and Gist, C.S. Observations of lizards and tree frogs in an irradiated Puerto Rican forest, in *A Tropical Rain Forest, a Study of Irradiation and Ecology at El Verde, Puerto Rico*, Odum, J.T., and R.F. Pigeon, Eds., TID-24270, U.S. Atomic Energy Division of Technical Information, Oak Ridge, TN, 1970, pp. E-25–E-49.
44. Hammerton, K.M., Jayasinghe, N., Jeffree, R.A., and Lim, R.P. Experimental study of blood lead kinetics in estuarine crocodiles (*Crocodylus porosus*) exposed to ingested lead shot, *Arch. Environ. Contam. Toxicol.*, 45, 390, 2003.
45. Bull, J.J. Sex determination in reptiles, *Q. Rev. Biol.*, 55, 4, 1980.
46. Gutze, W.H.N., Paukstis, G. L., and McDaniel, L.L. Skewed sex rations for adult and hatchling bull snakes, *Pituophis melanoleucus*, in Nebraska, *Copeia*, 1985, 649, 1985.
47. Gutze, W.H.N. and Crews, D. Embryonic temperature determines adult sexuality in a reptile, *Nature*, 332, 832, 1988.
48. Roosenburg, W. M. and Kelley, K.C. The effect of egg size and incubation temperature on growth in the turtle, *Malaclemys terrapin*, *J. Herpetol.*, 30, 198, 1996.
49. Burger, J. and Zappalorti, R. Nest site selection by pine snakes, *Pituophis melanoleucus*, in the New Jersey pine barrens, *Copeia*, 1986, 116, 1988.
50. Burger, J. Effects of incubation temperature on behavior of hatchling pine snakes: implications for reptilian distribution, *Behav. Ecol. Sociobiol.*, 28, 297, 1991a.
51. Burger, J. Response of hatchling pine snakes (*Pituophis melanoleucus*) to chemical cues of sympatric snakes, *Copeia*, 1990, 1160, 1990.
52. Burger, J. Anti-predator behavior of hatchling pine snakes: effects of incubation temperature and simulated predators, *Anim. Behav.*, 56, 547, 1998.

53. Burger, J. Effects of incubation temperature on hatchling pine snakes: implications for survival, *Behav. Ecol. Sociobiol.*, 43, 11, 1998.
54. Burger, J., Carruth-Hinchey, C., Ondroff, J., McMahon, M., Gibbons, J.W., and Gochfeld, M. Effects of lead on behavior, growth, and survival of hatchling slider turtles, *J. Toxicol. Environ. Health*, 55, 495, 1998.
55. Burger, J. and Gochfeld, M. Behavior effects of lead exposure on different days for herring gull (*Larus argentatus*) chicks, *Pharmacol. Biochem. Behav.*, 50, 97, 1995a.
56. Burger, J. Response to prey chemical cues by hatchling pine snakes (*Pituophis melanoleucus*): effects of incubation temperature and experience, *J. Chem. Ecol.*, 17, 1069, 1991b.
57. Burger, J. Trace element levels in pine snake hatchlings. Tissue and temporal differences, *Arch. Environ. Contam. Toxicol.*, 22, 209, 1992.
58. Burger, J. and Gochfeld, M. Behavioral impairment of lead-exposed herring gulls in nature, *Fund. Appl. Toxicol.*, 23, 553, 1995b.
59. Burger, J. and Gochfeld, M. Lead and neurobehavioral development in gulls: a model for understanding effects in the laboratory and the field, *Neurotoxicology*, 18, 279, 1997.
60. Burger, J. and Gochfeld, M. Trace element distribution in growing feathers: additional excretion in feather sheaths, *Arch. Environ. Contam. Toxicol.*, 23, 105, 1992.
61. Prechtl, J.C. and Bullock, T.H. Barbiturate sensitive components of visual ERPs in a reptile, *NeuroReport*, 3, 801, 1992.
62. Deckel, A.W. and Fuqua, L. Effects of serotonergic drugs on lateralized aggression and aggressive displays in *Anolis carolinensis*, *Behav. Brain Res.*, 95, 227, 1998.
63. Hornby, T.G., McDonagh, J.C., Reinking, R.M., and Stuart, D.G. Effects of excitatory modulation on intrinsic properties of turtle motorneurons, *J. Neurophysiol.*, 88, 86, 2002.
64. Belliure, J. and Clobert, J. Behavioral sensitivity to corticosterone in juveniles of the wall lizard, *Podarcis muralis*, *Physiol. Behav.*, 81, 121, 2004.
65. Yang, E.J., and Wilczynski, W. Interaction effects of corticosterone and experience on aggressive behavior in the green anole lizard, *Horm. Behav.*, 44, 281, 2003.
66. Knapp, R. and Moore, M.C. Male morphs in tree lizards have different testosterone responses to elevated levels of corticosterone, *Gen. Comp. Endocrinol.*, 107, 273, 1997.
67. Biczycki, M. An experimental study on the effects of the selected lead compounds on the hypothalamic secretory centres in sand lizards (*Lacerta agillis* L). I. Hypothalamic secretory centres after the acute intoxication with an organic lead compound lead nitrate, *Prace naukowe uniwersytetus slaskiego w Katowicach*, 1287, 9, 1992a.
68. Biczycki, M. An experimental study on the effects of the selected lead compounds on the hypothalamic secretory centres in sand lizards (*Lacerta agillis* L). II. Hypothalamic secretory centres after the acute intoxication with an organic lead compound lead nitrate, *Prace naukowe uniwersytetus slaskiego w Katowicach*, 1287, 26, 1992b.
69. O'Bryant, E.L. and Wade, J. Sexual dimorphism in a neuromuscular system regulating courtship in the green anole lizard: effects of season and androgen treatment, *J. Neurobiol.*, 40, 202, 1999.

70. Bergeron, J.M., Crews, D., and McLachian, J.A. PCBs as environmental estrogens: turtle sex determination as a biomarker of environmental contamination, *Environ. Health Perspect.*, 102, 780, 1994.
71. Dey, P.M., Burger, J., Gochfeld, M., and Reuhl, K.R. Developmental lead exposure disturbs expression of synaptic neural cell adhesion molecules in herring gull brains, *J. Toxicol.*, 146, 137, 2000.
72. Clotfelter, E.D., Bell, A. M., and Levering, E.R. The role of animal behaviour in the study of endocrine-disrupting chemicals, *Anim. Behav.*, 68, 665, 2004.

chapter 8

Immunotoxicology and Implications for Reptilian Health

Jennifer M. Keller, Margie M. Peden-Adams, **and**
A. Alonso Aguirre

Table of Contents

I. Introduction to the Reptilian Immune System

The immune system has evolved from simple barriers between self and nonself to complex cell-signaling pathways that recognize and destroy invading pathogens. Vertebrate immunity is a complex and intricate system that can detect and clear bacterial, viral, fungal, and parasitic infections, as well as tumor cells. These defenses are vital to the health and survival of individual animals, and when compromised by insults such as environmental contaminants, the organism may become more vulnerable to debilitating or deadly diseases.[1]

As a result of their location on the evolutionary tree, reptiles have more highly developed and complex immune organs than fish and anurans and they share many similarities to endothermic vertebrates (Figure 8.1; reviewed by Zapata and Amemiya[2] and Zapata and Cooper[3]). Reptiles possess the more evolved functions of bone marrow, including hematopoiesis and maturation of B-lymphocytes. These functions evolved first in the lungless salamander and began to replace the kidney as the main hematopoietic organ in reptiles. Reptiles have a functional thymus that serves as the site of T-lymphocyte maturation, and the reptilian thymus has structural similarities with all other jawed vertebrates. The spleen filters the blood, removing foreign antigens, dead cells, and cellular debris. It evolved in cartilaginous fish and became more histologically complex in reptiles, suggesting specialized immune functions in specific areas. Both B- and T-lymphocytes are present in the reptilian spleen. Reptiles are missing the Bursa of Fabricius, which is the site of B-lymphocyte development in birds. Although reptiles do not have true lymph nodes like birds, and mammals, gut-associated lymphoid aggregates (GALTs) are well developed, and lymphoid aggregates have been documented in the trachea and lungs of some species. More information on the evolution of the immune system from primitive chordates (tunicates) to tetrapods can be found in Zapata and Amemiya,[2] and a more thorough description of reptilian immunology is provided in Zapata and Cooper[3] and in Cooper and colleagues.[4]

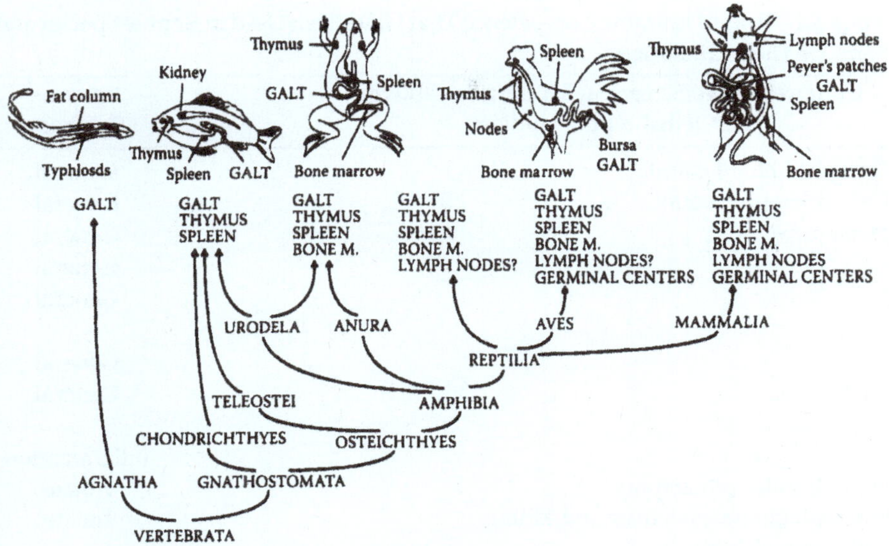

Figure 8.1 Evolution of the lymphoid systems. Bone M., bone marrow; GALT, gut-associated lymphoid tissue. (From Du Pasquier, L. and Flajnik, M., Origin and evolution of the vertebrate immune system, in *Fundamental Immunology*, 4th ed., Paul, W.E., Ed., Lippincott-Raven, 1999; 640. With permission.)

Although the immune organs of reptiles have been described, many cellular and molecular pathways that are known in mammalian immune responses are not well understood in reptiles. However, it is known that reptiles produce only two antibody or immunoglobulin classes (immunoglobulin M, or IgM, and IgY), whereas mammals produce five (IgM, IgG, IgA, IgD, and IgE).[5,6] The reptilian IgY is similar to mammalian IgG.[5,6] Reference texts, such as Janeway and colleagues[7] and Goldsby and colleagues,[8] provide details on cellular and molecular immune functions in mammals and may be useful for future comparative research using reptiles.

Reptiles possess all three categories of immune function: innate, cell mediated, and humoral (Table 8.1). Innate immunity comprises functions without specific recognition of individual or unique antigens, thereby no memory of previous exposure to a particular antigen is required. Innate functions include macrophage phagocytosis and natural killer cell activity. Cell-mediated and humoral functions are categorized in acquired immunity, which requires the recognition of individual antigens and retains memory of that exposure. Cell-mediated immunity can be measured using tissue graft rejection, mixed leukocyte reaction (MLR), or mitogen-induced T-lymphocyte proliferation. Humoral immune responses can be measured by using the

Table 8.1 Tests of Immune Competence That Have Been Used in Reptile Species and Some of Their Limitations

Comprehensive Screening Set for non-captive, non-lethal reptile studies[a]	Immune Category
Complete blood count	General
Cell differential count	General
Hematocrit	General
	General
	General
	General
	General
	Inflammation
Natural killer cell activity	Innate
Macrophage phagocytosis and killing	Innate
Lysozyme activity	Innate
	CMI
Mixed leukocyte response	CMI
	CMI
	CMI
	HI
	HI
Lymphocyte immunophenotyping (i.e., CD4+ cells)	General
Respiratory (oxidative) burst assay	Innate
	Innate
	Innate
Lymphocyte proliferation (T-cell mitogens)	CMI
	CMI
	CMI
	CMI
	CMI
Native immunoglobulin titers (IgM and IgY-like)	HI
Lymphocyte proliferation (B-cell mitogens)	HI
	HI
	HI
	Soluble mediators

Table 8.1 (continued) Tests of Immune Competence That Have Been Used in Reptile Species and Some of Their Limitations

Comprehensive Screening Set for non-captive, non-lethal reptile studies[a]	Immune Category
	Integrated
	Integrated
	Integrated
	HI
Compare diseased vs. non-diseased animals	Integrated

Natural cytotoxic cell, NCC; cell-mediated immunity, CMI; humoral immunity, HI.
[a] Correlative field studies can presume, but not prove, altered host resistance, such as Tangredi and Evans.[31]

mitogen-induced B-lymphocyte proliferation assay or by quantifying antibody production using an enzyme-linked immunosorbent assay (ELISA), the plaque-forming cell (PFC) response, or hemagglutination. All these assays have been performed in several reptilian species[20,23,25,29,32] (see additional references in Table 8.1).

The reptilian immune response is profoundly affected by ecologic factors, including population dynamics, stress, nutritional state, environmental temperature, seasonal variations, age, and infectious pathogens. The dramatic effects of seasonal changes and related steroid fluctuations have been the topic of many studies on the reptilian immune system.[33] For example, cell-mediated immunity in lizards and snakes, as measured by MLR or rejection of skin grafts, shows strong responses during certain seasons and a complete lack of response during other times of the year.[20,34] Seasonal responses vary depending on the reptilian species, but many studies have linked seasonal immunomodulation with changes in steroid concentrations (reviewed by Zapata and colleagues[33]). For example, the ocellated skink (*Chalcides ocellatus*) demonstrates peak immune functions, such as lymphocyte proliferation responses and PFC activity, in summer months when corticosterone and testosterone are low, whereas weaker immune responses are observed in winter months when the hormones are at higher concentrations.[35,36] In fact, injection of hydrocortisone or testosterone in these lizards during the summer results in decreased humoral and cell-mediated responses, including the PFC response to rat erythrocytes, skin allograft rejection, and T-lymphocyte proliferation.[35–37] Likewise, intraperitoneal injection of physiologically relevant concentrations of testosterone in the Caspian turtle (*Mauremys caspica*) significantly reduces the number of leukocytes in the blood, thymus, and spleen.[38] Therefore, because of seasonal fluctuations in immunity, it is critical that reptilian immunotoxicology studies standardize sampling to one season or account for seasonal changes in the study design with the inclusion of steroid hormones as additional cofactors.

II. Wildlife Immunotoxicology

The field of wildlife immunotoxicology is relatively new, and studies with reptiles are extremely limited. Previous studies with free-ranging birds and mammals provide a foundation for this field and call attention to population-level implications for the suppression of the immune system. Suppression of immune functions has been noted in Caspian terns (*Sterna caspia*) and herring gulls (*Larus argentatus*) inhabiting sites around the Great Lakes that are more heavily contaminated with polychlorinated biphenyls (PCBs) and organochlorine pesticides.[39,40] Similarly, a captive study with harbor seals (*Phoca vitulina*) described suppression of many immune functions after feeding seals fish from the more PCB- and pesticide-contaminated Baltic Sea compared with feeding them fish from the less contaminated Atlantic Ocean.[41] These studies provide evidence that exposure to environmentally relevant concentrations of contaminants can suppress immune function in wildlife and may lead to increased vulnerability to disease. In addition, the findings of the harbor seal study suggest that contaminant levels found in the wild could play a contributing role in viral epizootics that lead to mass mortalities of marine mammals. For example, 20,000 harbor seals died in Europe in 1988 during a phocid distempter virus-1 epizootic.[42] The seals that died from the infection had higher levels of organochlorine contaminants than those that survived. The same finding was observed with striped dolphins (*Stenella coeruleoalba*) that died during a dolphin morbillivirus epizootic in the Mediterranean Sea.[43] These studies, taken together, suggest that exposure to environmental contaminants in the wild may result in decreased host resistance, increased disease and mortality, and reduced population size.

III. Reptilian Immunotoxicology

Although many traditional immune assays have been used in reptilian species (Table 8.1), studies assessing toxic effects of contaminants on the reptilian immune system began only in the past 10 years. Studies available for review are so limited that past wildlife immunotoxicology reviews could not include reptilian sections.[1,44–46] The currently available studies focused on the effects of mainly organic contaminants, especially those with endocrine activity, on western fence lizards (*Sceloporus occidentalis*), box turtles (*Terrapene carolina carolina*), American alligators (*Alligator mississippiensis*), and two species of sea turtles.

A. Western Fence Lizards

Previous research on the effects of contaminants on reptiles has focused primarily on endocrine and reproductive alterations as discussed in Chapter 6. Studies have shown that testosterone and other steroid injections dramatically alter reptilian immune functions.[35–37] These studies emphasize the complex and strong interactions between the immune and endocrine systems and suggest that contaminants with endocrine activity (see Chapter 6) may also affect the reptilian immune system. This is highlighted by Burnham and colleagues[47] who showed that a single injection of the xenoestrogen, 17α-ethinylestradiol, in male western fence lizards resulted in decreased spleen leukocyte counts (at 0.1 mg/kg and 1.0 mg/kg), whereas 0.01, 0.1, and 1.0 mg/kg caused decreased peripheral blood leukocyte counts and increased MLR as compared with controls. This study, together with the steroid injection studies mentioned earlier, suggests that contaminants with endocrine activity (estrogenic and androgenic) may modulate reptilian immunity as has been shown or suggested in mammalian studies.[48–51]

B. Box Turtles

Only one study has investigated the link between environmental contaminants and immunomodulation in freshwater and terrestrial turtles. Tangredi and Evans[31] measured organochlorine pesticide concentrations in the liver of three wild eastern box turtles found on Long Island, NY, that had chronic bacterial infections and compared them with four wild healthy turtles that were in robust condition with accidental injuries. Two of the infected turtles had concentrations of oxychlordane and heptachlor epoxide that were more than two times higher than the mean of the control turtles (Figure 8.2). Although the sample size was small and one diseased turtle had low levels, this study suggests that environmental exposure of box turtles to chlordanes may be sufficient to suppress the immune system and lead to reduced host resistance and increased infection. This conclusion is supported by mammalian *in vitro* studies showing that 4100 μg/kg chlordane suppresses MLR, lymphocyte proliferation, and antibody secretory cell response to sheep red blood cells (SRBCs).[52] Similarly,

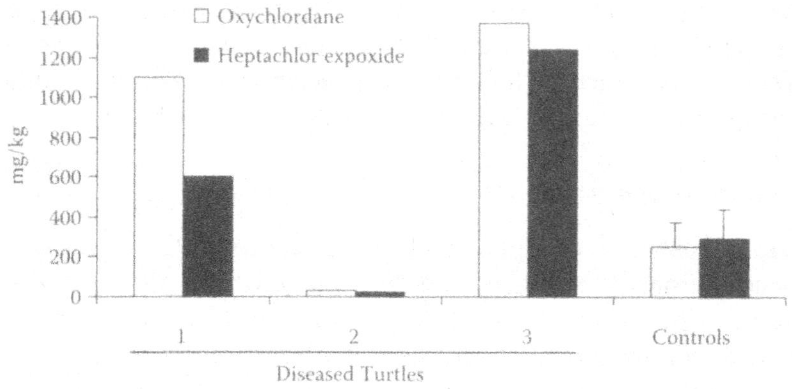

Figure 8.2 Liver concentrations of chlorinated pesticides in three diseased eastern box turtles from Long Island, NY, compared with the mean and standard deviation of four control, nondiseased box turtles. (Data from Tangredi and Evans.[31])

two separate studies found significant, negative associations in loggerhead sea turtles between total white blood cell (WBC) counts and blood oxychlordane concentrations[53] and between lymphocyte counts and chlordane blood concentrations.[9]

C. *American Alligators*

The effect of organic contaminants on the American alligator immune system has been the subject of several studies.[28,54–56] These studies were spurred from the toxicologic effects repeatedly observed in an alligator population in Lake Apopka, FL. This particular alligator population plummeted to 10% of its original size in the 1980s after a spill of the organochlorine pesticide dicofol, which included dichlorodiphenyltrichloroethane (DDT).[57] Numerous studies have associated the pesticide contamination in this lake to endocrine and reproductive abnormalities in alligators inhabiting it (see Chapter 6).[58,59] The rationale behind the alligator immunotoxicity studies reviewed here was two-fold: (1) because organochlorine contaminants are known to have endocrine activity in reptiles (Chapter 6) and (2) because hormones dramatically affect reptile immune functions.[47]

1. *DDT and Dicofol*

Male and female alligators from Lake Apopka were shown to have significantly higher mean plasma concentrations of the major DDT metabolite, 4,4′-dichlorodiphenyldichloroethylene (4,4′-DDE; 7.35 ng/ml in males, 17.98 ng/ml in females) compared with two reference lakes (0.92 ng/ml in males from both reference lakes; 1.28 ng/ml and

0.77 ng/ml in females from reference lakes).[60] Morphology of the spleen and thymus of alligators from Lake Apopka was compared with alligators from the two reference lakes.[55] Juvenile alligators captured in Lake Apopka had decreased thymic ratios (area of medulla:area of cortex) as a result of increased cortical areas, suggesting that T-lymphocyte maturation may be altered. Lake Apopka juvenile females exhibited the smallest lymphocyte sheath widths in the spleen, and Malpighian bodies of the spleen were also the smallest in juveniles from Lake Apopka. Reduced splenic lymphocyte sheath widths were also observed in an additional sample set of 3-year-old female alligators that were raised in captivity from eggs collected from Lake Apopka compared with those collected from a control lake.[55] These changes in both major immune organs suggest possible alterations in the populations and functions of T- and B-lymphocytes in alligators from this contaminated lake.

In another study, Gross and coworkers[28,56] microscopically examined the bone marrow, spleen, and thymus of neonatal alligators hatched from eggs collected from Lake Apopka and a reference lake. All three tissues in the Lake Apopka neonates were hypocellular, especially the bone marrow, compared with alligators from the reference lake. To address alterations in immune function, hatchlings were raised in captivity after incubating eggs collected from Lake Apopka and a reference lake. At 6 months of age, antibody responses after immunization with SRBCs were measured by hemagglutination. The average peak titers in the Lake Apopka alligators were significantly reduced by 86% compared with the controls. These findings indicated that humoral immunity was significantly suppressed in alligators from this contaminated lake.

2. *2,3,7,8-Tetrachlorodibenzo-p-Dioxin*

Peden-Adams[54] and Peden-Adams and coworkers[61] exposed 6-month-old juvenile alligators orally to 10 μg/kg tetrachlorodibenzo-*p*-dioxin (TCDD) every other day for 6 days. Alligators were challenged with SRBCs on the first day of TCDD exposure. After euthanasia on day 7, mitogen-induced splenocyte proliferation and antibody titers were assessed. Antibody titers, measured in the plasma by hemagglutination, were not significantly altered. The lack of observed effect on antibody production in this study is not surprising because another study found that it took 4 weeks before an antibody response was detected in 6-month-old alligators after a single injection of SRBCs.[28] B-lymphocyte proliferation, measured by stimulation of 1.0 ∞ 10^6 cells/well with 400 ng/ml phorbol 12,13-dibutyrate (PDB; final concentration in culture wells) for 72 h, was suppressed by 54%, but no significant changes were observed in T-lymphocyte proliferation stimulated with 10 μg/ml concanavalin A (ConA). These effects on lymphocyte proliferation are similar to effects observed in white leghorn chickens exposed *in ovo* after maternal injection of TCDD.[62] This study suggests that a potent immunosuppressive contaminant, TCDD, can suppress humoral immune responses in alligators, as noted by suppressed B-cell proliferation, but studies attempting to examine antibody production in reptiles need to optimize for the onset and time course of antibody production.

3. *DDE, TCDD, and Phytoestrogens*

Peden-Adams[54] and Peden-Adams and coworkers[63] also assessed the effects of mixtures of estrogenic, antiestrogenic, antiandrogenic, and androgenic compounds on the developing immune system and ethoxyresorufin-*O*-deethylase (EROD) activity of American alligators. Eggs collected from the Rockefeller Wildlife Refuge, Grand Chenier, LA, were injected with the following compounds: ethynyl estradiol, TCDD, 2,4′-DDE, 4,4′-DDE, coumestrol, and indole-3-carbinol either singly or in binary mixtures. At 21 days posthatch, mitogen-induced splenocyte proliferation was measured using the same method described for the 6-month-old alligators by Peden-Adams[54] using PDB as a B-lymphocyte mitogen and ConA as a T-lymphocyte mitogen.

TCDD exposure alone resulted in an apparent biphasic dose-response curve, although not statistically significant, for both B- and T-lymphocyte proliferation (Figure 8.3a). Lower doses stimulated proliferation and higher doses suppressed proliferation compared with the control. For example, the lowest dose of TCDD (0.02 μg/kg egg mass) nearly doubled T-lymphocyte proliferation compared with the control, whereas higher concentrations suppressed proliferation by more than 60% of the control mean. An injection of TCDD at 0.2 μg/kg egg mass significantly suppressed T-lymphocyte proliferation by 81% of the control. TCDD at 0.2 μg/kg and 0.6 μg/kg egg mass increased mean B-lymphocyte proliferation by 87% and 122%, respectively, whereas 2 μg/kg TCDD suppressed it by 69% from the control mean. Differences between the egg injection and 6-month-old exposure studies suggest that the sensitivity of the developing alligator immune system to TCDD exposure is dependent on the timing of exposure. B-lymphocyte proliferation was affected in the 6-month-old juveniles,[54,61] whereas T-lymphocyte proliferation was affected in the posthatchlings.[54,63] Notably lower TCDD concentrations were used in the *in ovo* experiment (0.02 μg/kg to 2.0 μg/kg egg mass) compared with the

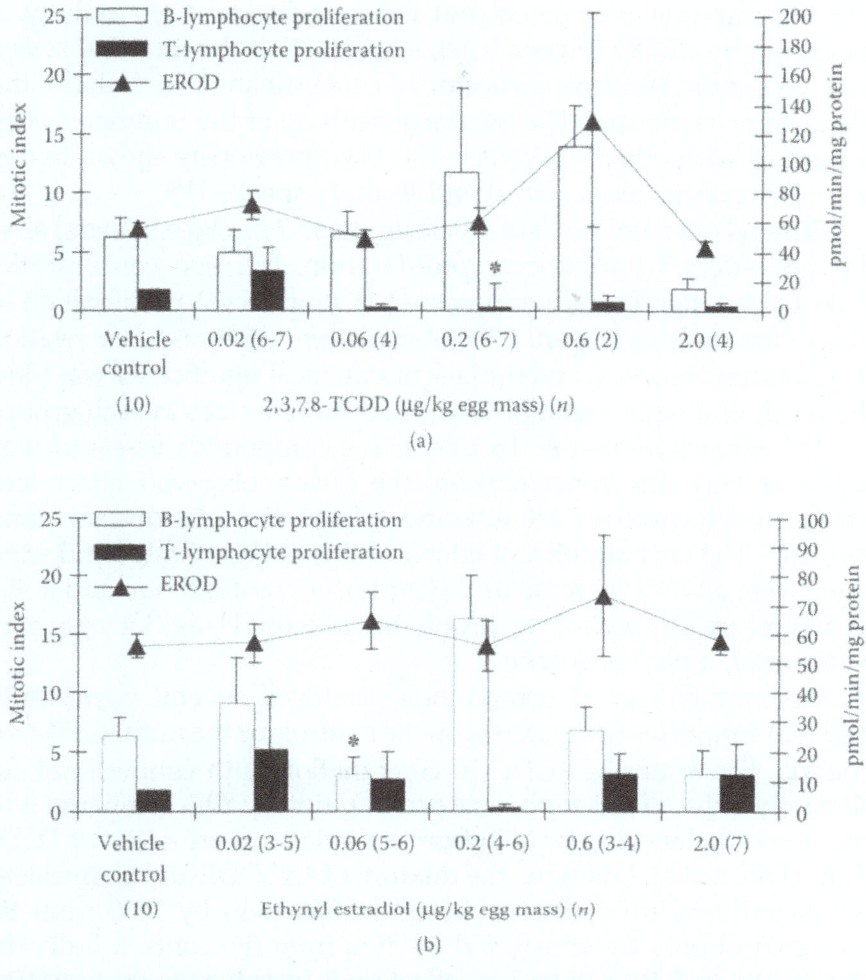

Figure 8.3 The effect of 2,3,7,8-TCDD (tetrachlorodibenzo-p-dioxin) (a) and ethynyl estradiol (EE) (b) on lymphocyte proliferation and ethoxyresorufin-O-deethylase (EROD) activity in 21-day-old American alligators exposed *in ovo* via yolk sac injections. The concentration of TCDD or EE injected is shown, and the number of animals in each group is indicated in parentheses along the x-axis. The vehicle control group received corn oil. Mitotic index is the unit for B- and T-lymphocyte proliferation and was calculated as counts per minute (cpm) of splenocytes stimulated with mitogen divided by cpm of unstimulated splenocytes. Units on the right y-axis describe the EROD activity. Data are shown as means ± one standard error. Significant differences from the vehicle control are indicated by an asterisk ($P < 0.05$). (Data from Peden-Adams.[54])

6-month-old oral experiment (10 mg/kg body mass). Because significant effects were observed for T-lymphocyte proliferation after *in ovo* exposure but not after the 6-month-old juvenile oral exposure, the embryonic immune system may be more sensitive than later developmental stages. The fact that the tested concentrations of TCDD affected immune parameters but not cytochrome P450 activity, as measured by EROD (Figure 8.3a), suggests that the immune system may be a more sensitive indicator of contaminant effects than other physiologic responses. The greater sensitivity of the immune system compared with other biomarkers has been previously shown in deer mice (*Peromyscus maniculatus*) and in other species.[64,65]

Ethynyl estradiol exposure (0.02 μg/kg to 2 μg/kg egg mass) alone did not affect T-lymphocyte proliferation, but one concentration (0.06 μg/kg) significantly suppressed B-lymphocyte proliferation by 51% of the controls (Figure 8.3b). A consistent dose-response relationship was not observed, and the lack of statistical significance was likely the result of a large variation and small sample sizes in each group.

The other potential endocrine active compounds assessed were tested at only one concentration, the lowest observed effect level based on cytochrome P450 alterations from previous dose-response studies.[66] The only significant effects of these single compound exposures were an 89% increase in B-lymphocyte proliferation and a 95% reduction in T-lymphocyte proliferation with 1 mg/kg egg mass coumestrol, a phytoestrogen.

Binary mixtures of compounds produced several significantly different responses compared with the controls or the individual compounds. For example, TCDD in combination with coumesterol significantly reduced B-lymphocyte proliferation by 80% compared with coumesterol alone and by 80% from a similar concentration of TCDD alone (Figure 8.4). Likewise, the mixture of 4,4′-DDE and coumesterol significantly reduced B-lymphocyte proliferation by 78% from the coumesterol-only treatment and by 58% from the control. Individually, TCDD or 2,4′-DDE had no effect on B-lymphocyte proliferation, but combined they caused suppression that was 75% lower than the control, 79% lower than 2,4′-DDE alone, and 87% lower than TCDD alone at a similar concentration. Mixtures of these compounds also affected T-lymphocyte proliferation in ways that differed from either the control or the individual compounds (Figure 8.5). These findings suggest that mixtures of xenobiotics can have a more pronounced effect on the immune system than individual compounds.

D. Sea Turtles

All species of sea turtles found in U.S. waters are protected by the U.S. Endangered Species Act with either an endangered or threatened status. Several sea turtle populations have declined or continue to decline because of various human impacts, including hunting, fisheries interactions, nesting habitat loss, and possibly chemical contamination.

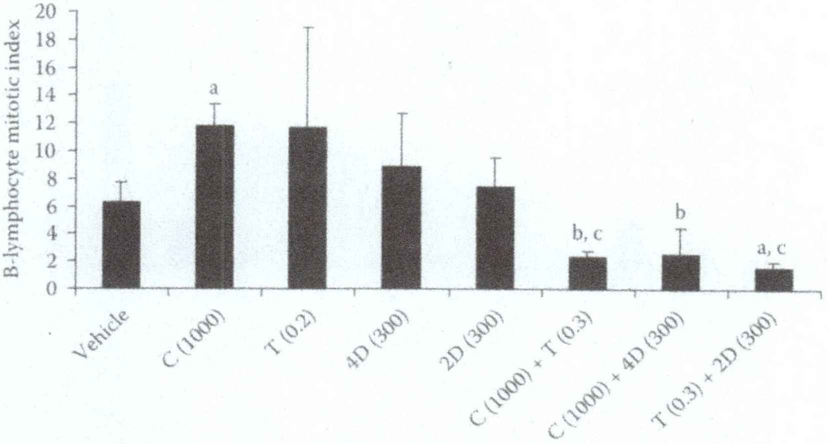

Figure 8.4 The effect of individual compounds and binary mixtures on B-lymphocyte proliferation in 21-day-old American alligators. Alligator eggs were injected with coumesterol (C), 2,3,7,8-TCDD (tetrachlorodibenzo-*p*-dioxin) (T), 4,4′-DDE (4D), 2,4′-DDE (2D), and binary mixtures at concentrations (μg/kg egg mass) shown in parentheses. Mitotic index was calculated as counts per minute (cpm) of splenocytes stimulated with mitogen (PDB, phorbol 12,13-dibutyrate) divided by cpm of unstimulated splenocytes. Sample sizes were 10 alligators for the vehicle control group, 5 for C, 7 for T, 4 for 4D, 5 for 2D, 6 for C+T, 3 for C+4D, and 5 for T+2D treatments. Data are shown as means ± one standard error. [a]indicates a significant difference from the vehicle control; [b]indicates a significant difference from C alone; and [c]indicates a significant difference from T alone ($P < 0.05$). (Data from Peden-Adams.[54])

In sea turtles, studies examining the effects of environmental contaminants are limited, but the immunotoxic effects of some environmental contaminants have been assessed. Because immunosuppressive contaminants have the potential to decrease survival of already threatened sea turtle populations, it is important to understand the risk of contaminants on their general health and immune functions.

1. *Petrochemicals*

Oil and tar exposure has been viewed as a major problem for certain sea turtle populations. For example, tar was found in the esophagus or stomach of 35% of posthatchling loggerhead sea turtles (*Caretta caretta*) captured in convergence zones off of eastern Florida, and more than 50% had visible tar in their jaws.[67] Of loggerhead sea turtles captured in the central Mediterranean Sea, 17% had visible signs of tar or oil exposure.[68] Chronic exposure to petroleum hydrocarbons was associated with poor body condition observed in two green sea turtles (*Chelonia mydas*) during the oil spill from Ixtoc in 1979.[69] In addition to causing direct mortality of embryos and developmental abnormalities,[70] sublethal effects of oil have been documented on the

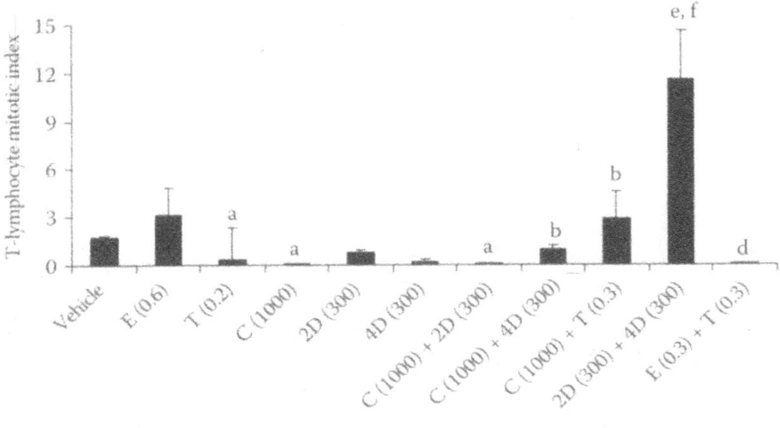

Figure 8.5 The effect of individual compounds and binary mixtures on T-lymphocyte proliferation in 21-day-old American alligators. Alligator eggs were injected with ethynyl estradiol (E), 2,3,7,8-TCDD (tetrachlorodibenzo-*p*-dioxin) (T), coumesterol (C), 2,4′-DDE (2D), 4,4′-DDE (4D), and binary mixtures at concentrations (μg/kg egg mass) shown in parentheses. Mitotic index was calculated as counts per minute (cpm) of splenocytes stimulated with mitogen (ConA) divided by cpm of unstimulated splenocytes. Sample sizes were 10 alligators for the vehicle control group, 4 for E, 7 for T, 5 for C, 5 for 2D, 4 for 4D, 3 for C+2D, 3 for C+4D, 6 for C+T, 2 for 2D+4D, and 4 for E+T treatments. Data are shown as means ± one standard error. [a]indicates a significant difference from the vehicle control; [b]indicates a significant difference from C alone; [c]indicates a significant difference from T alone; [d]indicates a significant difference from E alone; [e]indicates a significant difference from 4D alone; and [f]indicates a significant difference from 2D alone ($P < 0.05$). (Data from Peden-Adams.[54])

health and immune systems of juvenile loggerheads.[71] When loggerhead turtles were experimentally exposed to a thin film of South Louisiana crude oil for 96 h, the skin became inflamed and began to slough off and histologic examination showed extensive numbers of acute inflammatory cells, primarily heterophilic granulocytes.[71] Lesions in the integument may leave turtles vulnerable to infections because the skin is one of the first defenses against pathogens. Hematocrit also decreased by almost 50% of the controls after oil exposure. In addition, WBC counts significantly increased by the third day of exposure to four times the control. The increase in WBCs may have been caused by a stress reaction to the oil exposure, but stress hormones were not measured in the study and differential WBC counts were not reported. Limited studies with seabirds support the toxic effects of oil observed in the turtles. Hemolytic anemia and an increased heterophil:lymphocyte ratio, a common indicator of stress in birds, were noted in herring gulls (*Larus argentatus*) and Atlantic puffins (*Fratercula arctica*) exposed to Prudhoe Bay crude oil.[72] Decreased resistance to bacterial challenge was observed in mallards

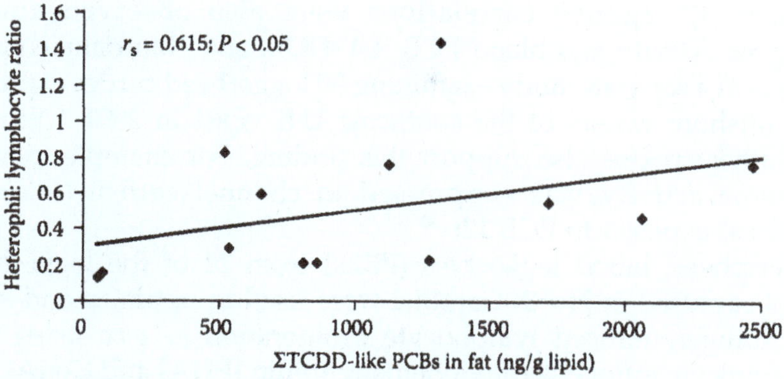

Figure 8.6 The heterophil: lymphocyte ratio of 13 loggerhead sea turtles compared with the concentrations of total TCDD (tetrachlorodibenzo-*p*-dioxin)-like polychlorinated biphenyls (PCBs; sum of PCB congeners 105, 156, 157, and 118) measured in their fat biopsies. r_s, Spearman rank correlation coefficient. (Data from Keller et al.[9])

(*Anas platyrhynchos*) exposed to 4 ml/kg of South Louisiana crude oil,[73] and tagging studies provide evidence that seabirds have lower survival and reproduction rates after being oiled, rehabilitated, and released (reviewed by Briggs and colleagues[74]).

2. *Organochlorine Contaminants*

Health assessment studies have evaluated juvenile loggerhead sea turtles along the southeast coast of the United States to better establish baseline reference values for health indicators, including plasma chemistries, WBC counts, mitogen-induced lymphocyte proliferation, and plasma lysozyme activity.[9,23,53,75–77] A major objective of these studies was to examine associations between the health indicators and organochlorine contaminant (OC) exposure. Blood and fat biopsies from 44 juvenile loggerhead turtles captured in Core Sound, NC, were analyzed for OCs, including PCBs, DDTs, and chlordane pesticides.[78] Total and differential WBC counts performed on a subset of the turtles significantly correlated with OC concentrations.[9] For example, blood and fat concentrations of chlordanes, mirex, DDTs, PCBs, and OC concentrations positively correlated with total WBC counts. The heterophil:lymphocyte ratio, a common indicator of stress, also positively correlated with concentrations of mirex and TCDD-like PCBs in the fat (Figure 8.6). These findings suggest that OCs may be modifying the number of immune cells in loggerhead turtles.

Plasma lysozyme activity, an indicator of innate immunity, was measured from 45 loggerhead sea turtles captured from Core Sound, NC, in the summer months of 2000 and 2001. Lysozyme activity was significantly and negatively correlated with 4,4′-DDE and chlordanes.[79]

Significant, negative correlations were also observed among lysozyme activity and blood PCB, 4,4′-DDE, and chlordane concentrations in a separate study examining 30 loggerhead turtles captured from offshore waters of the southeast U.S. coast in 2001.[76] Studies with other species also support this finding. For example, plasma lysozyme activity was suppressed in channel catfish (*Ictalurus punctatus*) exposed to PCB 126.[80]

Peripheral blood leukoctyes (PBLs) from 24 of the loggerhead turtles captured in North Carolina were used to optimize and measure mitogen-induced lymphocyte proliferation *ex vivo* using two T-lymphocyte mitogens (phytohemagglutinin [PHA] and ConA) and two B-lymphocyte mitogens (lipopolysaccharide [LPS] and PDB).[23] Significant correlations were seen between blood OC concentrations and lymphocyte proliferation.[77,81] Specifically, positive correlations were observed between the following blood OC concentrations and lymphocyte proliferation using the following mitogens: PCBs, DDTs, chlordanes, dieldrin, and OCs with LPS; PCBs and OCs with PHA; and PCBs, DDTs, mirex, and OCs with PDB (see Figure 8.7a for PCBs vs. PDB and Figure 8.8a for 4,4′-DDE vs. PDB). Most of the significant correlations seen with blood OC concentrations were also seen with OC concentrations measured in fat biopsies. All the significant correlations with lymphocyte proliferation had positive slopes, indicating that OCs may have an immunostimulatory effect. Initially, these results were unexpected, but findings from other studies with birds and rodents provide support for this suggested immunoenhancement. For example, adult male American kestrels (*Falco sparverius*) that were fed a mixture of three technical PCB products (Aroclor 1248, 1254, and 1260) demonstrated an increased PHA-skin response.[82] Juvenile herring gulls from heavily contaminated sites in the Great Lakes had higher lymphocyte-proliferation responses than did gulls from reference sites.[83] Higher brain concentrations of DDTs in warbler chicks (*Prothonotaria citrea*) significantly correlated with stronger T-lymphocyte proliferation.[84] White-footed mice (*Peromyscus leucopus*) exposed to Aroclor 1254 *in utero* exhibited an increase in mitogen-induced proliferation of thymocytes and splenocytes.[85,86] These examples show that immunoenhancement by OCs observed in the loggerhead turtles is often seen in other wildlife species. Furthermore, it is important to note that immunoenhancement is not necessarily a healthy outcome because it can lead to hypersensitivity responses and autoimmune diseases.[87]

To further explore the positive correlations observed between lymphocyte proliferation and OC concentrations in loggerhead sea

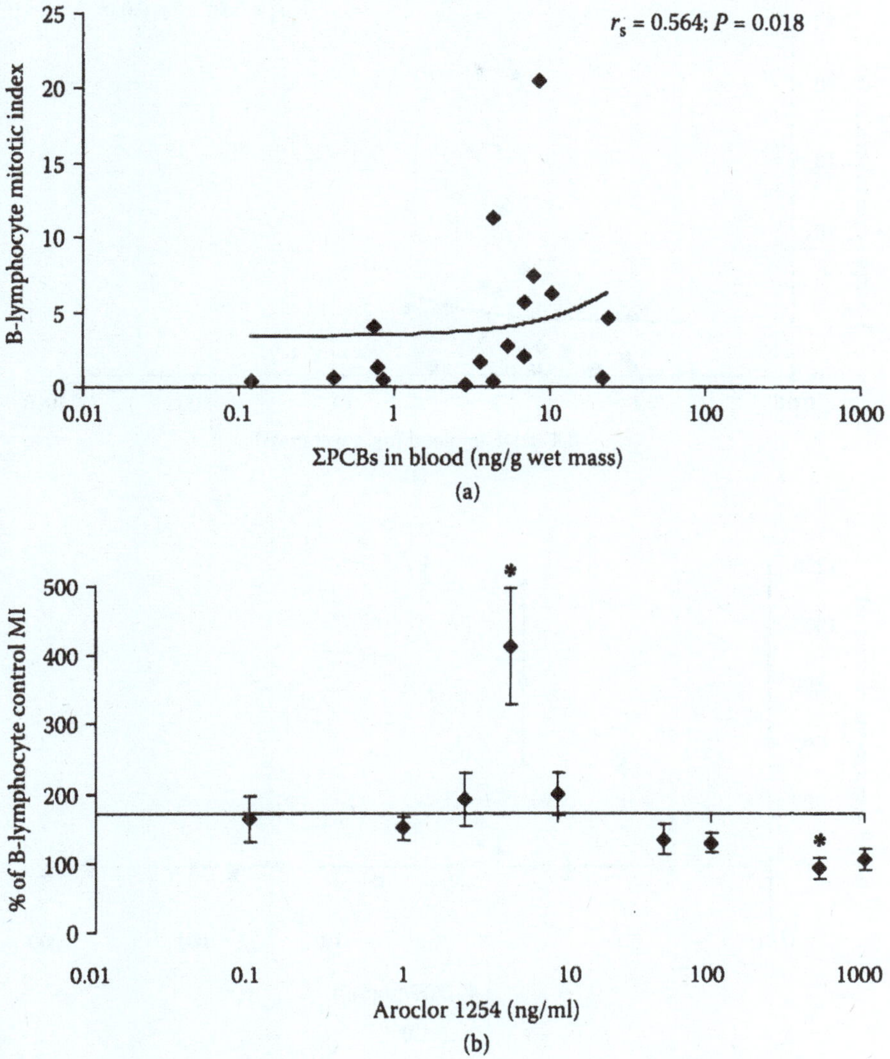

Figure 8.7 The effect of polychlorinated biphenyls on loggerhead sea turtle mitogen-induced lymphocyte proliferation. (a) PCB concentrations measured in the blood of individual turtles compared with their B-lymphocyte proliferation mitotic index (MI) using 800 ng/ml PDB (phorbol 12,13-dibutyrate) as the mitogen. MI was calculated as counts per minute (cpm) of lymphocytes stimulated with mitogen divided by cpm of unstimulated lymphocytes. The trendline indicates a linear regression; note the x-axis is on a logarithmic scale. r_s, Spearman rank correlation coefficient. (b) The effect of *in vitro* exposure of peripheral blood leukocytes from 16 loggerhead turtles to a mixture of PCBs (polychlorinated biphenyls) (Aroclor 1254) on B-lymphocyte proliferation by using 200 ng/ml PDB. Data are shown as mean of the percentage of the MI measured in the control (no vehicle and no Aroclor 1254) for each turtle ± one stand ad error. Error bars are the standard error. The x-axis crosses the y-axis at the vehicle control mean (172%), which had a standard error of 39%. Significant differences from the vehicle control are indicated by an asterisk ($P < 0.05$). (Data from Keller et al.[81])

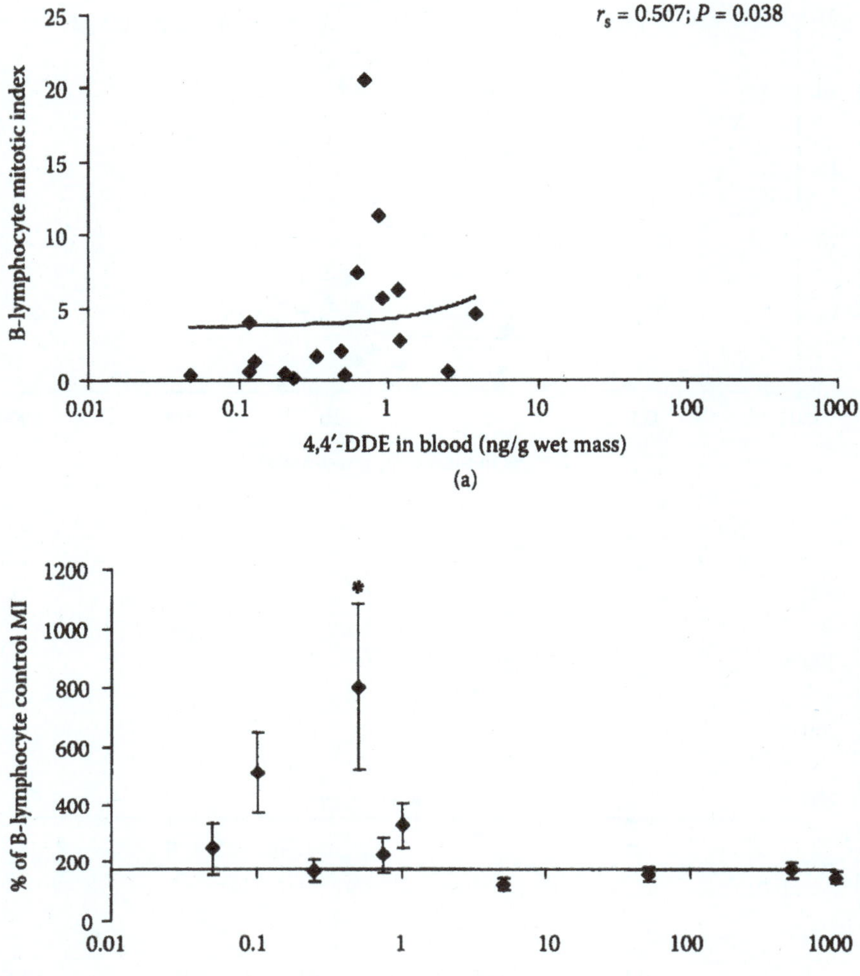

Figure 8.8 The effect of 4,4′-DDE (1,1′-(dichloroethenylidiene) bis 4-chlorobenzene) on loggerhead sea turtle mitogen-induced lymphocyte proliferation. (a) 4,4′-DDE concentrations measured in the blood of individual turtles compared with their B-lymphocyte proliferation mitotic index (MI) by using 800 ng/ml PDB as the mitogen. MI was calculated as counts per minute (cpm) of lymphocytes stimulated with mitogen divided by cpm of unstimulated lymphocytes. The trendline indicates a linear regression; note the x-axis is on a logarithmic scale. r_s, Spearman rank correlation coefficient. The linear trendline is shown. (b) The effect of *in vitro* exposure of peripheral blood leukocytes from 16 loggerhead turtles to 4,4′-DDE on B-lymphocyte proliferation by using 200 ng/ml PDB (phorbol 12,13-dibutyrate). Data are shown as mean of the percentage of the MI measured in the control (no vehicle and no Aroclor 1254) for each turtle ± one standard error. The x-axis crosses the y-axis at the vehicle control mean (172%), which had a standard error of 39%. Significant differences from the vehicle control are indicated by an asterisk ($P < 0.05$). (Data from Keller et al.[81])

turtles, PBLs collected from the buffy coat of 16 loggerhead blood samples were exposed *in vitro* to a PCB mixture (1 ng/ml to 13,500 ng/ml Aroclor 1254).[81] No tested concentration of Aroclor 1254 significantly affected T-lymphocyte proliferation, but all concentrations resulted in a general increase over the vehicle control. A biphasic dose-response curve was observed with B-lymphocyte proliferation, as 5 ng/ml Aroclor 1254 increased proliferation and 498 ng/ml decreased proliferation (Figure 8.7b). It is interesting that PCB concentrations measured in the blood of loggerhead turtles ranged from 0.121 ng/g to 23.9 ng/g (Figure 8.7a), which falls primarily within the early, ascending phase of the dose-response curve. Likewise, *in vitro* exposure of loggerhead PBLs to 4,4′-DDE (0.1 ng/ml to 13,400 ng/ml) resulted in an enhancement of T- and B-lymphocyte proliferation. T-lymphocyte proliferation was significantly increased at 48 ng/ml and 992 ng/ml 4,4′-DDE. B-lymphocyte proliferation was significantly increased at 0.5 ng/ml 4,4′-DDE (Figure 8.8b). Loggerhead blood concentrations of 4,4,′-DDE ranged from 0.0473 ng/g to 3.8 ng/g, and the turtle with the strongest B-lymphocyte proliferation response had a blood concentration of 0.675 ng/g 4,4′-DDE (Figure 8.8a), which was similar to 0.5 ng/ml that produced the peak B-lymphocyte proliferation response *in vitro* (Figure 8.8b). The immunoenhancing trends observed in the *in vitro* exposure experiment supported the positive correlations seen between blood OC levels and lymphocyte proliferation responses assessed *ex vivo*.[81] The similarities between the correlative field study and the *in vitro* exposure experiment suggest that environmental exposure to PCBs and 4,4′-DDE may be enhancing sea turtles' proliferative immune responses.

Several studies have performed *in vitro* OC exposures with mammalian lymphocytes. De Guise and colleagues[88] found suppressed proliferation in beluga whale (*Delphinapterus leucas*) splenocytes after exposure to a mixture of 5 μg/ml each of PCB congeners 138, 153, and 180 (15 μg/ml total PCBs). *In vitro* exposure of B6C3F1 mouse splenocytes to 10 μg/ml or greater of Aroclor 1242 inhibited B-lymphocyte proliferation.[89] Rat splenocytes, however, exposed to a lower concentration of Aroclor 1254 (0.01 μg/ml), exhibited increased proliferation.[90] The mammalian studies that demonstrated suppression by PCBs used concentrations near the high end of the range used in the loggerhead turtle study (13.5 μg/ml), which killed a significant percentage of the loggerhead PBLs. Enhancement, however, was observed in the rat study at a similar concentration (0.01 μg/ml) that enhanced loggerhead proliferation (0.005 μg/ml). These comparisons suggest a hormetic dose-response curve with immunoenhancement at lower concentrations and immunosuppression at higher concentrations.

Hormetic dose-response relationships are commonly encountered in a variety of toxicologic studies,[91] and an interesting, but purely speculative, mechanism linking endocrine disruption with immunotoxicity may explain the immunoenhancing effects of OCs observed on loggerhead turtle lymphocyte proliferation. PCBs, DDTs, and chlordane, all measurable in sea turtle tissues, are known to possess estrogenic and antiandrogenic activity in reptiles (Chapter 6).[92–96] Additionally, it is known that females of many species generally exhibit stronger immune responses than do males, and sex steroids are partly the cause of this sexual dimorphism (reviewed by Olsen and Kovacs[97]). Concentrations slightly higher than endogenous estrogen levels enhance many immune functions, including lymphocyte proliferation, but even higher estrogen concentrations suppress immune functions.[98] Together these three pieces of information suggest that low-level, chronic exposure to estrogenic OCs may enhance immune functions through estrogenic mechanisms. The OC concentrations in sea turtles are much lower than those in harbor seals and Caspian terns,[78] which may explain why turtles exhibited immunoenhancement whereas the latter species demonstrated immunosuppression in relation to OCs.[40,41] It is important to note, however, that the differences in immunotoxic effects observed between these species may also be explained simply by species differences in immune responses or the use of different immune function tests. Nevertheless, the complex interactions along the neuro–endocrine–immune axis are only now beginning to be understood in traditional laboratory species, and their relationship in most wildlife species is completely unknown. Further studies are needed to examine the mechanistic links between endocrine disruption and immunotoxicity.

To summarize the findings of the sea turtle responses to OCs, concentrations of certain OCs in loggerhead turtle blood correlated positively with total WBC counts, the heterophil:lymphocyte ratio, and both T- and B-lymphocyte proliferation, whereas negative correlations were observed between lysozyme activity and certain OCs. Loggerhead PBLs exposed *in vitro* to PCBs and 4,4′-DDE exhibited enhanced lymphocyte proliferation at concentrations found in sea turtle blood. Taken together these studies suggest that current exposure of loggerhead sea turtles to OCs may be enhancing their acquired immune functions, which could lead to hypersensitivity and autoimmune disorders.

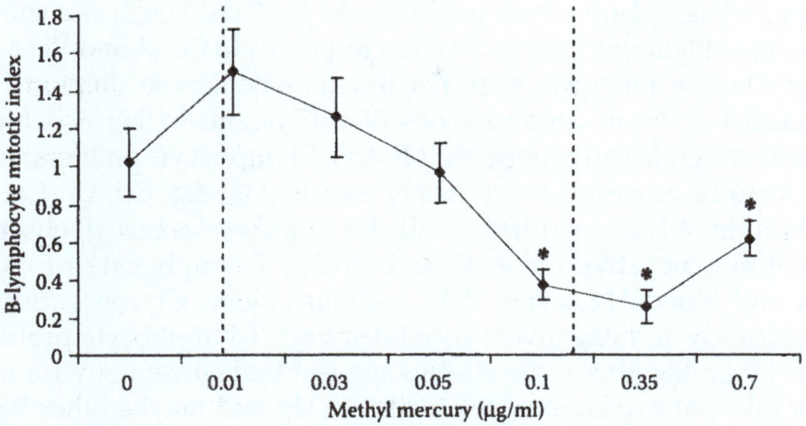

Figure 8.9 The effect of *in vitro* exposure of peripheral blood leukocytes from 20 loggerhead turtles to methyl mercury on B-lymphocyte proliferation. Mitotic index was calculated as counts per minute (cpm) of peripheral blood leukoctyes (PBLs) stimulated with mitogen (200 ng/ml PDB, or phorbol 12,13-dibutyrate) divided by cpm of unstimulated PBLs. Data are shown as means ± one standard error. Significant differences from the control are indicated by an asterisk ($P < 0.05$). Vertical dotted lines indicate the concentrations of total mercury measured in loggerhead turtle blood. (Data from Heesemann et al.[100] and Day et al.[101])

3. *Heavy Metals*

Associations between total mercury (Hg) in juvenile loggerhead blood and measured health indicators have been investigated.[99,100] Blood Hg concentrations measured in loggerhead turtles captured from coastal waters between northeast Florida and northeast South Carolina were significantly and negatively correlated with B-lymphocyte proliferation using PDB.[99] Blood Hg concentrations were also significantly negatively correlated with total WBC counts.[99] These data suggest that current environmental Hg exposures may suppress loggerhead immunity. *In vitro* exposure of loggerhead PBLs to methyl mercury (MeHg) corroborated the correlation observed in the field study (Figure 8.9).[100] B-lymphocyte proliferation exhibited a decreasing trend from 0.01 μg/ml to 0.35 μg/ml MeHg, and the three highest concentrations (0.1 μg/ml to 0.7 μg/ml) significantly suppressed proliferation compared with the control. These results support the negative correlations observed with measured concentrations of total Hg in loggerhead blood (0.01 μg/g to 0.2 μg/g) over a similar concentration range.[99–101] Furthermore, these results are supported by studies with other species. Both B- and T-lymphocyte proliferation using red drum (*Sciaenops ocellatus*) PBLs were inhibited by *in vitro* exposure to 10 μM $HgCl_2$ (~6 μg/ml) or higher concentrations.[102]

Beluga whale splenocytes exposed *in vitro* to 10 μM $HgCl_2$ also showed suppressed lymphocyte proliferation responses to ConA and PHA, but B-lymphocyte mitogens were not tested.[103] Bottlenose dolphin PBLs exposed *in vitro* to concentrations of 0.03 μg/ml MeHg and higher exhibited significantly suppressed B- and T-lymphocyte proliferation.[100]

A study assessing seven heavy metals (Hg, Ag, Cd, Cr, Cu, Pb, and Zn) in 36 Kemp's ridley sea turtles (*Lepidochelys kempii*) observed significant, negative correlations between T-lymphocyte proliferation and blood Hg levels.[104] In addition, blood Cr concentrations significantly and negatively correlated with T-lymphocyte proliferation. Taken together these studies suggest that current environmentally relevant exposure of sea turtles to Hg and maybe other heavy metals may suppress their immune functions. Effects observed in the *in vitro* Hg experiment over the same concentration ranges support this conclusion.

4. *Fibropapillomatosis*

The previous sea turtle immunotoxicity studies have focused primarily on healthy turtles. Our laboratories are currently examining immune function and contaminant concentrations in diseased sea turtles.[105] One disease of significant importance affecting sea turtles is fibropapillomatosis (FP; Figure 8.10). FP is a metaplastic disease first recorded in the 1930s in green turtles (*Chelonia mydas*),[106] and since the mid-1980s FP has reached epidemic proportions[107] with prevalence as high as 92% in Kaneohe Bay, Oahu, HI.[108] Turtles with FP are found in tropical and subtropical coastal waters worldwide, and the disease appears to be linked to environmentally stressed

Figure 8.10 Fibropapillomatosis in a green sea turtle. (Picture taken by Chris Johnson.)

habitats (i.e., high human use, low water turnover rates).[109] This severely debilitating disease is characterized by mostly benign tumors of the skin, periocular tissues, carapace, and plastron as well as tumors in all internal organs.[109–111] Large tumors can severely, mechanically hamper the ability of the animal to swim and dive; locate, capture, and swallow food; and avoid predators and may ultimately prove fatal. Internal tumors may also interfere with systemic function. It is generally accepted that the etiology of FP is related to a virus; however, primary etiologic agents and the natural transmission of the disease remain unknown.[109,112–114] Further details on the possible link of genotoxicants to FP are reviewed in Chapter 9.

Several studies have demonstrated that green turtles afflicted with FP have suppressed immune functions. Captive turtles with FP exhibited significantly reduced T- and B-lymphocyte proliferation as compared with captive turtles without the disease.[115] The diseased turtles also had lower lymphocyte numbers and albumin:globulin ratios. Sposato and colleagues[116] observed reduced lymphocyte proliferation, lower lymphocyte counts, and higher heterophil counts in diseased wild turtles with FP compared with nondiseased turtles from three habitats in East Florida. Similarly, Hawaiian green turtles with the most advanced severity of FP tumors exhibited significantly increased heterophil:lymphocyte ratios, reduced T-lymphocyte proliferation, altered biochemical enzyme activities, and decreased plasma total protein and globulin concentrations.[117–119] Aguirre and colleagues[119] examined the effect of captivity stress on wild-caught green turtles with and without FP. After 24 h in captivity both groups exhibited increased plasma corticosterone levels and increased heterophil:lymphocyte ratios. Both of these stress responses were significantly elevated in the FP-afflicted turtles compared with healthy turtles.[119] These studies suggest that turtles afflicted with FP are immunosuppressed and chronically stressed; however, it is not known if immunosuppression is the cause or the consequence of the disease.

Although Work and colleagues[117] suggested that the suppression of T-lymphocyte proliferation in sea turtles with FP may be a secondary effect and could be related to viral or parasitic infection, the connection between FP and contaminants is unclear and warrants further research. Only two studies have examined tissues of sea turtles afflicted with FP for trace metals and organic contaminants,[120,121] and both of these studies had several limitations. In Aguirre and colleagues,[120] 14 metals were detected in some turtles, but no PCBs or pesticides (organochlorine, carbamate, or organophosphate) were detected because the detection limits were much higher than concentrations generally detected in adipose tissue of sea turtles.[78] Mercury, a well-documented immunosuppressive metal, was not measured, and a control group of turtles large enough for statistical comparison with the diseased group was not included. Miao and colleagues[121] detected PCBs in the liver and adipose tissue of three green sea turtles, but statistical comparisons were not possible because only one turtle per disease category was analyzed (no tumors, moderate, and severe affliction).

Future studies should more thoroughly investigate the potential role of immunosuppressive contaminants in the susceptibly of turtles to FP because (1) the immune system of green turtles with FP has been shown to be suppressed[115–117,119] and (2) studies with loggerhead and Kemp's ridley turtles have observed correlations between contaminants and immune function.[81,99,100,104] Moreover, studies with other species have shown that contaminant exposure can increase disease susceptibility. For example, salmon (*Oncorhynchus tshawytscha*) exposed to PCBs or polycyclic aromatic hydrocarbons (PAHs) exhibit a reduced PFC response and increased susceptibility to *Vibrio anguillarum*.[122] Benzo[a]pyrene, a carcinogenic and immunotoxic PAH, was shown to reduce host resistance against a bacterial infection in Japanese medaka (*Oryzias latipes*).[123] Notably, mercury increased the severity of a herpes virus infection in mice,[124] which is an important finding because mercury has been shown to reduce immune function in sea turtles[99,100] and a herpes virus is a putative cause of FP.[113,114]

IV. Future Directions

Immunotoxicologic biomarkers are well suited for assessing a variety of stressors and are classified as "Silver Standard" markers, meaning they are affected by a broad range of chemicals.[125] There is profound potential for the use of immunotoxicologic tests in assessing both general health and contaminant effects in reptilian species. In fact, it is recognized that assays of immune function are often more sensitive than other toxicologic endpoints.[65] Immune assays can indicate (1) exposure to immunotoxic compounds, (2) susceptibility of particular species to toxicants, and (3) actual health effects that can lead to a disease state.[65] Thus, immune function tests can be used as biomarkers to indicate exposure to contaminants and as biomarkers of effects indicating the health status of a population as discussed in Chapter 4. The field of reptilian immunotoxicology is in its infancy and future directions need to focus on (1) assessing the effects of various contaminant classes on reptilian immunity, (2) identifying reptilian model species, and (3) improving immune assessments tools (i.e., assay development, optimization, and validation). These will each be discussed in the following paragraphs.

A. Testing Additional Contaminant Classes

Few contaminant classes have been tested for immunotoxicity in reptiles. Past studies have focused mainly on organochlorine contaminants (PCBs, dioxin, and pesticides), many of which have been restricted or banned from use. Additional contaminants need to be screened for immunotoxicity in reptiles, such as metals, organometals (i.e., methylmercury and tributyltin), PAHs, and other halogenated organic contaminants (i.e., perflourinated compounds [PFCs] and polybrominated diphenyl ethers [PBDEs]). All these classes are known to affect the immune system of other animal species.[65,126–129] Studies might focus on current-use compounds that are increasing in the environment. For example, PFCs are currently used in textiles as stain and soil repellents, in fire-fighting foams, and as polymerization aids. The levels of PFCs have increased in human blood samples over the past 25 years.[130] PBDEs are another class of compounds of emerging concern. They are used as flame retardants in many consumer products, such as textiles and electronic plastics, and have also been increasing in wildlife and human tissues over the past three decades.[131]

B. Choosing Reptilian Model Species

Currently available immunotoxicity studies using reptiles are limited in their scope. The studies have focused primarily on alligators and turtles, with only one study using lizards and none using snakes. The choice of appropriate reptilian models for assessing immunotoxicity is clearly needed but is a complex issue. Models are chosen based on a variety of needs often specific to the study design. For instance, a researcher might choose a small, easily maintained species for routine screening of different compounds for reptile immunotoxicity testing, whereas a larger, slow-growing species may better suit a study examining chronic environmental exposure in field situations. On occasion a surrogate species may be needed to represent a species of concern that cannot be subjected to captive exposure studies. The diamondback terrapin (*Malaclemys terrapin*), an estuarine turtle, may be a good surrogate species for sea turtle toxicity studies. Additional types of model species include indicator species, which are chosen to represent a group of species of concern, and sentinel species, which are usually more sensitive and can predict impending harm or change to other species or the environment. Species used in the laboratory setting, such as surrogates, should be easy to maintain and breed in captivity, exhibit minimum captivity-related stress, be receptive to a battery of immune tests (including host-resistance studies using pathogens common to reptiles), and represent the particular reptilian order of interest. Determining which reptilian orders or species are most and least sensitive will be important in deciding on good laboratory model species as well as in choosing indicator and sentinel species to assess the risk of contaminant effects in wild reptiles.

Developing a single-standard reptilian immunotoxicologic model, similar to the choice of B6C3F1 mice for mammals, would be highly advantageous for comparative purposes. However, choosing a reptilian model will not be as clear-cut as using a single species. Reptilian immunotoxicologic studies have diverse objectives and require the consideration of the physiologic and life history differences between reptilian orders and species. Some potentially useful model species have been suggested. For instance, the western fence lizard has been suggested as a model species for immunotoxicity[47] and for terrestrial wildlife ecotoxicology.[132] A model snake species may be important because snakes, being carnivorous, feed high on the terrestrial food chain and have the ability to bioaccumulate certain contaminants.[133] The red-eared slider turtle (*Trachemys scripta*) might serve as a good laboratory model species because of its use in previous endocrine toxicity studies.[93] For field studies, aquatic reptiles that might serve as model or indicator species include the American alligator and the snapping turtle (*Chelydra serpentina*). These candidates are important because of the increased risk of exposure in aquatic food webs compared with terrestrial food webs and because of their previous use in toxicology studies.[28,54,55,134,135] In addition to its possible use as a surrogate for sea turtles, the diamondback terrapin may also serve as a sentinel or indicator for estuarine health because of its strong site fidelity and large geographic range.

C. *Improving Immune Assessment Tools*

The scope of previous immunotoxicologic assessments in reptiles is limited by the study of only a few immune parameters: primarily histology, lymphocyte proliferation, and hemagglutination. Assessing immune function should incorporate measures of all branches of the immune system, including innate and acquired (cell-mediated and humoral) immunity, to achieve a better understanding of the function of the system as a whole. Luster and colleagues[136] proposed a two-tiered testing scheme that is currently used by the National Toxicology Program to screen xenobiotics for immunotoxicity using rodents

(Table 8.2). Tier I was designed to detect immunotoxic compounds, whereas Tier II uses tests to detect mechanisms of immunotoxicity. Modifications of this testing scheme were then suggested by Weeks and colleagues[1] to include a three-tiered approach for wildlife risk assessment (Table 8.2). The endpoints described in this approach were indicated for studies with fish, birds, and mammals but can be adapted to reptiles.

Many traditional immune function assays have been used with reptiles (Table 8.1), but the tools available are still greatly limited. Sensitive and predictive assays still need to be developed to measure contaminant effects in a variety of reptilian species. The greatest need may be species-specific reagents, such as monoclonal antibodies to immunoglobulins and cell surface markers, to allow more modern and informative molecular and mechanistic analysis. Having more tools available will improve our ability to test contaminant effects as well as advance our basic understanding of reptilian immunity.

1. *General Considerations*

Controlled laboratory experiments or localized field studies where the animals can be tracked, recaptured, and euthanized should take advantage of a battery of available immune parameters, including tests to assess the functions of innate, cell-mediated, and humoral immunity. In situations where the test species can be euthanized, a battery of tests can be selected from Table 8.2. If suppression is observed in a number of these parameters, host-resistance studies should be undertaken if possible. Additionally, if stimulation of immune functions is observed, it may be useful to determine serum levels of antinuclear antibodies or antibodies to single-stranded or double-stranded DNA to screen for autoimmunity.

Often when reptilian wildlife is studied, the animals cannot be immunized, recaptured, or euthanized. In these situations many of the parameters suggested by Luster and colleagues[136] and Weeks and colleagues[1] are not possible. Table 8.1 lists the limitations of several assays. In these situations, we recommend a smaller, yet informative, Comprehensive Screening Set that can be performed in reptiles non-lethally without requiring captivity, recapture, or injection of the animals with antigens (Table 8.2).

Optimization, standardization, and validation of immunologic endpoints in reptiles are critical to the progress of reptilian immunotoxicology. Validation of each assay for each species of interest should be the ultimate goal because it promotes uniformity throughout the field and permits ease of comparison among studies. Validated assays

Table 8.2 Tiered Approaches Previously Recommended for Immunotoxicity Testing with Rodents (Luster et al.[136]) and with Wildlife (Weeks et al.[1]) and a New Recommended Comprehensive Screening Set for Reptilian Studies Requiring Noncaptive, Nonlethal Techniques

	Test recommended by Luster et al.[136]	Test recommended by Weeks et al.[1]
Tier I	Complete blood count	Complete blood count
	Cell differential count	Cell differential count
		Hematocrit
		Leukocrit for fish only
	Organ weights (body, spleen, thymus, kidney, liver)	Organ weights (spleen, thymus, bursa, lymph nodes, somatopleure pronephros, etc.)
		Wound healing
	Natural killer cell activity	Natural killer cell activity
		Macrophage phagocytosis and killing
		Lysozyme activity
	Lymphocyte blastogenesis to ConA	
	Mixed leukocyte response	
		Chemiluminescence
		Graft rejection
	Plaque-forming cell assay for IgM production to SRBCs	
		Agglutination assay
Tier II	Quantitation of splenic B- and T-cells	Immune cell quantitation (Surface markers and flow cytometry)
		NBT reduction
	Macrophage function-quantitation of resident peritoneal cells	
	Macrophage phagocytic ability (basal and activated by MAF)	Macrophage responses (melanin accumulation; chemotaxis, pinocytosis)
		Lymphocyte blastogenesis (T-cell mitogens)
		Melanomacrophage centers
	Cytotoxic T-lymphocyte cytolysis	Cytotoxic T-cell (leukocyte) activity
	Delayed hypersensitivity response	Delayed-type hypersensitivity response
		Mixed leukocyte response
	Enumation of IgG response to SRBCs	
		Plaque-forming cell assay
		Lymphokine quantitation
	Host resistance challenge (to tumors, bacteria, viruses, or parasites)	
Tier III		Host resistance challenge
		Bacteremia/viremia/parasitemia/tumor quantitation and duration
		Mortality
		Specific antibody quantitation

Table 8.2 (continued)

CMI, cell-mediated immunity; HI, humoral immunity; SRBCs, sheep red blood cells; NBT, nitrotetrazolium blue; MAF, macrophage activating factor.

[a] The Comprehensive Screening Set is not suggested as a tiered approach. Tests listed in this column are aligned with those recommended by Luster et al.[136] and Weeks et al.[1] for simplicity.

Test names from Luster et al.[136] and Weeks et al.[1] were taken directly from their respective tables. Tests aligned on the same row are considered the same (or very similar) test(s) even if they are shown with different names.

are those that have undergone peer review and have been performed in several laboratories yielding similar results. Certain criteria must be met to validate immune biomarkers, including (1) reproducibility within and among labs, (2) specificity for the type of immune function being assessed, (3) sensitivity to measure both normal and abnormal function, and (4) measurability of altered function caused by exposure to known immunotoxicants.[65] Validation is a daunting task but one that is needed to yield quality data and to elevate reptilian immunotoxicologic studies into the ranks with fish, bird, and mammal studies.

2. *Hematology Tools*

When assessing immunotoxicity in reptiles, traditional hematology parameters (i.e., differential WBC counts) can provide information on general stress to the immune system. For instance, an increase in the heterophil:lymphocyte ratio is a common response to stress in reptiles[137,138] and correlates with disease in reptiles.[117,119,139] Contaminant exposure has also been associated with an increase in this ratio in juvenile herring gulls from the Great Lakes and in loggerhead sea turtles,[9,140] suggesting that a differential WBC count should be performed in reptilian immunotoxicity studies. It is important to note, in mammals at least, that simple hematologic assessments, such as leukocyte counts, are not thought to be extremely predictive of reduced host resistance, but more advanced counts of different lymphocyte populations using their cell-surface markers (i.e., CD4 and CD8 on T-lymphocytes) are more predictive.[141]

The fields of reptilian immunology and immunotoxicology would benefit greatly from improved methods of identifying, separating, and collecting certain cell types. The current criteria for microscopic identification of different leukocyte types in reptilian blood are inconsistent and need to be standardized.[11,12] Species-specific reagents that detect cell-surface markers are lacking but could be used to count different cell types using automated methods of flow cytometry if they were available. Techniques to isolate and collect particular cell types are important for certain immune function assays. A method to partially separate peripheral blood mononuclear cells (which include lymphocytes, monocytes, and macrophages) from other blood cells has been described for loggerhead sea turtles.[142]

3. *Innate Immunity Tools*

Innate immunity appears to be an extremely important defense in reptiles, but it has not been a common tool of reptilian immunotoxicology studies. Lysozyme activity is the only innate immune parameter used in a reptilian immunotoxicology study.[76,79] Merchant and colleagues[143] observed just how strong innate immunity is in alligators. Alligator serum rapidly and effectively inhibited gram-negative and gram-positive bacterial growth. The evidence provided from this study suggests that the complement system is responsible for this surprisingly strong immune defense and is likely activated through the "alternative" innate pathway instead of the "classical" humoral pathway. Other tests of innate immunity, such as macrophage phagocytosis or respiratory (oxidative) burst, have been measured in reptiles[16,18] but have not been applied to immunotoxicity studies. In rodents, a very predictive endpoint of innate function is natural killer cell function. It is not fully known if reptiles possess true natural killer cells similar to mammals or if they are more similar to the evolutionary precursors found in fish (natural cytotoxic cells). Natural killer cell–like activity was successfully measured from the spleen and thymus of the Caspian turtle to investigate seasonal changes,[14,15] yet an attempt to measure this activity was unsuccessful with green sea turtle blood leukocytes using three different types of target cells.[25] In both species, however, the antibody-dependent cell-mediated cytotoxicity (ADCC) response was detected. Because innate immunity appears to be both strong and important to immune defenses in reptiles, more assays addressing this branch of immunity should be developed and explored for use in immunotoxicology studies.

4. *Acquired Immunity Tools*

Acquired immunity involves both the cell-mediated and humoral responses. Cell-mediated immunity has been used most frequently in reptilian immunotoxicology studies. T-lymphocytes are the main cell types in cell-mediated immunity, and at least three types of T-cells (helper, cytotoxic, and regulatory) are known to play different roles in mammalian immune functions and to express different cell surface receptors (T-cell receptors). Mammalian helper T-cells express the cell-surface marker CD4, whereas cytotoxic T-cells express CD8. The number of CD4+ cells is a predictive measure of contaminant-mediated immunosuppression.[141] Cell-surface markers for reptilian lymphocytes have been investigated in a few studies and have been suggested to be similar to mammals.[15] Species-specific molecular markers for different cell types would allow separation of individual cell types to test their individual functions and help explain reptilian immune functions in more detail.

Measures of cell-mediated immunity, such as MLR, skin allograft rejection, and mitogen-induced lymphocyte proliferation, have been commonly used in reptiles to assess basic immunology, but the lymphocyte proliferation assay has been the primary tool used in immunotoxicology studies. Lymphocyte proliferation measures only a single, early step in a complex immune response and therefore has little functional meaning to the overall immune system. However, T-lymphocyte proliferation along with MLR and delayed-type hypersensitivity (DTH) are predictive of compromised host resistance.[141] The latter two assays provide data on more complex immune functions. Although lymphocyte proliferation has limited explanatory power for mammals, it appears to be important for reptilian studies because it has been shown to be suppressed in diseased green turtles[117] and it correlated with organochlorine contaminants and mercury in sea turtles.[81,99,104] These studies suggest that this relatively simple assay may be important in reptilian immunotoxicity studies.

When performing cell-culture experiments such as MRL or lymphocyte proliferation, incubation and culture conditions must be optimized for each species. For example, it is important in studies with ectotherms to use the appropriate physiologic temperature, which is dictated by the preferred water or air temperature, or to optimize assays for the best temperature. Studies measuring reptilian immune functions *ex vivo* have chosen to use incubation temperatures that vary from 27°C to 37°C.[15,23,25,115] The temperature 37°C, which is used in mammalian studies, may not be a normal physiologic temperature for reptiles and should at least be tested and compared with lower temperatures during optimization of assays. One study provides a detailed explanation of optimization methods for lymphocyte proliferation in red-eared slider turtles (*Trachemys scripta*).[24] Additional optimization studies have been performed for the mitogen-induced lymphocyte proliferation assay using loggerhead sea turtles,[23] American alligators,[26] and green sea turtles.[25,144]

Very few tools have been used to measure contaminant alterations on reptilian humoral immunity. Antibody titers after immunization have been measured in reptilian immunotoxicology studies using hemagglutination, which does not require species-specific reagents.[28,54]

As mentioned previously, one important consideration is the length of time needed after immunization before antibody production is detectable. It is typical in mammalian studies to test antibody production (IgM) at only 4 to 7 days after the first immunization. However, Gross and colleagues[28] noted that alligator antibody titers were undetectable until 4 weeks after first immunization. Detection of IgY production in green sea turtles was found to take between 5 weeks and 9 months after several immunizations.[27] This comparison suggests that the reptilian antibody response is considerably slower than that of mammals. Few studies have developed monoclonal antibodies for specific reptilian immunoglobulins. Monoclonal antibodies have been developed for green sea turtle IgY and IgM,[27] for desert tortoise (*Gopherus agassizii*) IgY,[145] and for Mediterranean tortoise (*Testudo graeca* and *T. hermanni*) IgY.[146] A polyclonal antibody was also developed for American alligator IgY.[147] As more species-specific antibodies are produced, immunotoxicology studies testing reptiles will benefit greatly.

A final note on measuring humoral immunity concerns the PFC assay. This assay is considered to be one of the most predictive tests of immunotoxic effects because it measures the complete and integrated T-lymphocyte–dependent humoral response.[141] It requires macrophages to present antigens to helper T-cells, which initiate antibody production by B-cells. This assay has been performed with splenocytes from the Caspian turtle to investigate seasonal changes in immunity.[29] Although this test is highly recommended because of its sensitivity in immunotoxicity testing, it requires captive animals, antigen immunization, and euthanasia to harvest splenocytes. Modifications of this assay, however, eliminate these limitations by allowing it to be performed in mammals without using prior challenge with an antigen.[148] In this modified technique, lymphocytes can be exposed simultaneously *in vitro* to the antigen and to a xenobiotic.[149] This is an unlikely option for reptiles because their lymphocytes could not be maintained alive in culture long enough for antibody production, if the long duration observed in alligators and green turtles[27,28] is required for most reptile species.

5. *Host-Resistance Tools*

Determining whether host resistance is compromised is an important goal of immunotoxicity studies and can be tested by exposing animals to bacteria or other pathogens and directly assessing the rate or degree of infection or mortality. Host-resistance studies examine the extent to which the integrated immune system, which has built-in reserve and redundancy capacities, can fight an infection or disease. Studies that correlate contamination with disease in free-ranging animals begin to address this aspect of immunotoxicology but usually cannot conclusively define a cause-and-effect relationship because of the inherent nature of the correlative relationship. Examples of such studies include those with marine mammals and box turtles in which higher contaminant tissue burdens were observed in diseased compared with nondiseased animals.[31,42,43] Laboratory host-resistance experiments with reptiles are extremely limited. One study immunized mountain leopard tortoises (*Geochelone pardalis*) with homogenates of ticks, and on subsequent tick challenge the animals exhibited resistance to tick infestations.[30] A host-resistance model in juvenile alligators with a *Mycoplasma* species has also been developed and used in immunotoxicology studies.[150] Because host resistance is one of the most important aspects of immunotoxicity, these assays need further development in reptiles.

V. Conclusions

The field of reptilian immunotoxicology should (1) develop and validate a standard set of immune assays to test all categories of reptilian immunity, (2) test a variety of reptilian species and orders to assist researchers in choosing appropriate reptilian model species, and (3) determine the sensitivity of reptiles to many classes of immunotoxic compounds. More *in vivo* experimental exposures are needed, such as those performed with alligators and fence lizards,[47,54] as well as *in vitro* studies using cells from reptilian species, such as sea turtles, that cannot be studied in captivity or in an invasive manner.

The immune system is sensitive to alterations caused by all classes of environmental contaminants,[151] and suppression of immune functions may lead to increased disease in reptiles as suggested by Tangredi and Evans[31] and Gross and colleagues.[28] If disease prevalence increases sufficiently, then large-scale reductions in survival or reproduction can lead to population declines, as exemplified by mass mortality events of marine mammals.[41] Globally, reptiles have experienced population declines as a result of a number of human-imposed threats (see Chapter 2). A great need exists to better understand the effects of environmental contaminants on reptilian health, immunity, and ultimately survival.

Acknowledgments

The authors thank Deborah Keil, Rusty Day, Michelle Lee, Charlie Rice, Paul Becker, John Kucklick, Michele Schantz, Steve Wise, and Katherine Sharpless for their helpful review of this chapter and Jackie EuDaly for formatting assistance.

Literature Cited

1. Weeks, B.A. et al., Immunological biomarkers to assess environmental stress, in *Biomarkers: Biochemical, Physiological, and Histological Markers of Anthropogenic Stress*, Huggett, R.J. et al., Eds., Lewis Publishers, Boca Raton, FL, 1992, 211–234, Chap. 5.
2. Zapata, A. and Amemiya, C.T., Phylogeny of lower vertebrates and their immunological structures, in *Origin and Evolution of the Vertebrate Immune System*, Du Pasquier, L. and Litman, G.W., Eds., Springer, New York, 2000, 67–107.
3. Zapata, A.G. and Cooper, E.L., *The Immune System: Comparative Histophysiology*, John Wiley & Sons, New York, 1990.
4. Cooper, E.L., Klempau, A.E., and Zapata, A.G., Reptilian immunity, in, *Biology of the Reptilia*, Gans, C., Billett, F., and Maderson, P.F.A., Eds., John Wiley & Sons, New York, 1985, 599–678, Chap. 8.
5. Bengtén, E. et al., Immunoglobulin isotypes: structure, function, and genetics, in *Origin and Evolution of the Vertebrate Immune System*, Du Pasquier, L. and Litman, G.W., Eds., Springer, New York, 2000, 189–219.
6. Warr, G.W., Magor, K.E., and Higgins, D.A., IgY: clues to the origins of modern antibodies, *Immunol. Today*, 16, 392, 1995.
7. Janeway, C.A. et al., *Immunobiology 5: The Immune System in Health and Disease*, 5th ed., Garland Publishing, New York, 2001.
8. Goldsby, R.A., Kindt, T.J., and Osborne, B.A., *Kuby Immunology*, 4th ed., W.H. Freeman, New York, 2000.
9. Keller, J.M. et al., Associations between organochlorine contaminant concentrations and clinical health parameters in loggerhead sea turtles from North Carolina, USA, *Environ. Health Perspect.*, 112, 1074, 2004.
10. Bolten, A.B., Jacobson, E.R., and Bjorndal, K.A., Effects of anticoagulant and autoanalyzer on blood biochemical values of loggerhead sea turtles (*Caretta caretta*), *Am. J. Vet. Res.*, 53, 2224, 1992.
11. Work, T.M. et al., Morphologic and cytochemical characteristics of blood cells from Hawaiian green turtles, *Am. J. Vet. Res.*, 59, 1252, 1998.
12. Alleman, A.R., Jacobson, E.R., and Raskin, R.E., Morphologic and cytochemical characteristics of blood cells from the desert tortoise (*Gopherus agassizii*), *Am. J. Vet. Res.*, 53, 1645, 1992.
13. Leceta, J. and Zapata, A., Seasonal changes in the thymus and spleen of the turtle *Mauremys caspica* a morphometric light microscopic study, *Dev. Comp. Immunol.*, 9, 653, 1985.
14. Muñoz, F.J. and De la Fuente, M., The immune response of thymic cells from the turtle *Mauremys caspica*, *J. Comp. Physiol. B*, 171, 195, 2001.
15. Muñoz, F.J. and De la Fuente, M., The effect of the seasonal cycle on the splenic leukocyte functions in the turtle *Mauremys caspica*, *Physiol. Biochem. Zool.*, 74, 660, 2001.
16. Mondal, S. and Rai, U., Dose and time-related *in vitro* effects of glucocorticoid on phagocytosis and nitrite release by splenic macrophages of wall lizard *Hemidactylus flaviviridis*, *Comp. Biochem. Physiol. C*, 132, 461, 2002.
17. Ingram, G.A. and Molyneux, D.H., The humoral immune response of the spiny-tailed agamid lizard (*Agama caudospinosum*) to injection with *Leishmania agamae* promastigotes, *Vet. Immunol. Immunopathol.*, 4, 479, 1983.

18. Pasmans, F. et al., Induction of the respiratory burst in turtle peritoneal macrophages by *Salmonella muenchen*, *Dev. Comp. Immunol.*, 25, 159, 2001.
19. El Masri, M. et al., Seasonal distribution and hormonal modulation of reptilian T cells. *Immunobiology*, 193, 15, 1995.
20. Saad, A.H. and El Ridi, R., Mixed leukocyte reaction, graft-versus-host-reaction, and skin allograft rejection in the lizard, *Chalcides ocellatus*, *Immunobiology*, 166, 484, 1984.
21. Cope, R.B. et al., Resistance of a lizard (the green anole, *Anolis carolinensis*; polychridae) to ultraviolet radiation-induced immunosuppression, *Photochem. Photobiol.*, 74, 46, 2001.
22. Farag, M.A. and El Ridi, R., Functional markers of the major histocompatibility gene complex of snakes, *Eur. J. Immunol.*, 20, 2029, 1990.
23. Keller, J.M. et al., Mitogen-induced lymphocyte proliferation in loggerhead sea turtles: comparison of methods and effects of gender, plasma testosterone concentration, and body condition on immunity, *Vet. Immunol. Immunopathol.*, 103, 269, 2005.
24. Ulsh, B.A. et al., Culture methods for turtle lymphocytes, *Methods Cell Sci.*, 22, 285, 2001.
25. McKinney, E.C. and Bentley, T.B., Cell-mediated immune response of *Chelonia mydas*, *Dev. Comp. Immunol.*, 9, 445, 1985.
26. Cuchens, M.A. and Clem, L.W., Phylogeny of lymphocyte heterogeneity. III. Mitogenic responses of reptilian lymphocytes, *Dev. Comp. Immunol.*, 3, 287, 1979.
27. Herbst, L.H. and Klein, P.A., Monoclonal antibodies for the measurement of class-specific antibody responses in the green turtle, *Chelonia mydas*, *Vet. Immunol. Immunopathol.*, 46, 317, 1995.
28. Gross, D.A. et al., Potential contaminant-induced immuno-suppression in neonatal alligators from contaminated and control lakes in Central Florida, presented at Soc. Environ. Toxicol. Chem. 18th Annual Meeting, San Francisco, Nov. 16–20, 1997, 174.
29. Leceta, J. and Zapata, A., Seasonal variations in the immune response of the tortoise *Mauremys caspica*, *Immunology*, 57, 483, 1986.
30. Tembo, S.D. and Kiwanuka, A., Acquisition of protective immunity in *Geochelone pardalis* against *Amblyomma marmoreum* (Acari:Ixodidae) nymphal ticks, *Onderstepoort J. Vet. Res.*, 64, 1, 1997.
31. Tangredi, B.P. and Evans, R.H., Organochlorine pesticides associated with ocular, nasal, or otic infection in the eastern box turtle (*Terrapene carolina carolina*), *J. Zool. Wildl. Med.*, 28, 97, 1997.
32. Farag, M.A. and El Ridi, R., Proliferative responses of snake lymphocytes to concanavalin A, *Dev. Comp. Immunol.*, 10, 561, 1986.
33. Zapata, A.G., Varas, A., and Torroba, M., Seasonal variations in the immune system of lower vertebrates, *Immunol. Today*, 13, 142, 1992.
34. Farag, M.A. and El Ridi, R., Mixed leucocyte reaction (MLR) in the snake *Psammophis sibilans*, *Immunology*, 55, 173, 1985.
35. Saad, A.H., Khalek, N.A., and El Ridi, R., Blood testosterone level: a season-dependent factor regulating immune reactivity in lizards, *Immunobiology*, 180, 184, 1990.
36. Saad, A.H. and El Ridi, R., Endogenous corticosteroids mediate seasonal cyclic changes in immunity of lizards, *Immunobiology*, 177, 390, 1988.

37. Saad, A.H. et al., Effect of hydrocortisone on immune system of the lizard *Chalcides ocellatus* III. Effect on cellular and humoral immune responses, *Dev. Comp. Immunol.*, 10, 235, 1986.
38. Saad, A.H. et al., Testosterone induces lymphopenia in turtles, *Vet. Immunol. Immunopathol.*, 28, 173, 1991.
39. Grasman, K.A. and Fox, G.A., Associations between altered immune function and organochlorine contamination in young Caspian terns (*Sterna caspia*) from Lake Huron, 1997–1999, *Ecotoxicology*, 10, 101, 2001.
40. Grasman, K.A. et al., Organochlorine-associated immunosuppression in prefledgling Caspian terns and herring gulls from the Great Lakes: an ecoepidemiological study, *Environ. Health Perspect.*, 104 (Suppl. 4), 829, 1996.
41. Ross, P. et al., Contaminant-induced immunotoxicity in harbour seals: wildlife at risk?, *Toxicology*, 112, 157, 1996.
42. Hall, A.J. et al., Organochlorine levels in common seals (*Phoca vitulina*) which were victims and survivors of the 1988 phocine distemper epizootic, *Sci. Total Environ.*, 115, 145, 1992.
43. Aguilar, A. and Borrell, A., Abnormally high polychlorinated biphenyl levels in striped dolphins (*Stenella coeruleoalba*) affected by the 1990–1992 Mediterranean epizootic, *Sci. Total Environ.*, 154, 237, 1994.
44. Keller, J.M. et al., Assessment of immunotoxicology in wild populations: review and recommendations, *Rev. Toxicol.*, 3, 167, 1999/2000.
45. Zelikoff, J.T., Biomarkers of immunotoxicity in fish and other non-mammalian sentinel species: predictive values for mammals?, *Toxicology*, 129, 63, 1998.
46. Luebke, R.W. et al., Aquatic pollution-induced immunotoxicity in wildlife species, *Fund. Appl. Toxicol.*, 37, 1, 1997.
47. Burnham, D.K. et al., Effects of 17α-ethinylestradiol on immune parameters in the lizard *Sceloporus occidentalis*, *Environ. Toxicol.*, 18, 211, 2003.
48. Hong, C.C. et al., Effect of endocrine disrupting chemicals on lipopolysaccharide-induced tumor necrosis factor-alpha and nitric oxide production by mouse macrophages, *Biol. Pharm. Bull.*, 27, 1136, 2004.
49. Iwata, M. et al., The endocrine disruptors nonylphenol and octylphenol exert direct effects on T cells to suppress Th1 development and enhance Th2 development, *Immunol. Lett.*, 94, 135, 2004.
50. Forawi, H.A., Tchounwou, P.B., and McMurray, R.W., Xenoestrogen modulation of the immune system: effects of dichlorodiphenyltrichloroethane (DDT) and 2,3,7,8-tetrachlorodibenzo-*p*-dioxin (TCDD), *Rev. Environ. Health*, 19, 1, 2004.
51. Ahmed, S.A. et al., Linking environmental agents and autoimmune diseases, *Environ. Health Perspect.*, 107 (Suppl. 5), 681, 1999.
52. Johnson, K.W., Kaminski, N.E., and Munson, A.E., Direct suppression of cultured spleen cell responses by chlordane and the basis for differential effects on *in vivo* and *in vitro* immunocompetence, *J. Toxicol. Environ. Health*, 22, 497, 1987.
53. Peden-Adams, M.M. et al., Relationship of lymphoproliferation and clinical blood parameters to contaminants in loggerhead turtles, presented at Soc. Environ. Toxicol. Chem. 23rd Annual Meeting, Salt Lake City, Nov. 16–20, 2002, 175.
54. Peden-Adams, M.M., Evaluation of xenobiotic-induced immunotoxicity and CYP450 activity in wildlife species, Ph.D. thesis, Clemson University, Clemson, 1999.

55. Rooney, A.A., Bermudez, D.S., and Guillette, L.J., Jr., Altered histology of the thymus and spleen in contaminant-exposed juvenile American alligators, *J. Morphol.*, 256, 349, 2003.
56. Gross, D.A. et al., Characterization of potential contaminant-induced clinical manifestations in neonatal alligators from contaminated and control lakes in central Florida, presented at Soc. Environ. Toxicol. Chem. 17th Annual Meeting, Washington, Nov. 17–21, 1996, 213.
57. Woodward, A.R. et al., Low clutch viability of American alligators on Lake Apopka, *Fla. Sci.*, 56, 52, 1993
58. Crain, D.A. et al., Sex steroid and thyroid hormone concentration in juvenile alligators (*Alligator mississippiensis*) from contaminated and reference lakes in Florida, USA, *Environ. Toxicol. Chem.*, 17, 446, 1998.
59. Guillette, L.J., Jr. et al., Reduction in penis size and plasma testosterone concentrations in juvenile alligators living in a contaminated environment, *Gen. Comp. Endocrinol.*, 101, 32, 1996.
60. Guillette, L.J., Jr. et al., Serum concentrations of various environmental contaminants and their relationship to sex steroid concentrations and phallus size in juvenile American alligators, *Arch. Environ. Contam. Toxicol.*, 36, 447, 1999.
61. Peden-Adams, M., et al., 2,3,7,8-TCDD effects on immune function in the American alligator (*Alligator mississippiensis*): a scoping study, presented at Soc. Environ. Toxicol. Chem. 17th Annual Meeting, Washington, Nov. 17–21, 1996, 265.
62. Peden-Adams, M. et al., Effects of environmentally relevant concentrations of 2,3,7,8-TCDD on domestic chicken immune function and CYP450 activity: F_1 generation and egg injection studies, *Chemosphere*, 37, 1923, 1998.
63. Peden-Adams, M.M. et al., *In-ovo* effects of endocrine disrupting chemicals on immune function and CYP450 induction in juvenile American alligators, *Toxicologist*, 42, 340, 1998.
64. Dickerson, R.L. et al., Toxicological foundations of ecological risk assessment: biomarker development and interpretation based on laboratory and wildlife species. *Environ. Health Perspect.*, 102 (Suppl. 12), 65, 1994.
65. National Research Council, *Biologic Markers in Immunotoxicology*, National Academy Press, Washington, 1992.
66. Taylor, M. et al., Preliminary evaluation of cytochrome P450 activity in American alligators (*Alligator mississippiensis*) following *in ovo* exposure, presented at Soc. Environ. Toxicol. Chem. 17th Annual Meeting, Washington, Nov. 17–21, 1996, 139.
67. Witherington, B.E., Flotsam, jetsam, post-hatchling loggerheads, and the advecting surface smorgasbord, in *Proc. 14th Annual Symposium on Sea Turtle Biology and Conservation*, Bjorndal, K.A. et al., compilers, NOAA Technical Memorandum NMFS-SEFSC-351, U.S. Department of Commerce, Miami, FL, 1994, 166–168.
68. Gramentz, D., Involvement of loggerhead turtle with the plastic, metal, and hydrocarbon pollution in the central Mediterranean, *Mar. Pollut. Bull.*, 19, 11, 1988.
69. Hall, R.J., Belisle, A.A., and Sileo, L., Residues of petroleum hydrocarbons in tissues of sea turtles exposed to the Ixtoc I oil spill, *J. Wildl. Dis.*, 19, 106, 1983.
70. Fritts, T.H. and McGehee, M.A., Effects of petroleum on the development and survival of marine turtle embryos, Report FWS/OBS-82/37, U.S. Fish and Wildlife Service, Belle Chasse, LA, 1982.

71. Lutcavage, M.E. et al., Physiologic and clinicopathologic effects of crude oil on loggerhead sea turtles, *Arch. Environ. Contam. Toxicol.*, 28, 417, 1995.
72. Leighton, F.A., Clinical, gross, and histological findings in herring gulls and Atlantic puffins that ingested Prudhoe Bay crude oil, *Vet. Pathol.*, 23, 254, 1986.
73. Rocke, T.E., Yuill, T.M., and Hinsdill, R.D., Oil and related toxicant effects on mallard immune defenses, *Environ. Res.*, 33, 343, 1984.
74. Briggs, K.T., Yoshida, S.H., and Gershwin, M.E., The influence of petrochemicals and stress on the immune system of seabirds, *Regul. Toxicol. Pharmacol.*, 23, 145, 1996.
75. Segars, A. et al., Hematology and plasma chemistry reference values from free-ranging loggerhead sea turtles along the southeastern U.S. coast, presented at 25th Annual Symp. Sea Turtle Biol. Conserv., Savannah, Jan. 18–22, 2005.
76. Peden-Adams, M.M. et al., unpublished data, 2004.
77. Keller, J.M., Occurrence and effects of organochlorine contaminants in sea turtles, Ph.D. thesis, Duke University, Durham, 2003.
78. Keller, J.M. et al., Organochlorine contaminants in sea turtles: correlations between whole blood and fat, *Environ. Toxicol. Chem.*, 23, 726, 2004.
79. Keller, J.M. et al., unpublished data, 2004.
80. Burton, J.E., et al., Circulating lysozyme and hepatic CYP1A activities during a chronic dietary exposure to tributyltin (TBT) and 3,3′,4,4′,5-pentachlorobiphenyl (PCB-126) mixtures in channel catfish, *Ictalurus punctatus*, *J. Toxicol. Environ. Health A*, 65, 589, 2002.
81. Keller, J.M., Kucklick, J.R., and Peden-Adams, M.M., Comparison of mitogen-induced lymphocyte proliferation of loggerhead sea turtles after *in vivo* and *in vitro* exposure to PCBs and 4,4′-DDE, presented at Soc. Environ. Toxicol. Chem. 25th Annual Meeting, Portland, Nov. 14–18, 2004, 259.
82. Smits, J.E. et al., Thyroid hormone suppression and cell-mediated immunomodulation in American kestrels (*Falco sparverius*) exposed to PCBs, *Arch. Environ. Contam. Toxicol.*, 43, 338, 2002.
83. Croisant, E.T. and Grasman, K.A., Altered lymphocyte mitogenesis in fish-eating birds of the Great Lakes, presented at Soc. Environ. Toxicol. Chem. 23rd Annual Meeting, Salt Lake City, Nov. 16–20, 2002, 139.
84. Peden-Adams, M.M. et al., Evaluation of the lymphoproliferative response as a biomarker for ecological risk assessment in feral juvenile prothonotary warblers following DDT and Hg exposure, presented at Soc. Environ. Toxicol. Chem. 17th Annual Meeting, Washington, Nov. 17–21, 1996, 264.
85. Segre, M. et al., Immunological and physiological effects of chronic exposure of *Peromyscus leucopus* to Aroclor 1254 at a concentration similar to that found at contaminated sites, *Toxicology*, 174, 163, 2002.
86. Wu, P.J. et al., Immunological, hematological, and biochemical responses in immature white-footed mice following maternal Aroclor 1254 exposure: a possible bioindicator, *Arch. Environ. Contam. Toxicol.*, 36, 469, 1999.
87. Burns, L.A., Meade, B.J., and Munson, A.E., Toxic responses of the immune system, in *Casarett & Doull's Toxicology: The Basic Science of Poisons*, Klaassen, C.D., Ed., McGraw-Hill, New York, 1996, 355–402, Chap. 12.
88. De Guise, S. et al., Effects of *in vitro* exposure of beluga whale leukocytes to selected organochlorines, *J. Toxicol. Environ. Health Part A*, 55, 479, 1998.
89. Smithwick, L.A. et al., Inhibition of LPS-induced splenocyte proliferation by *ortho*-substituted polychlorinated biphenyl congeners, *Toxicology*, 188, 319, 2003.

90. Snyder, C.A. and Valle, C.D., Lymphocyte proliferation assays as potential biomarkers for toxicant exposures, *J. Toxicol. Environ. Health*, 34, 127, 1991.
91. Calabrese, E.J. and Baldwin, L.A., The hormetic dose-response model is more common than the threshold model in toxicology, *Toxicol. Sci.*, 71, 246, 2003.
92. Guillette L.J., Jr., Vonier, P.M., and McLachlan, J.A., Affinity of the alligator estrogen receptor for serum pesticide contaminants, *Toxicology*, 181–182, 151, 2002.
93. Willingham, E. and Crews, D., The red-eared slider turtle: an animal model for the study of low doses and mixtures, *Am. Zool.*, 40, 421, 2000.
94. Matter, J.M., et al., Development and implementation of endocrine biomarkers of exposure and effects in American alligators (*Alligator mississippiensis*), *Chemosphere*, 37, 1905, 1998.
95. Vonier, P.M., Interaction of environmental chemicals with the estrogen and progesterone receptors from the oviduct of the American alligator, *Environ. Health Perspect.*, 104, 1318, 1996.
96. Bergeron, J.M., Crews, D., and McLachlan, J.A., PCBs as environmental estrogens: turtle sex determination as a biomarker of environmental contamination, *Environ. Health Perspect.*, 102, 780, 1994.
97. Olsen, N.J. and Kovacs, W.J., Gonadal steroids and immunity, *Endocr. Rev.*, 17, 369, 1996.
98. Hall, N.R. and Goldstein, A.L., Endocrine regulation of host immunity, in *Immune Modulation Agents and Their Mechanisms*, Fenichel, R.L. and Chirigos, M.A., Eds., Marcel Dekker, Inc., New York, 1984, 533–563, Chap. 26.
99. Day, R., Mercury in loggerhead sea turtles, *Caretta caretta*: developing monitoring strategies, investigating factors affecting contamination, and assessing health impacts, Master's thesis, University of Charleston, Charleston, 2003.
100. Heesemann, L.M. et al., Exposure to methylmercury (MeHg) *in vitro* alters lymphocyte proliferation in loggerhead turtles and bottlenose dolphin blood leukocytes, presented at Soc. Environ. Toxicol. Chem. 25th Annual Meeting, Portland, Nov. 14–18, 2004, 258.
101. Day, R. et al., Monitoring mercury in the loggerhead sea turtle, *Caretta caretta*, *Environ. Sci. Technol.*, 39, 437, 2005.
102. MacDougal, K.C., Johnson, M.D., and Burnett, K.G., Exposure to mercury alters early activation events in fish leukocytes, *Environ. Health. Perspect.*, 104, 1102, 1996.
103. De Guise, S. et al., Effects of *in vitro* exposure of beluga whale splenocytes and thymocytes to heavy metals, *Environ. Toxicol. Chem.*, 15, 1357, 1996.
104. Peden-Adams, M.M. et al., Relationship of lymphoproliferation and clinical blood parameters to heavy metals in Kemp's ridley sea turtles, presented at Soc. Environ. Toxicol. Chem. 24th Annual Meeting, Austin, Nov. 9–13, 2003, 240.
105. Norton, T.M. et al., Debilitated loggerhead turtle (*Caretta caretta*) syndrome along the southeastern U.S. coast: incidence, pathogenesis, and monitoring, presented at 25th Annual Symp. Sea Turtle Biol. Conserv., Savannah, Jan. 18–22, 2005.
106. Smith, G.M. and Coates, C.W., Fibro-epithelial growths of the skin in large marine turtles *Chelonia mydas* (Linneaus), *Zoologica (NY)*, 23, 93, 1938.
107. Williams, E.H., Jr. et al., An epizootic of cutaneous fibropapillomas in green turtles *Chelonia mydas* of the Caribbean: part of a panzootic?, *J. Aquat. Anim. Health*, 6, 70, 1994.

108. Balazs, G.H., Current status of fibropapillomas in the Hawaiian green turtle, *Chelonia mydas*, in *Research Plan for Marine Turtle Fibropapilloma*, Balazs, G.H. and Pooley, S.G., Eds., NOAA Technical Memorandum NMFS-SWFSC-156, U.S. Department of Commerce, Honolulu, HI, 1991, 47–48.
109. Aguirre, A.A. and Lutz, P.L., Marine turtles as sentinels of ecosystem health: is fibropapillomatosis an indicator? *EcoHealth*, 1, 275, 2004.
110. Aguirre, A.A. et al., Low-grade fibrosarcomas in green turtles (*Chelonia mydas*) from the Hawaiian Islands, presented at 21st Annual Symp. Sea Turtle Biol. Conserv., Philadelphia, Feb. 24–28, 2001.
111. Norton, T.M., Jacobson, E.R., and Sundberg, J.P., Cutaneous fibropapillomas and renal myxofibroma in a green turtle, *Chelonia mydas*, *J. Wildl. Dis.*, 26, 265, 1990.
112. Aguirre, A.A. et al., Monitoring the health and conservation of marine mammals and sea turtles and their ecosystems, in *Conservation Medicine: Ecological Health in Practice*, Aguirre, A.A. et al., Eds., Oxford University Press, New York, 2002, 79–94.
113. Lu, Y.N. et al., Identification of a small, naked virus in tumor-like aggregates in cell lines derived from a green turtle, *Chelonia mydas*, with fibropapillomas, *J. Virol. Methods*, 86, 25, 2000.
114. Herbst, L.H. et al., Experimental transmission of green turtle fibropapillomatosis using cell-free tumor extracts, *Dis. Aquat. Org.*, 22, 1, 1995.
115. Cray, C. et al., Altered *in vitro* immune responses in green turtles (*Chelonia mydas*) with fibropapillomatosis, *J. Zool. Wildl. Med.*, 32, 436, 2001.
116. Sposato, P.L., Lutz, P.L., and Cray, C., Immunosuppression and fibropapilloma disease in wild green sea turtle populations (*Chelonia mydas*), in *Proc. 20th Annual Symposium on Sea Turtle Biology and Conservation*, Mosier, A., Foley, A., and Brost, B., compilers, NOAA Technical Memorandum NMFS-SEFSC-477, U.S. Department of Commerce, Miami, FL, 2002, 152–153.
117. Work, T.M. et al., Immune status of free-ranging green turtles with fibropapillomatosis from Hawaii, *J. Wildl. Dis.*, 37, 574, 2001.
118. Aguirre, A.A. and Balazs, G.H., Blood biochemistry values of green turtles, *Chelonia mydas*, with and without fibropapillomatosis, *Comp. Haematol. Int.*, 10, 132, 2000.
119. Aguirre, A.A. et al., Adrenal and hematological responses to stress in juvenile green turtles (*Chelonia mydas*) with and without fibropapillomas, *Physiol. Zool.*, 68, 831, 1995.
120. Aguirre, A.A. et al., Organic contaminants and trace metals in the tissues of green turtles (*Chelonia mydas*) afflicted with fibropapillomas in the Hawaiian Islands, *Mar. Pollut. Bull.*, 28, 109, 1994.
121. Miao, X-S. et al., Congener-specific profile and toxicity assessment of PCBs in green turtles (*Chelonia mydas*) from the Hawaiian Islands, *Sci. Total Environ.*, 281, 247, 2001.
122. Arkoosh, M.R. and Collier, T.K., Ecological risk assessment paradigm for salmon: analyzing immune function to evaluate risk. *Hum. Ecol. Risk Assess.*, 8, 265, 2002.
123. Carlson, E.A., Li, Y., and Zelikoff, J.T., Exposure of Japanese medaka (*Oryzias latipes*) to benzo[a]pyrene suppresses immune function and host resistance against bacterial challenge, *Aquat. Toxicol.*, 56, 289, 2002.
124. Christensen, M.M. et al., Influence of mercuric chloride on resistance to generalized infection with herpes simplex virus type 2 in mice, *Toxicology*, 114, 57, 1996.

125. Peakall, D.B., Biomarkers: the way forward in environmental assessment, *Toxicol. Ecotoxicol. News*, 1, 55, 1994.
126. Kuriyama, S. and Chahoud, I., Maternal exposure to low dose 2,2′,4,4′,5 pentabromodiphenyl ether (PBDE 99) impairs male reproductive performance in adult rat offspring, *Organohalogen Compounds*, 61, 92, 2003.
127. Yang, Q. et al., Further evidence for the involvement of inhibition of cell proliferation and development in thymic and splenic atrophy induced by the peroxisome proliferator perfluoroctanoic acid in mice, *Biochem. Pharmacol.*, 62, 1133, 2001.
128. Darnerud, P.O., and Thuvander, A., Studies on immunological effects of polybrominated diphenyl ether (PBDE) and polychlorinated biphenyl (PCB) exposure in rats and mice, *Organohalogen Compounds*, 70, 415, 1998.
129. Zelikoff, J.T., Fish Immunotoxicology, in *Immunotoxicology and Immunopharmacology*, 2nd ed., Dean, J.H. et al., Eds., Raven Press, New York, 1994, 71–95, Chap. 5.
130. Harada, K. et al., The influence of time, sex and geographic factors on levels of perfluorooctane sulfonate and perfluorooctanoate in human serum over the last 25 years, *J. Occup. Health*, 46, 141, 2004.
131. deWit, C.A., An overview of brominated flame retardants in the environment, *Chemosphere*, 46, 583, 2002.
132. Janz, D.M. et al., Development of a reptile model for terrestrial wildlife ecotoxicological studies, presented at Soc. Environ. Toxicol. Chem. 23rd Annual Meeting, Salt Lake City, Nov. 16–20, 2002, 128.
133. Bishop, C.A. and Rouse, J.D., Chlorinated hydrocarbon concentrations in plasma of the Lake Erie water snake (*Nerodia sipedon insularum*) and Northern water snake (*Nerodia sipedon sipedon*) from the Great Lakes Basin in 1998, *Arch. Environ. Contam. Toxicol.*, 39, 500, 2000.
134. de Solla, S.R., Bishop, C.A., and Brooks, R.J., Sexually dimorphic morphology of hatchling snapping turtles (*Chelydra serpentina*) from contaminated and reference sites in the Great Lakes and St. Lawrence River basin, North America, *Environ. Toxicol. Chem.*, 21, 922, 2002.
135. Crain, D.A. et al., Alterations in steroidogenesis in alligators (*Alligator mississippiensis*) exposed naturally and experimentally to environmental contaminants. *Environ. Health Perspect.*, 105, 528, 1997.
136. Luster, M.I. et al., Development of a testing battery to assess chemical-induced immunotoxicity: National Toxicology Program's guidelines for immunotoxicity evaluation in mice, *Fundam. Appl. Toxicol.*, 10, 2, 1988.
137. Lance, V.A. and Elsey, R.M., Plasma catecholamines and plasma corticosterone following restraint stress in juvenile alligators, *J. Exp. Zool.*, 283, 559, 1999.
138. Morici, L.A., Elsey, R.M., and Lance, V.A., Effects of long-term corticosterone implants on growth and immune function in juvenile alligators, *Alligator mississippiensis*, *J. Exp. Zool.*, 279, 156, 1997.
139. Knotek, Z. et al., Renal disease haemogram and plasma biochemistry in green iguana. *Acta Vet. Brno*, 71, 333, 2002.
140. Grasman, K.A., Scanlon, P.F., and Fox, G.A., Geographic variation in hematological variables in adult and prefledgling herring gulls (*Larus argentatus*) and possible associations with organochlorine exposure, *Arch. Environ. Contam. Toxicol.*, 38, 244, 2000.
141. Luster, M.I. et al., Risk assessment in immunotoxicology II. Relationships between immune and host resistance tests, *Fundam. Appl. Toxicol.*, 21, 71, 1993.

142. Harms, C.A., Keller, J.M., and Kennedy-Stoskopf, S., Use of a two-step Percoll® gradient for separation of loggerhead sea turtle peripheral blood mononuclear cells, *J. Wildl. Dis.*, 36, 535, 2000.
143. Merchant, M.E. et al., Antibacterial properties of serum from the American alligator (*Alligator mississippiensis*), *Comp. Biochem. Physiol. B*, 136, 505, 2003.
144. Work, T.M. et al., Assessing humoral and cell-mediated immune response in Hawaiian green turtles, *Chelonia mydas*, *Vet. Immunol. Immunopathol.*, 74, 179, 2000.
145. Schumacher, I.M. et al., Detection of antibodies to a pathogenic mycoplasma in desert tortoises (*Gopherus agassizii*) with upper respiratory tract disease, *J. Clin. Microbiol.*, 31, 1454, 1993.
146. Origgi, F.C. et al., Enzyme-linked immunosorbent assay for detecting herpesvirus exposure in Mediterranean tortoises (spur-thighed tortoise [*Testudo graeca*] and Hermann's tortoise [*Testudo hermanni*]), *J. Clin. Microbiol.*, 39, 3156, 2001.
147. Brown, D.R. et al., Detection of antibodies to a pathogenic mycoplasma in American alligators (*Alligator mississippiensis*), broad-nosed caimans (*Caiman latirostris*), and Siamese crocodiles (*Crocodylus siamensis*), *J. Clin. Microbiol.*, 39, 285, 2001.
148. Mishell, R. and Dutton, R., Immunization of disassociated spleen cell cultures from normal mice, *J. Exp. Med.*, 126, 423, 1967.
149. Harper, N. et al., Halogenated aromatic hydrocarbon-induced suppression of the plaque-forming cell response in B6C3F1 splenocytes cultured with allogenic mouse serum: Ah receptor structure activity relationships, *Toxicology*, 99, 199, 1995.
150. Gross, T.S., personal communication, 2004.
151. Peakall, D.B., *Animal Biomarkers as Pollution Indicators*, Chapman & Hall, New York, 1992.

chapter 9

Reptilian Genotoxicity

Apolonia Novillo, Noppadon Kitana, Emily Marquez, and Ian P. Callard

Table of Contents

I. Introduction

A. Historical Perspective of Genotoxicity and Genetic Ecotoxicology

The beginnings of genotoxicity as an area of research may be traced to Muller's studies of gene mutability using radiation,[1] followed 20 years later by the studies of Auerbach and Robson[2] using chemicals. Subsequently, several authors demonstrated genetic changes in different organisms induced by chemicals and radiation, but genotoxicity was not recognized as a discipline until 1969 when the Environmental Mutagen Society was founded. The recognition of this discipline reflected a growing concern with the potential genetic impact associated with the proliferation of manmade chemicals on species other than man.

Genotoxicity can be defined in broad terms as a discipline concerning the ability of chemicals or radiation to interact with DNA and cellular apparatus that regulate the fidelity of genomic replication.[3] A chemical is considered to be mutagenic if it is capable of inducing heritable changes (mutations) in the genotype of a cell as a consequence of alterations to, or loss of, genes, chromosomes, or part of chromosomes; and the chemical is considered to be carcinogenic if it induces cancer.[4] Because pollution of the environment has become a major concern of our society, a definition of genetic ecotoxicology was adopted in 1994 after the Napa conference on Genetic and Molecular Ecotoxicology: "The study of chemical- or radiation-induced changes in the genetic material of natural biota. Thus changes can be direct alterations in genes and gene expression or selective effects of pollutants on gene frequencies." However, it also includes epigenetic effects and changes in gene pools attributable to chemical exposures. The relevance of this new discipline in relation to pollution, and from an individual and ecologic point of view, in particular for reptilian communities, is that the long-term effects of environmental genotoxic substances, as well as the effects of increasing ultraviolet-B (UV-B) radiation as a consequence of ozone depletion, are largely unknown.[5] Until the risks associated with the exposure of reptilian populations to genotoxic chemicals have been evaluated and investigated in detail, it is important to adopt a cautious approach in managing the exposure of reptilian species to genotoxic chemicals or radiation, given their importance in terrestrial and aquatic ecosystems.

Researchers in genotoxicity have begun to capitalize on the rapidly advancing technology of molecular biology; in human health, the consequences of exposure to genotoxic substances have been intensively studied for decades. In contrast, the number of studies of animal populations, and in particular reptiles, has been fragmented at best. Given the extent of environmental degradation and exposure of both terrestrial and aquatic reptiles to toxic insult, it is important to assess genotoxicity in this important vertebrate group.

Unlike human studies of genotoxicity where the concern is the individual, ecologic studies deal with genetic effects in natural populations as influenced by the exposure. The most important negative outcomes of concern after the exposure to genotoxic substances are as follows: gamete loss as a result of cell death; embryo mortality (lethal mutations); abnormal development; neoplasia; and heritable mutations, which can cause changes in genetic diversity and gene expression that affect Darwinian fitness (such as altered growth rates, reproductive output, and viability of offspring).[5,6]

B. *Genotoxic Compounds and Agents in the Environment*

In Chapter 3, the uptake, accumulation, and distribution of xenobiotics are discussed. Here we will discuss the sources and types of genotoxic compounds in the environment to understand how reptilian populations may be exposed to genotoxic agents.

A wide array of synthetic chemicals are produced by man: (1) as pharmaceutical products (used as antitumor agents, narcotics, contraceptive hormones, antibiotics, anesthetics); (2) as pesticides; (3) as additives (food, plastic, others); and (4) as by-products of industry (used in the industry or occurring in the environment) such as heavy metals, alkylating agents, organic solvents, oil, and so on. These substances are released into the environment — "unintended," accidentally, or as disposal of wastes and deliberate application of biocides — on the order of megatons per year.[7] In addition, more than 500 new compounds appear in the market every year. A subset of these compounds is classified as genotoxic based on their ability to damage the DNA, produce cancer, and induce heritable changes. Thus, present-day contamination is characterized by long-term exposure of organisms to low doses of complex chemical mixtures from wastes, soil leachates, and atmospheric deposition.

It is difficult, if not impossible, to compile an exhaustive list of genotoxic compounds and to propose a general classification of genotoxic compounds. Most are substances of anthropogenic origin that are synthesized by man used in industry, or occurring in the environment as by-products, but some are naturally occurring genotoxic substances (such as alkaloids, products of bacterial microorganisms). A tentative list of genotoxic compounds (in particular carcinogenics) is available in the International Agency for Research on Cancer

(IARC) Monographs on the Evaluation of Carcinogenic Risks to Humans (http://www.iarc.fr/). Suspected agents are categorized by IARC as carcinogenic (Group 1), probably carcinogenic (Group 2A), or possibly carcinogenic (Group 2B) to humans. Compounds for which there is inadequate evidence are classified as Group 3 and Group 4 (probably not carcinogenic). This categorization reflects the strength of evidence relative only to the human, but it does not indicate carcinogenetic potency. According to supplementary IARC Monographs, some suspected carcinogenic agents in the environment are as follows: (1) metals (arsenic, nickel, cadmium, and lead); (2) benz[a]anthracene, benzo[a]pyrene, anthracene, and chrisene; (3) chlorinated organics, such as polychlorinated biphenyls, DDT (dichlorodiphenyltrichloroethane), 1,2-dichloroethane, lindane, dieldrin, aldrin, and trichloroethylene; and (4) oil dispersants and hydrocarbon solvents.

To this list of chemicals, we have to add the genotoxic potential of water- and airborne industrial wastes and effluents. These undesirable by-products of economic development and technologic advancement represent more than 384 million tons of waste annually,[8] and approximately 275 million metric tons are classified as hazardous. Houk[9] reviewed the data available on the genotoxicity of industrial wastes and effluents and provides an assessment of the genotoxic burden that industrial wastes place on the environment. Diffuse air and water pollution consists of an omnipresent complex mixture of pollutants that is emitted from many widely dispersed sources such as automobile traffic, industries, households, energy plants, waste incinerators, and agriculture. Water- and airborne pollutants can be deposited in relatively remote areas as a result of long-range transport.

Over the century, another group of genotoxicant agents such as radioactive materials has been generated. They have been used in production of weapons, generation of electricity, industry, research, and so on, and huge amounts of radioactive waste have been created. As a consequence thousands of cities in the United States have been contaminated (reviewed by Wolbarst and colleagues[10]). In the United States the current plans for permanent disposal of radioactive waste have some uncertainties[11] principally with concern about the effect of low-level radiation on natural biota.[12] In a wide scale, contamination from the Chernobyl accident has generated growing concern about the presence of radiation in the environment.[13]

C. *Linking the Mechanism of Genotoxicity with Its Consequences*

From an ecologic point of view, the potential for exposure to genotoxic substances is of serious concern.[3] The evaluation of genotoxicity is particularly important because of the delayed appearance of the genotoxic effects that may require months or years to be fully manifest and because later genotoxic events may be crucially important at the population and community level. A unique aspect of genotoxicity in somatic cells is the potential to lead to cellular dysfunction and eventually cell death. Further, tumors may appear, in which case the genotoxic substance is also carcinogenic.[14] In contrast, genotoxic effects in germ cells can be passed on to future generations; thus, genotoxicity can result in rapid alterations in gene frequencies (relative to normal evolutionary rates) in natural populations. The ecologic consequences are unknown or poorly understood but are likely serious.[15–18]

Exposure of an organism to a genotoxic agent usually disrupts normal cellular processes and can result in direct interactions of the toxic agent with DNA, inducing structural modifications of the DNA. If the agent is a physical entity such as UV light or ionizing radiation, the possibility exists for immediate and structural damage to the DNA molecule, inducing breaks, adducts, mutation, and dimers.[16] If the genotoxicant is a chemical, it must first be made available to the organism. If the cellular mechanisms of detoxification (e.g., biotransformation, bioactivation, excretion, sequestering, etc.) are successful, no DNA damage may be observed. The cellular barriers to toxicity of the chemical of its metabolites may be circumvented, however, resulting in DNA changes. Structural modification to DNA does not always lead to damaging sequelae. Thus, damage will have no further consequences if it is repaired or if it is provoked in regions of DNA that are not functionally important and therefore cannot produce adverse effects.

Some genotoxic agents do not interact directly with the DNA, but exposure of an organism to these compounds can result in a genotoxic effect. This mechanism is named indirect mechanism of genotoxicity.[19] So far these indirect mechanisms are thought to involve aneugenic activity, oxidative stress, inhibition of enzymes involved in DNA synthesis such as topoisomerase or cytotoxicity, lipid peroxidation, and protein adducts. Recent research in this area is focused on inhibition of repair enzymes (e.g., 8 oxoguanine DNA glycolsylase (OGG1), xeroderma pigmentosus (XPD)), cell-cycle control proteins (e.g. p53, retinoblastoma, cyclins), apoptosis-related gene products (e.g., p53, bax, bcl-2), nuclear lamins, oxidative defense proteins (glutathione), metabolizing enzymes, and tubulins of the mitotic and meiotic spindle.[19] It is accepted that genotoxic effects via indirect mechanisms demonstrate a threshold, an important issue for regulatory agencies.[20]

The exposure of an organism to genotoxic chemicals may induce a cascade of events at the individual and population levels (see Figure 9.1, adapted from references 16–18, 21), which are not always easy to detect and correlate with the causative agent. In summary, at the individual level, (1) initially, the genotoxicant or its metabolites can interact with DNA-targets (by direct mechanisms) and non-DNA targets (by indirect mechanism), (2) followed subsequently by the appearance of cytogenetic alterations and pathology in the organism.

At the population level, three types of genotoxic syndrome can be identified: the presence of malignant tumors; decreased reproductive success; and altered genotypic diversity, resulting in long-term consequences for the population, which could range from alteration in genotypic diversity, to possible population extinction.[16–18,21]

Thus, the organism functions as an integrator of toxic exposure, accounting for abiotic and physiologic factors that modulate the dose of genotoxicant. The inability of the organism to cope with this stress and to maintain structural and functional integrity of DNA provides the investigator with the opportunity to test the severity of the exposure and use various events as biomarkers that can be used to test for the genotoxicity of agents in the environment.[16,22] In addition to structural changes in DNA, the detection and quantification of several events in this cascade (see Figure 9.1) can be used as biomarkers of exposure and effects in organisms and populations environmentally exposed to genotoxic agents.[17]

II. The Use of Reptilian Species in the Study of Aquatic and Terrestrial Genotoxicity

Reptiles are an important vertebrate class and comprise a large percentage of the faunal biomass in many terrestrial and aquatic ecosystems. The estimated number of species is 6000.[23] They are predators and prey of vertebrates and invertebrates, and their unique life stories make their roles in food webs diverse and important. In relation to biomonitoring, reptiles are particularly suitable because of their persistence in a variety of habitats; wide geographic distribution; longevity; and, in many cases, site fidelity.[24] Furthermore, reptiles exhibit a sensitivity to contaminants similar to that reported for birds and mammals,[25–27] and they bioaccumulate and biomagnify contaminants to levels equal to or greater than that reported for birds and mammals.[26,28–33]

Besides these helpful characteristics, the literature shows that reptiles are perhaps the least investigated group with regard to environmental

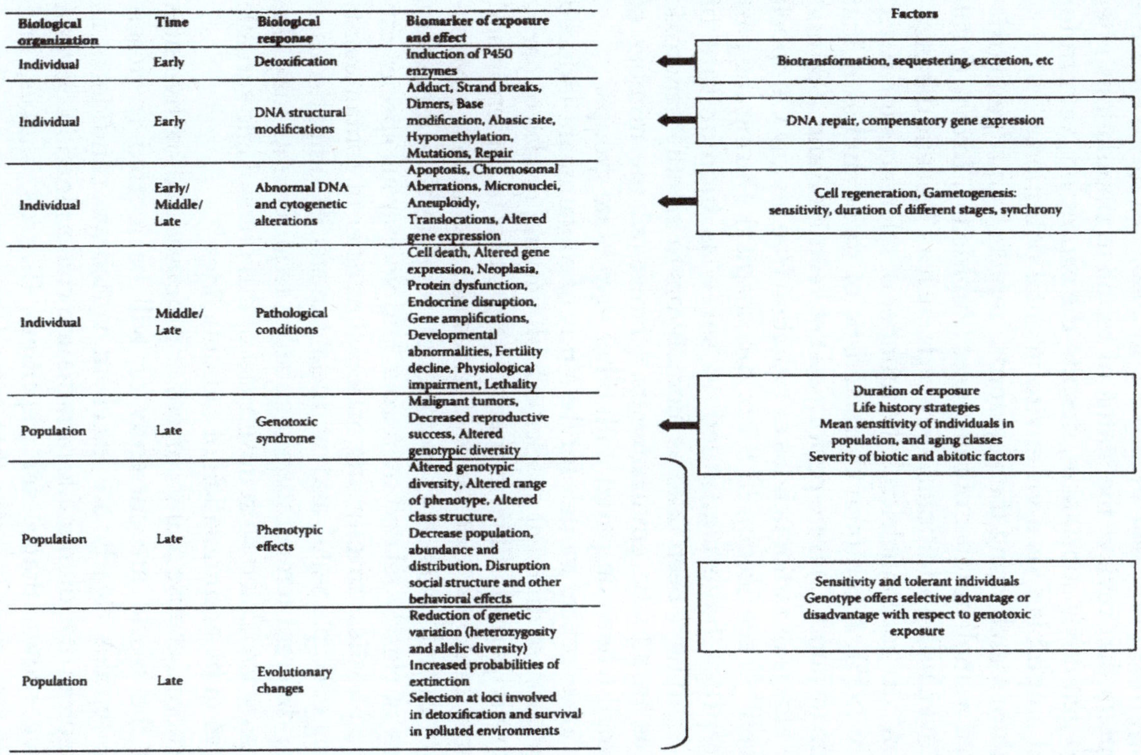

Figure 9.1 Schematic representation of a possible temporal sequence of steps leading from the exposures of organism to genotoxicants to the expression of evolutionary effects and description of potential factors that will determine the consequences of exposure to genotoxicants. (Figure adapted from Shugart,[16] Anderson et al.,[17] Bickman and Smolen,[18] and Llorente.[21])

contamination studies (see review in Campbell and Campbell[33]). The first-ever session devoted exclusively to reptile toxicology was held at the meeting of the Society of Environmental Toxicology and Chemistry's (SETAC) 2000 meeting in Nashville (TN, USA), following which a state-of-the-art review and synthesis of amphibian and reptile ecotoxicology was published by SETAC in 2000.[30] In this book there are no chapters that review the status of reptilian populations after exposure to genotoxic substances, despite the expressed concern for reptiles and amphibian vis á vis environmental pollution.

A literature search using the keywords "reptiles" and "genotoxicity" retrieved a small number of references (PubMed: 1, Toxnet: 6; Fall 2004). However, using additional keywords such as "genetic damage" or "radiation" or "radionucleids," the number of references increased considerably (>35). A common characteristic of these studies is that most of them (around 22) were performed between 1960 and 1980 and concern the effects of radionuclides or radiation (see review Campbell and Campbell[31–33]; see Section III.B.2 of this chapter). During the past 20 years reptiles have continued to serve as biomonitors of radiation[12,34] and have been used as biomonitors to identify genotoxic contaminants in aquatic environments.[35] However, the number of studies is small when compared with other vertebrate groups.

Reptiles are excellent models for the study of contaminant-induced genotoxicity because different species exhibit varying modes of parity (oviparity or viviparity), differ in thermal regulation (ectotherms and poikilotherms), inhabit a wide range of terrestrial and aquatic environments, have varied diets, and experience considerable longevity. Comparisons of closely related oviparous and viviparous species can elucidate complexities surrounding maternal transfer of contaminants by comparing yolk, placental, and oviductal transfer.[24] Comparisons of reptiles with different temperature regulatory requirements may elucidate processes associated with the impact of this physiologic variable on differences in bioaccumulation of radionuclides.[36]

Furthermore, because many current or proposed radioactive sites throughout the world are currently or will be located in desert ecosystems, reptiles clearly are excellent vertebrate candidates as ecologic receptors of radionuclide or radiation contamination. Several authors have shown that desert reptilians (lizards, snakes, and tortoises) are abundant in areas that were formerly used for above- and below-ground nuclear testing. Considering they have small home ranges that these species exhibit, they may be continually exposed to radiation and potentially are excellent models for ecologic risk assessment relative to radiation exposure.[12,33,36]

A. Biomarkers of Exposure and Effect in Reptilian Populations

In Chapter 4, the current status of tools for assessing contaminant exposure and effects on reptile are reviewed in detail. However, because exposure to genotoxic agents may cause unique and distinctive effects on DNA, we will describe the array of techniques currently available to assess DNA damage and discuss the limitations of these approaches. Because another outcome of exposure to genotoxic chemicals is carcinogenesis, the incidence of neoplasia in reptiles in relation to contamination will be discussed.

1. DNA Damage as a Biomarker of Exposure

Some environmental chemicals and physical agents are classified as genotoxicants because they have the ability to interact with and damage the structure of DNA, often with negative outcomes for the cell and organism.[22,37,38] Thus the specific type of structural DNA damage is considered a biomarker of exposure. To detect DNA damage, there is an array of techniques available, for which sensitivity and applicability to different species from mammals to fish to invertebrates have been discussed by Shugart[16] and Hebert and Luiker[39] and are summarized in Table 9.1.

Table 9.1 Current Markers and Techniques Available to Analyze DNA Damage

Marker Type	Biomarker	Technique
Structural	Adducts	^{32}P post labeling
		Immunoassays: RIA, ELISA, USRA
		HPLC combined with fluorescence
		Gas chromatography
	Base changes	Hypomethylation
	Strand breaks	Alkaline unwinding assays, comet assay, alkaline elution, DNA precipitation, Electrophoresis agarose
Irreversible events	Chromosomal aberrations	Flow cytometry, FISH, CGH Micronucleus assay
	Sister chromatid exchanges	FISH
	Micronuclei	
	Aneuploidy	
	Gene mutations	Mutation rates
		Oncogene activation
		RAPD, AFLP, RFLPs, SSR

USRA: ultra sensitive radioimmunoassay; FISH: Fluorescence in situ Hybridization, CGH: Comparative Genomic Hybridization, RAPD: Random Amplified Polymorphic DNA, AFLP: Amplified fragment Length Polymorphism, RFLPs: Restriction Fragment Length, Polymorphism, RIA: Radioimmunoassay, ELISA: Enzyme-Linked Sorbent Assay, SSR: Single Sequence Repeats

Table 9.2 DNA Damage as Biomarker of Exposure and Effect in Studies with Reptilians

Biomarker	Technique	Species	Tissue	Environment	Reference
DNA strand break	Alkaline unwinding assay	Trachemys scripta, Chelydra serpentina	Liver	Reservoir	Meyers-Schöne et al.[35]
Chromosomal aberrations	Flow cytometry	Pseudemi scripta	Blood	Reservoir	Bickham et al.[34]
DNA content Aneuploid mosaicism	Flow cytometry	Trachemys scripta	Blood	Reservoir	Lamb et al.[12,40]
Chromosomal translocation	FISH	Trachemys scripta	Lymphocytes	Laboratory	Ulsh et al.[41,54,55]
Micronuclei	Micronucleus assay	Emys orbicularis	Several tissue	Reservoir	Swartz et al.[57]

FISH, fluorescence *in situ* hybridization.

In studies with reptiles, some of these techniques (flow cytometry, alkaline unwinding assay, fluorescence *in situ* hybridization [FISH], and micronucleus assay) have been developed and used by different authors (summarized in Table 9.2) for detecting exposure to and in some cases adverse effects of radionuclides in reptilian populations.[12,34–35,40,41] The approach used in these studies was to measure genetic damage in the form of aneuploidy, DNA content, DNA strand break, micronucleus, chromosomal aberrations, and chromosome translocations (Table 9.2). It should be recognized that few investigations, with approaches similar to those listed here, have been initiated to assess the environmental impact of genotoxins in reptilian populations. The salient features of these studies are summarized in Table 9.2 and consist of the following: (1) only freshwater environments were investigated, (2) different changes in the DNA damage were measured that included both structural and irreversible events, and (3) sensitive and selective analytical methodologies were used to detect these changes in DNA integrity. These studies will be discussed in more detail later in this chapter.

Despite limitations, DNA damage is considered a good biomarker of exposure[16] and has been used in biomonitoring studies in different species. For this reason, developments in molecular biology during the past years have focused on new possibilities for detecting DNA damage. The more classical and modern approach to population genetic analysis is to examine protein polymorphism (allozyme analysis). The revolution of DNA sequencing and polymerase chain reaction (PCR) has revolutionized genetic toxicology. Examples of such new techniques include restriction fragment length polymorphism (RFLPs), single sequence repeats (SSR), amplified fragment length polymorphism (AFLP), or random amplified polymorphic DNA (RAPD). The RAPD technique may reveal differences in the DNA fingerprints of individuals from control and polluted sites, providing a useful alternative biomarker assay for detection of genotoxic effects.[42,43] Another set of new approaches allows the detection of irreversible changes and can detect single genes and centromeres by using FISH to directly identify a specific DNA region. A variation of the FISH assay is the comparative genomic hybridization (CGH) technique. This approach can potentially demonstrate all regions of amplification and deletion in the genome and allows comparison of one genome with another. The principles of these techniques can be found elsewhere (see Kallioniemi and colleagues[44]).

As we indicated earlier, promising protocols for the detection of DNA damage, using reptilian species, are flow cytometry (FCM) and micronucleus assay techniques. The FCM procedure has been demonstrated to be a sensitive indicator of clastogenic damage in both laboratory and field studies using reptilians (see review in Lamb and colleagues[40]). In contrast, micronucleus assay is simple and allows the detection of both clastogens and aneuploidy-inducing chemicals. This procedure has been considered to be useful as a new genotoxicity assay.[45,46] In particular, these new techniques potentially signal new ways to monitor the exposure of reptilian populations to genotoxicants.

2. *Neoplasia as a Biomarker of Effect*

Neoplasia in fish and aquatic mammals has been strongly associated with polluted environments,[43,47] and a casual relationship between pollution and neoplasia in marine invertebrates is indicated by some authors.[48] In the case of reptiles, the scarcity of studies makes it almost impossible to correlate pollution with the appearance of neoplasia. Campbell and Campbell[31] indicated that snakes that have been exposed to high levels of radionuclide developed necrosis of the liver and pancreas. Another controversial type of neoplasia is the fibropapillomas of green sea turtles. Some authors relate these neoplasias to herpesvirus, and others ascribe them to indirect consequences of pollution (see Section III.A.3 and Chapter 8).

The problem of identifying the causative agents of neoplasia in reptiles has yet to be overcome. Reviews of reptile neoplasias have been published,[49] but many of these cases are in captive animals, and viruses have been implicated as causative agents. Done[49] suggests that neoplasia may be developed in any organ system and in all reptilian species, although snakes appear to be more susceptible.[50]

III. Reptilian Models for Studies of Genotoxicity

A. Chelonia

Animals in the order *Chelonia* (or *Testudine*) are shelled reptiles commonly known as turtles, tortoises, and terrapins. Chapter 2 covers the diversity and global status of these species in further detail. The potential for exposure and reactions to genotoxic agents depends on life history and pattern of movement of a species in different habitats.[23] Most turtles have a life history characterized by slow growth, late maturity, and long life, allowing long-term bioaccumulation of pollutants, which places them at great risk in the current chemical environment of the world.[52] The life history traits of turtles make them well suited as sentinel species based on the characteristics suggested by the National Research Council.[53] These include measurable response to the agent in question, territory or home range that overlaps the area to be monitored, relatively easily enumerated and captured, and of sufficient population size and density to permit enumeration. The uptake and transport of contaminants is more fully addressed in Chapter 3.

Turtles have been used as model species for genotoxicity in several ways, ranging from studies of direct DNA damage to alterations in reproductive and immune systems, as discussed in the following examples.

1. Direct DNA Damage

Biodosimetry or an estimation of doses by determining the frequency of radiation-induced chromosome aberrations has been developed using the red-eared slider turtle (*Trachemys scripta*) as a model. *In vitro* dosimetry was established in an embryonic fibroblast cell line using a FISH painting probe. In regard to induction of chromosome interchange aberrations, the dose-response curve for ^{137}Cs gamma irradiation showed that red-eared slider fibroblasts were ~1.7 times less sensitive than human fibroblasts.[41] However, using cultured lymphocytes, whole-genome spontaneous background level of symmetric translocations in turtle cells was 6 to 25 times less than human lymphocytes. This relatively low background would be a significant advantage for resolution of effects at low doses and rates. A dose-response curve of *in vivo* dosimetry using cultured lymphocytes from turtles chronically exposed to radiation and whole-chromosome FISH painting probe was established.[54] It was suggested that the frequency of symmetric chromosome translocations in peripheral blood lymphocytes is suitable as a biomarker of cumulative radiation exposure, and because symmetric chromosome translocations may affect reproductive success, this could prove to be ecologically relevant as well.[55]

Genetic effects of low-level radiation on the natural population of turtles were studied in red-eared sliders living in catchment basins at the Savannah River Plant (SRP), a nuclear materials production facility of the U.S. Department of Energy in South Carolina, which is also included on the National Priority List (Superfund site) by U.S. Environmental Protection Agency. Turtles that lived in seepage basins with histories of radioactive and nonradioactive influent for 7 to 28 years showed high body burdens of ^{90}Sr and ^{137}Cs. Because radioactive strontium is usually deposited in calcareous skeletal tissue as a calcium analog, the turtle shell enhances the retention rate of radionuclides and prolongs radiation exposure.[56] Using FCM to detect variation in DNA content of cell nuclei, significantly higher variation in DNA content in red blood cells was found in turtles from these basins compared with turtles from a control population. It was concluded that radiation or some unidentified chemical in seepage basins causes chromosomal rearrangements leading to deletions and duplications that have the effects of increasing DNA content variation in blood cells.[34] These results were confirmed by additional FCM analysis on spleen cells by using multiple-tissue assay in seepage basin turtles.[40] Similarly, greater variation of DNA content in blood cells was also found in turtles living in radioactive reservoirs that had lower radiation levels, and the contamination is strictly radiologic without effects of nonradioactive pollutants.[12]

The integrity of DNA as a biological marker of exposure to genotoxic agents was examined in populations of the common snapping turtle (*Chelydra serpentina*) and the red-eared slider turtle (*T. scripta*) living at White Oak Lake, a settling basin for low-level radioactive (^{90}Sr, ^{137}Cs, ^{60}Co, ^{3}H) and nonradioactive (Hg) wastes generated at the Oak Ridge National Laboratory of the U.S. Department of Energy. Turtles from a contaminated area showed a significantly higher body burden of radionuclides and mercury as well as significantly increased single-stranded breaks in liver DNA, indicating a more severe genotoxic stress than in turtles from a reference site. It was concluded that both turtle species were effective monitors of the genotoxic contaminants.[22,35]

Micronuclear assay, one of the most common assays for genotoxicity, was performed in the European pond turtles (*Emys orbicularis*) from the industrialized zone of Sumgayit, Republic of Azerbaijan. The area suffered widespread contamination through spill, discharge, run-off, and aerial dispersal of several potential genotoxicants, including mercury, by chemical factories. Pond turtles from the contaminated site showed a significantly higher level of 11 chemicals or chemical classes in tissues than did turtles from the reference site. Although the elevated micronuclear number in turtle blood smears was not statistically different from the reference site, elevated micronuclear number in contaminated site turtles was significantly correlated with the elevated tissue levels of mercury, heptachlor, dichlorodiphenyldichloroethone (DDD), hexachlorobenzene (HCB), and transnonachlor.[57]

2. *Potential Reproductive Impairments*

The painted turtle (*Chrysemys picta*) has been used to study potential impacts of contaminants originating from the Massachusetts Military Reservation (MMR), a Superfund site on Cape Cod, MA. Assessment of contamination in the areas showed several genotoxic agents at above ecotoxicology benchmarks, including volatile organic compounds (trichloroethylene, tetrachloroethylene, ethylene dibromide), semivolatile organic compounds (benzo(a) pyrene), pesticides (DDT), polychlorinated biphenyls (PCBs), and heavy metals (As, Be, Cd, Pb, Ni) in ground water, surface water, and sediments.[58] Using hepatic enzymes involved in the two-phase process of xenobiotic biotransformation as biomarkers, it was found that turtles from the affected site showed significantly higher activities of ethoxyresorufin-*O*-deethylase (EROD) and glutathione-S-transferase (GST) as well as higher expression of cytochrome P4501A protein,[59] indicating exposure to organic contaminants[60,61] (see also Chapter 4 and Chapter 5). The affected-site animals also showed a higher concentration of cadmium in liver, kidney, and reproductive tract as well as a higher level of metallothionein-like protein, indicating exposure to heavy metal toxicants.[59] Further investigations provided cause-and-effect association with significant reproductive impairment of animals from the affected site, including lower Gonadosomatic Index (GSI) and sperm number in males and lower oviduct weight, follicle number, levels of plasma estradiol, and vitellogenin in females.[62,63] In additional microcosm studies, neonate red-eared sliders native to Cape Cod environments exposed to sediment and water from the affected or reference site for 1 year showed that gonial proliferation was lower and apoptosis was much higher in male turtles raised under the affeced-site environment than turtles maintained under reference site conditions. Although there are no differences in proliferation or cell death in females between sites, some females raised under affected-site conditions showed a gross gonadal abnormality. The results indicate that the lower sperm and oocyte number found in adult turtles may be the result of slow gonial progression and an enhanced rate of apoptosis during gametogenesis as a consequence of low-level exposure to environmental xenobiotic mixtures.[63] However, there is no current direct evidence of genotoxicity in these studies.

The potentially genotoxic effect of cadmium on gonadal development was studied using red-eared slider embryos as a model. Cadmium is a candidate toxin in the study because female painted turtles from the affected area near MMR showed a higher concentration of cadmium in the reproductive tract.[59] A study on distribution of isotopic cadmium in adult female *C. picta* after 8 days also showed a significant amount of cadmium in reproductive endocrine tissue (follicular wall and corpora lutea, 0.47% of the total injected dose), yolk (0.23%), oviductal wall (1.34%), and oviductal egg (0.45%), indicating a possible role of maternal cadmium transfer.[64] After 10 days of *in ovo* exposure to cadmium chloride at approximately 62.9 ppb of total cadmium, the total number of germ cells in the genital ridge of early stage red-eared slider embryos was significantly reduced in the treated embryos compared with the germ cell population in the control embryos. The results suggest that exposures to an environmentally relevant concentration of cadmium may reduce proliferation or delay migration of germ cells into the genital ridge, which may cause reproductive impairment later on in adult life,[63] possibly through early developmental genotoxicity.

3. *Potential Immune Suppression*

In higher vertebrates, DNA adduct formation (binding of pollutants to DNA) is thought to be the early event leading to neoplasia and malignancy; in general, neoplasia or malignant tumors are uncommon in lower vertebrates.[15] However, the incidence of fibropapillomatosis, characterized by single to multiple histologically benign fibroepithelial tumors, has been reported in several marine turtle species. These lesions are generally called green turtle fibropapillomatosis (GTFP) because of their presumed origin, the highest incidence being observed in this species[52] (see also Chapter 8). GTFP was first reported in 1938 in captive green turtles (*Chelonia mydas*)[65] and subsequently observed in other marine turtle species, including olive ridleys (*Lepidochelys olivacea*), flatbacks (*Natator depressus*), and loggerheads (*Caretta caretta*).[66] Although the GTFP is not cancerous, its excessive growth internally and externally is life threatening to turtles.

Experimental results suggest a filterable infectious agent, probably a virus, as the primary cause of GTFP.[67] The papillomas have never been reported in the youngest juveniles of the pelagic phase but increase in populations in the larger size class living in near-shore waters.[52] Several field studies also suggest association of high GTFP prevalence with marine habitats that have been affected by agricultural, industrial, or urban development. This has led to speculation that environmental contaminants may play a role in GTFP pathogenesis. Current hypotheses for environmental contaminant effects in GTFP include cocarcinogenesis (induction of latent virus infections) and contaminant-induced immune suppression.[68] However, no evidence presently supports a cause-and-effect association.[52]

B. *Squamates*

Squamates are an extremely diverse group of reptiles, constituting 96% of extant reptile species.[23] The order Squamata is composed of amphisbaenids, lizards, and snakes, and Chapter 2 covers the diversity and global status of these species in further detail. Amphisbaenids, lizards, and snakes demonstrate either oviparity or viviparity, and viviparity is exclusive to extant species of squamates.[73] The diversity of lifestyle strategies makes squamates important candidates for genotoxic studies. However, this same diversity confounds any attempt to use a single reptile species as a model. Different life strategies could affect length of exposure to a genotoxic substance, the nature of the effects elicited, and thus the ability of an individual species to adapt to the presence of genotoxic substances. Additionally, the diversity of life strategies must be considered when evaluating potential negative effects of genotoxic substances on reptile species.

1. *Tumorigenesis*

Tumorigenesis in squamates has not been extensively studied. Although environmental contaminants have been suggested as possible carcinogens, there are no studies examining the relationship between environmental contaminants and tumors in squamates. Viruses can cause neoplasia in lizards, as in the case of a papovavirus causing papillomatosis in the European green lizard (*Lacerta viridis*) and a poxvirus causing brown papules in a tegu (*Tempinambis teguexin*).[77] In snakes, C-type virus particles have been found in the tumor tissue of a chrondrosarcoma in a corn snake (*Elaphe guttata*).[78] Viruslike eosinophilic intranuclear inclusions were found in a lymphosarcoma in a California king snake (*Lampropeltis getulus californiae*).[79] Up to now the focus has been on tumorigenesis in individuals kept in zoos or as pets; however, future studies should also examine wild populations located in sites contaminated with potential carcinogens.

2. *Radiation Studies*

Eleven studies were conducted with lizards continuously exposed to ^{137}Cs (gamma radiation) in the Mojave Desert at the U.S. Department of Energy Nevada Test Site. Exposure to radiation caused sterilization in female side-blotched lizards (*Uta stansburiana*), longnose leopard lizards (*Gambelia wislizenii*), zebra-tailed lizards (*Callisaurus draconoides*), desert horned lizards (*Phrynosoma coronatum*), and western whiptails (*Cnemidophorus tigris*). Initially no effects were found; however, longer term studies indicated that females and some males in all of the species became sterile over time.[80–82] All female longnose leopard, zebra-tailed, and desert horned lizards became sterile, often with hypertrophied fat bodies and an absence of ovarian tissue.[80,81,83,84] Some female side-blotched lizards and western whiptails became sterile; usually older females were affected in this manner.[80,81,83,84] The species where all the females became sterile were longer lived, attained sexual maturity later in life, and had low fecundity, whereas the species less affected were shorter lived, had earlier sexual maturity, and high reproductive capacity. The longer-lived species became extinct from the area the study was conducted. Side-blotched lizards maintained their population because year-old lizards were able to successfully reproduce before they accumulated dosages of radiation, whereas the older lizards of the same species had a longer exposure period.[80]

Two laboratory studies have been conducted examining the effects of X-irradiation on side-blotched lizards. In one, lizards were irradiated in the laboratory and returned to a study area, where a decreased number of young were produced a year later, resulting in a decrease in the population.[85] Sterility occurred at doses lower than the $LD_{50/30}$ of experiments conducted in the laboratory.[86]

3. *Suggestions for Future Models*

Using a single model for squamates is especially problematic as a result of the diversity of their life histories. The results from the studies conducted at the Nevada test site, where only females relatively longer-lived of five species of lizard were negatively affected by radiation, demonstrate this fact.[80,81,83,84] Future models would therefore have to be chosen with regard for the wide diversity of life strategies among squamates, especially in the case of field studies. The type of sex determination could also result in variable responses to environmental contaminants. Although exposure to estradiol during incubation results in a female-biased sex ratio in red-eared slider turtles and the leopard gecko, these experiments have been conducted in species with temperature sex determination. It is likely that species where sex ratio responds to environmental cues such as temperature could also be more susceptible to environmental contaminants, whereas species with genetic sex determination could be more resistant to environmental contaminant exposure. Finally, the longevity of a species can also affect the data obtained in field studies, as demonstrated by the Nevada test site studies. Length of exposure may be the major determinant in whether a population is able to sustain itself. For instance, populations where the majority of animals are young and sexually mature, as in the case of the side-blotched lizard, may be able to reproduce and sustain their population after exposure to environmental contaminants. Species that take longer to attain sexual maturity will have a longer exposure time and more opportunities to accumulate harmful dosages of environmental contaminants, which could in turn negatively affect their ability to produce viable offspring.

C. *Crocodylia*

Modern crocodilians consist of 25 species in three families distributed throughout the world's tropics, subtropics, and slightly in temperate zones. Chapter 2 covers the diversity and global status in more detail.

Compared with freshwater turtles, which are found in similar habitats, crocodilians not only share a common route of exposure to genotoxic substance but also are better models for biological magnification in ecosystem because they are long -lived, are top predators, and consume primarily aquatic prey.[52,53] This character was assessed in American alligators (*Alligator mississippiensis*) living in the Florida Everglades by analyzing bioconcentration factor (BCF), the ratio of concentration of chemicals in the organism to its concentration in the surrounding media. BCFs for mercury in alligator liver and kidney compared with value in water column were found to be very high: $39.9 \infty 10^{7}$ and $32.9 \infty 10^{7}$ in adults and $10.5 \infty 10^{7}$ and $9.34 \infty 10^{7}$ in juveniles.[87] However, for obvious reasons (availability, handling, housing, etc.) adult alligators are not suited as models, though they added drama to the field. Laboratory-reared hatchlings, however, may be expected to yield important data on the responses of this group when exposed to genotoxic substances. Furthermore, because most of the crocodilians are listed as threatened species by the World Conservation Union (IUCN; 13 of 25 species[88]) and are protected by the U.S. Endangered Species Act (20 of 25 species[89]), invasive techniques for an extensive genotoxic study are unlikely. Most if not all previously published research on contamination of potential genotoxic substance in crocodilians are dealing with contaminant tissue burden, with only few on measurable effects of exposure on biological functions (see review in Bishop and Martinovic[90]). Among these studies, *A. mississippiensis* is the most popular model as a result of availability of specimens because it is a regionally harvested game animal. Several noninvasive techniques to detect contaminant tissue burden have been developed and validated in this species. Examples of these studies include tissue mercury concentration in dermal scutes,[91] serum organochlorine pesticides and PCBs,[92] organochlorine pesticides and PCBs in eggs,[93–95] and organic contaminants in the chorioallantoic membrane of the egg.[93,96,97] These nonlethal techniques are of importance in potential assessment of bioaccumulation of genotoxic substances in other crocodilians. Some of these techniques were successfully used to study organic contaminants in eggs and chorioallantoic membranes of an endangered Morelet's crocodile (*Crocodylus moreletii*).[96,98] To further provide cause-and-effect association of crocodilian genotoxicity, application of nonlethal biomarker approach to assess genetic and epigenetic effects (e.g., blood cell DNA content variation by FCM, micronucleus assay) should be incorporated.

IV. Summary

It is obvious that the small number of studies available makes it difficult to conclude whether reptilian populations are in danger as a consequence of exposure to genotoxicant agents. Nonetheless, the studies discussed here illustrate that sensitive and selective techniques are available that may be used to monitor and investigate the status of reptilian populations. Furthermore, these studies illustrate the existence of gaps and the need for future research. Environmental genotoxicology is a new discipline, and its application to reptilian communities is still in its infancy. DNA alterations that induce posterior damage may be different depending on the genotoxic agent and the background levels of DNA adducts, strand breaks, mutations, and other DNA alterations that occur as a result of natural phenomena; also, the level of background changes in DNA can vary among species and among different tissues.[16,48] Therefore the ability to measure contaminant-induced DNA damage is directly dependent on accurate background levels of such alterations, and these data are missing in reptiles. It is anticipated that as genomes of reptiles are added to those of other vertebrate groups, the application of genomic analysis will lead to a better understanding of genotoxic effects caused by environmental agents in this important vertebrate group.

Acknowledgments

This work was supported by NIHES 07381 to Ian P. Callard and Postdoctoral Fellowship from Ministry of Science and Technology of Spain to A. Novillo.

Literature Cited

1. Muller, H.J., Artificial transmutation of the gene. *Science*, 66, 84, 1927.
2. Auerbach, C. and Robson, J.M., Chemical production of mutations. *Nature* (London), 157, 302, 1946.
3. Wogan, G.N. and Gorelick, N.J., Chemical and biochemical dosimetry to exposure to genotoxic chemicals. *Environ. Health Perspect.*, 62, 5, 1985.
4. ATLA, *Genotoxicity and Carcinogenicity*, ATLA, 30, Suppl 1, 83, 2002.
5. Anderson, S.L. and Wild, G.C., Linking genotoxic responses and reproductive success in ecotoxicology. *Environ. Health Perspect.*, 102, Suppl 12, 9, 1994.
6. Depledge, M.H., Genetic ecotoxicology: an overview. *J. Exp. Mar. Biol. Ecol.*, 200, 57, 1996.
7. Daughton, C.G. and Ternes, T.A., Pharmaceuticals and personal care products in the environment: agents of subtle change? *Environ. Health Perspect.*, 107, Suppl 6, 907, 1999.
8. U.S. Environmental Protection Agency, *The Waste System, Office of Solid Waste, and Emergency Response*, USEPA, Washington, 1998.
9. Houk, V.S., The genotoxicity of industrial wastes and effluents. *Mutation Res. Rev. Genet. Toxicol.*, 277, 91, 1992.
10. Wolbarst, A.B., et al., Sites in the United States contaminated with radioactivity. *Health Phys.*, 77, 247, 1999.
11. Campbell, J.E. and Cranwell, R.M., Performance assessment of radioactive waste repositories. *Science*, 239, 1389, 1988.
12. Lamb, T., et al., Genetic damage in a population of slider turtles (*Trachemys scripta*) inhabiting a radioactive reservoir. *Arch. Environ. Contam. Toxicol.*, 20, 138, 1991.
13. Clark, M.J. and Smith, F.B., Wet and dry deposition of Chernobyl releases. *Nature*, 332, 245, 1988.
14. Harvey, R.C., Polycyclic hydrocarbons and cancer. *Am. Sci.*, 70, 386, 1982.
15. Depledge, M.H., Genotypic toxicity: implications for individuals and populations. *Environ. Health Perspect.*, 102, Suppl 12, 101, 1994.
16. Shugart, L.R., DNA damage as a biomarker of exposure. *Ecotoxicology*, 9, 329, 2000.
17. Anderson, S., et al., Genetic and molecular ecotoxicology: a research framework. *Environ. Health Perspect.*, 102, Suppl 12, 3, 1994.

18. Bickham, J.W. and Smolen, M.J., Somatic and heritable effects of environmental genotoxins and the emergence of evolutionary toxicology. *Environ. Health Perspect.*, 102, 25, 1994.
19. Kirsch-Volders, M., et al., Indirect mechanisms of genotoxicity. *Toxicol. Lett.*, 140–141, 63, 2003.
20. Pratt, I.S. and Barron, T., Regulatory recognition of indirect genotoxicity mechanisms in the European Union. *Toxicol. Lett.*, 140–141, 53, 2003.
21. Llorente, M.T., *Valoración de alteraciones citogenéticas en estudios medioambientales: desarrollo de protocolos metodológicos utilizando citometría de flujo*, PhD Thesis, Facultad de Ciencias Biologicas, Universidad Complutense de Madrid, Madrid, Spain, 2002.
22. Shugart, L.R. and Theodorakis, C., Environmental genotoxicity: probing the underlying mechanisms. *Environ. Health Perspect.*, 102, Suppl 12, 13, 1994.
23. McDiarmid, R.W. and Mitchell, J.C., Diversity and distribution of amphibians and reptiles, in *Ecotoxicology of Amphibians and Reptiles*, Sparling, D.W., Linder, G. Bishop, C.A., Eds., SETAC Press, Pensacola, FL, 2000, 15.
24. Crain, D.A. and Guillette, L.J.J., Reptiles as models of contaminant-induced endocrine disruption. *Anim. Reprod. Sci.*, 53, 77, 1998.
25. Hall, R.J. and Clark, D.R.J., Response of the iguanid lizard *Anolis carolinensis* to four organophosphorus pesticides. *Environ. Pollut. Series A*, 28, 45, 1982.
26. Hall, H.J. and Henry, P.F.P., Assessing effects of pesticides on amphibians and reptiles: status and needs. *Herpetological J.*, 2, 65, 1992.
27. Pauli, B.D. and Money, S., Ecotoxicology of pesticides in reptiles, in *Ecotoxicology of Amphibians and Reptiles*, Sparling, D.W., Linder, G. Bishop, C.A., Eds., SETAC Press, Pensacola, FL, 2000, 269.
28. Olafsson, P.G., et al., Snapping turtles—a biological screen for PCBs. *Chemosphere*, 12, 1525, 1983.
29. Bryan, A.M., Stone, W.B., and Olafsson, P.G., Disposition of toxic PCB congeners in snapping turtle eggs: expressed as toxic equivalents of TCDD. *Bull. Environ. Contam. Toxicol.*, 39, 791, 1987.
30. Sparling, D.W., Linder, G., and Bishop, C.A., Eds., *Ecotoxicology of Amphibians and Reptiles*, SETAC Technical publication series. SETAC Press, Pensacola, FL, 2000, 904.
31. Campbell, K.R. and Campbell, T.S., Lizard contamination data for ecological risk assessment. *Rev. Environ. Contam. Toxicol.*, 165, 39, 2000.
32. Campbell, K.R. and Campbell, T.S., The accumulation and effects of environmental contaminants on snakes: a review. *Environ. Monit. Assess.*, 70, 253, 2001.
33. Campbell, K.R. and Campbell, T.S., A logical starting point for developing priorities for lizard and snake ecotoxicology: a review of available data. *Environ. Toxicol. Chem.*, 21, 894, 2002.
34. Bickham, J.W., et al., Flow cytometric analysis of the effects of low-level radiation exposure on natural populations of slider turtles (*Pseudemys scripta*). *Arch. Environ. Contam. Toxicol.*, 17, 837, 1988.
35. Meyers-Schöne, L., et al., Comparison of two freshwater turtle species as monitors of radionuclide and chemical contamination: DNA damage and residue analysis. *Environ. Toxicol. Chem.*, 12, 1487, 1993.
36. Meyers-Schöne, L., Ecological risk assessment of reptiles, in *Ecotoxicology of Amphibians and Reptiles*, Sparling, D.W., Linder, G., Bishop, C.A., Eds., SETAC Press, Pensacola, FL, 2000, 793.

37. Shugart, L.R. and Theodorakis, C., New trends in biological monitoring: application of biomarkers to genetic ecotoxicology. *Biotherapy*, 11, 119, 1998.
38. Shugart, L.R., Quantitation of chemically induced damage to DNA of aquatic organisms by alkaline unwinding assay. *Aquatic Toxicol.*, 13, 43, 1998.
39. Hebert, P.D.N. and Luiker, M.M., Genetic effects of contaminant exposure—towards an assessment of impacts on animal populations. *Sci. Total Environ.*, 191, 23, 1996.
40. Lamb, T., et al., The slider turtle as an environmental sentinel: multiple tissue assays using cytometric analysis. *Ecotoxicology*, 4, 5, 1995.
41. Ulsh, B.A., et al., Chromosome translocation in turtles: a biomarker in a sentinel animal for ecological dosimetry. *Radiat. Res.*, 153, 752, 2000.
42. Savva, D., Use of DNA fingerprinting to detect genotoxic effects. *Ecotoxicol. Environ. Saf.*, 41, 103, 1998.
43. Castano, A., Becerril, C., and Llorente, M.T., Fish cells used to detect aquatic carcinogens and genotoxic agents, in *In Vitro Methods in Aquatic Toxicology*, Mothersill, C. and Austin, B., Eds., Praxis Publishing Ltd, Chichester, 2003, Chap. 2.
44. Kallioniemi, A., et al., Comparative genomic hybridization for molecular cytogenetic analysis of solid tumours. *Science*, 258, 818, 1992.
45. Kirsch-Volders, M.E., Cundari, A.E., and Van Hummelen, P., The in vitro micronucleus test: a multi-endpoint assay to detect simultaneously mitotic delay, apoptosis, chromosome breakage, chromosome loss and non-disjunction. *Mutation Res.*, 392, 19, 1997.
46. Kirsch-Volders, M., et al., Report from the in vitro micronucleus assay working group. *Environ. Molecular Mutagenesis*, 35, 167, 2000.
47. Wester, P.W., et al., Aquatic toxicology: opportunities for enhancement through histopathology. *Environ. Toxicol. Pharmacol.*, 11, 289, 2002.
48. Depledge, M.H., The ecotoxicological significance of genotoxicity in marine invertebrates. *Mutation Res.*, 399, 109, 1998.
49. Done, L., Neoplasia, in *Reptile Medicine and Surgery*, Mader, D.R., Ed., 1996, Saunders, Philadelphia, 1996, 125.
50. Crawshaw, G.J., Diseases and pathology of amphibians and reptiles, in *Ecotoxicology of Amphibians and Reptiles*, Sparling, D.W., Linder, G., Bishop, C.A., Eds., SETAC Press, Pensacola, FL, 2000, 199.
51. Kuchling, G., *The Reproductive Biology of the Chelonian*, Springer-Verlag, Berlin, 1999.
52. Zug, G.R., Vitt, L.J., and Caldwell, J.P., *Herpetology: An Introductory Biology of Amphibians and Reptiles*, Academic Press, San Diego, 2001.
53. National Research Council, *Animals as Sentinels of Environmental Health Hazards*, National Academy Press, Washington, 1991.
54. Ulsh, B.A., et al., Chromosome translocations in *T. scripta:* the dose-rate effect and in vivo lymphocyte radiation response. *Radiat. Res.*, 155, 63, 2001.
55. Ulsh, B.A., et al., Environmental biodosimetry: a biologically relevant tool for ecological risk assessment and biomonitoring. *J. Environ. Radioact.*, 66, 121, 2003.
56. Scott, D.E., Whicker, F.W., and Gibbons, J.W., Effect of season on the retention of 137Cs and 90Sr by the yellow-bellied slider turtle (*Pseudemys scripta*). *Can. J. Zool.*, 64, 2850, 1986.
57. Swartz, C.D., et al., Chemical contaminants and their effects in fish and wildlife from the industrial zone of Sumgayit, Republic of Azerbaijan. *Ecotoxicology*, 12, 509, 2003.

58. U.S. Air Force Center for Environmental Excellence (AFCEE). Massachusetts Military Reservation Plume Response Program; Final Ecological Quarterly Data Summary Report: Summer 1996, 1998.
59. Rie, M.T., *Assessment of the effects of groundwater pollution on a sentinel species, Chrysemys picta, on Cape Cod, MA: tissue contaminant levels and hepatic and reproductive bioindicators*, PhD Thesis, Biology Department, Boston University, Boston, 2000.
60. Yawetz, A., Benedek-Segal, M., and Woodin, B.R., Cytochrome P4501A immunoassay in freshwater turtles and exposure to PCBs and environmental pollutants. *Environ. Toxicol. Chem.*, 16, 1802, 1997.
61. Yawetz, A., Woodin, B.R., and Stegeman, J.J., Cytochrome P450 in liver of the turtle *Chrysemys picta picta* and the induction and partial purification of CYP1A-like proteins. *Biochim. Biophys. Acta*, 1381, 12, 1998.
62. Rie, M.T., et al., Reproductive endocrine disruption in a sentinel species (*Chrysemys picta*) on Cape Cod, Massachusetts. *Arch. Environ. Contam. Toxicol.*, 48, 217, 2005.
63. Kitana, N., *Potential environmental impacts on endocrine responses and gonadal development in freshwater turtles*, PhD Thesis, Biology Department, Boston University, Boston, 2004.
64. Rie, M.T., Lendas, K.A., and Callard, I.P., Cadmium: tissue distribution and binding protein induction in the painted turtle, *Chrysemys picta. Comp. Biochem. Physiol. C Toxicol. Pharmacol.*, 130, 41, 2001.
65. Smith, G.M. and Coates, C.W., Fibro-epithelial growths of the skin in large marine turtles *Chelonia mydas (L). Zoologica (NY)*, 23, 93, 1938.
66. Herbst, L.H., Fibropapillomatosis of marine turtles. *Annu. Rev. Fish. Dis.*, 4, 389, 1994.
67. Herbst, L.H., et al., Experimental transmission of green turtle fibropapillomatosis using cel-free tumor extracts. *Dis. Aquatic Org.*, 22, 1, 1995.
68. Herbst, L.H. and Klein, P.A., Green turtle fibropapillomatosis: challenges to assessing the role of environmental cofactors. *Environ. Health. Perspect.*, 103, Suppl 4, 27, 1995.
69. Estes, R., de Quieroz, K., and Gauthier, J.A., Phylogenetic relationships within Squamata, in *Phylogenetic Relationships of the Lizard Families, Essays Commemorating Charles L. Camp.*, Estes, R., de Quieroz, K., Gauthier, J.A., Eds., Stanford University Press, Stanford, NJ, 1988, 119.
70. Dunham, A.E., Miles, D.B., and Reznick, D.N., Life history patterns in squamate reptiles, in *Biology of the Reptilia*, Gans, C. and Huey R.B., Eds., Alan R. Liss, New York, 1988, 421.
71. Tinkle, D.W., Wilbur, H.M., and Tilley, S.G., Evolutionary strategies in lizard reproduction. *Evolution*, 24, 55, 1970.
72. Bull, J.J., *Evolution of Sex-Determining Mechanisms.* Benjamin/Cummings Publishing Co., Menlo Park, CA, 1983.
73. Lee, M.S. and Shine, R., Reptilian viviparity and Dollo's Law. *Evolution*, 52, 1441, 1998.
74. Blackburn, D.G., Evolutionary origins of viviparity in the Reptilia. I. Sauria. *Amphibian-Reptilia*, 3, 185, 1982.
75. Blackburn, D.G., Vitt, L.G., and Beuchat, C.A., Eutherian-like reproductive specializations in a viviparous reptile. *Proc. Nat. Acad. Sci.*, 81, 4860, 1984.
76. Thompson, M.B. and Speake, B.K., Energy and nutrient utilization by embryonic reptiles. *Comp. Biochem. Physiol. Part A*, 133, 529, 2003.

77. Cooper, J.E., Gschmeissner, S., and Holt, P.E., Viral particles in a papilloma from a green lizard (*Lacerta viridis*). *Lab. Anim.*, 16, 12, 1982.
78. Lunger, P.D., Hardy, W.D.J., and Clark, F.L.F., C-type virus particles in a reptilian tumor. *J. Natl. Cancer Inst.*, 52, 1231, 1974.
79. Jacobson, E.R., Seely, J.C., and Novilla, M.N., Lymphosarcoma associated with viruslike intranuclear inclusions in a California king snake (Colubridae, Lampropeltis). *J. Natl. Cancer Instit.*, 65, 577, 1980.
80. Turner, F.B., et al., Effects of continuous irradiation on animal populations. *Adv. Radiat. Biol.*, 5, 83, 1975.
81. Turner, F.B. and Medica, P.A., Sterility among female lizards (*Uta stansburiana*) exposed to continuous gamma radiation. *Radiat. Res.*, 70, 154, 1977.
82. French, N.R. Chronic low-level gamma irradiation of a desert ecosystem for five years, in *Actes du Symposium International de Radioecologie, Centre d'Etudes Nucleaires de Cadarache*, France, 1970.
83. Turner, F.B., et al. Radiation-induced sterility in natural populations of lizards (*Crotaphytus wislizenii* and *Cnemidophorous tigris*), in Proceedings of the 3rd National Symposium on Radioecology, Oak Ridge, TN, 1971.
84. Medica, P.A., Turner, F.B., and Smith, D.D., Effects of radiation on a forced population of horned lizards. *J. Herpetol.*, 7, 75, 1973.
85. Tinkle, D.W., Effects of radiation on the natality, density, and breeding structure of a natural population of lizards, *Uta stansburiana. Health Physiol.*, 11, 1595, 1965.
86. Dana, S.W. and Tinkle, D.W., Effects of X-radiation on the testes of the lizard, *Uta stansburiana stejnegeri. Int. J. Radiat. Biol.*, 9, 67, 1965.
87. Khan, B. and Tansel, B., Mercury bioconcentration factors in American alligators (*Alligator mississippiensis*) in the Florida Everglades. *Ecotoxicol. Environ. Saf.*, 47, 54, 2000.
88. IUCN, IUCN red list of threatened species, in *IUCN Species Survival Commission*, Gland, Switzerland. 2000
89. Levell, J.P., *A Field Guide to Reptiles and the Law*, Serpent's Tale Natural History Book Distributors, MN, 1997, 20.
90. Bishop, C.A. and Martinovic, B., Guideline and procedure for toxicological field investigations using amphibians and reptiles, in *Ecotoxicology of Amphibians and Reptiles*, Sparling, D.W., Linder, G., Bishop, C.A., Eds., SETAC Press, Pensacola, FL, 2000, 697.
91. Yanochko, G.M., Jagoe, C.H., and Brisbin, I.L.J., Tissue mercury concentration in alligators (*Alligator mississippiensis*) from the Florida Everglades and the Savannah River Site, South Carolina, *Arch. Environ. Contam. Toxicol.*, 32, 323, 1997.
92. Guillette, L.J.J., et al., Serum concentrations of various environmental contaminants and their relationship to sex steroid concentrations and phallus size in juvenile *American alligators. Arch. Environ. Contam. Toxicol.*, 36, 447, 1999.
93. Cobb, G.P., Wood, P.D., and O'Quinn, M., Polychlorinated biphenyls in eggs and chorioallantoic membranes of American alligators (*Alligator mississippiensis*) from coastal South Carolina. *Environ. Toxicol. Chem.*, 16, 1456, 1997.
94. Cobb, G.P., Houlis, P.D., and Bargar, T.A., Polychlorinated biphenyl occurrence in American alligators (*Alligator mississippiensis*) from Louisiana and South Carolina. *Environ. Pollut.*, 118, 1, 2002.

95. Heinz, G.H., Percival, H.F., and Jennings, M.L., Contaminants in American alligator eggs from Lake Apopka, Lake Griffin, and Lake Okeechobee, Florida. *Environ. Monit. Assess.*, 16, 277, 1991.
96. Cobb, G.P., et al., Using chorioallantoic membranes for non-lethal assessment of persistent organic pollutant exposure and effect in oviparous wildlife. *Ecotoxicology*, 12, 31, 2003.
97. Bargar, T.A., et al., Relative distribution of polychlorinated biphenyls among tissues of neonatal American alligators (*Alligator mississippiensis*). *Arch. Environ. Contam. Toxicol.*, 37, 364, 1999.
98. Wu, T.H., et al., Organochlorine contaminants in Morelet's crocodile (*Crocodylus moreletii*) eggs from Belize. *Chemosphere*, 40, 671, 2000.

chapter 10

Reptile Ecotoxicology: Studying the Effects of Contaminants on Populations

Kyle W. Selcer

Table of Contents

I. Populations: Theory and Background

A. Ecotoxicology and the Importance of Population Studies

Toxicology is the study of the effects of contaminants on individuals. The term *ecotoxicology* is commonly used to refer to the study of the effects of contaminants on organizational levels above the individual, including populations, communities, and ecosystems. Previous chapters in this book have provided a comprehensive review of reptile toxicology, documenting how contaminants affect a number of organism-level parameters. This chapter focuses on reptile ecotoxicology, with particular emphasis on contaminant effects at the population level. This chapter is not intended as a review of the reptile ecotoxicologic literature, various aspects of which have been well reviewed in the past.[1–9] Rather, this chapter is intended to provide an overview of the major theoretical considerations involved in understanding the effects of contaminants on reptile populations and to address some of the challenges to be faced in studying this topic in the future.

There are pragmatic and theoretical reasons for emphasizing populations as a starting point for ecotoxicology. Toxicologic data largely focus on individual-level effects, many of which cannot easily be translated into higher level effects (e.g., titers of cellular or molecular markers). However, some individual-level effects do translate well to the population level. These include hatchling size and quality, age and size at sexual maturity, egg size and quality, egg number, sperm counts and viability, among others. Certain individual parameters are also used to calculate population-level rates, the most important of which are fecundity and mortality. These rates are important variables in mathematic modeling of population-levle and higher level responses. Thus, a variety of population-level parameters are traceable to effects of contaminants at the individual level and yet are useful in predicting responses at the community and ecosystem levels. Projecting ecologic responses to contaminant exposure is one goal of ecologic risk assessors, and population-level data provide a framework for translating observed individual effects into predicted ecologic responses. As Rose and colleagues[9] stated, "Population-level responses bridge the gap between exposure and toxicological response on the one hand and risk assessment on the other." Additionally, populations are a unit of biological organization that people can identify with, including lawmakers. Environmental policies often reflect this bias by focusing preservation of certain species at the population level. Clark and colleagues[10] summed up the need for population data in toxicologic studies as follows: "As one goes from the population to higher levels or from population to lower levels, it becomes increasingly difficult and complex to assess contaminant effects in terms that are relevant to societal values and policy directives."

Despite the relevance of population information in toxicologic studies, there are a limited number of examples where it has been conclusively demonstrated that contaminants have adversely affected vertebrate populations. This is in large part a result of the focus of toxicologic studies on the organism level and to a corresponding lack of information at the ecologic levels.[5,11] There is compelling evidence for contaminant-induced reductions in populations of fish-eating birds from the Great Lakes as a result of polychlorinated biphenyl (PCB) exposure, the depletion of lake trout in the Great Lakes as a result of complex mixtures of toxicants, and decline of raptors in the United States as a result of dichlorodiphenyltrichloroethane (DDT) exposure (cases reviewed by Fairbrother and colleagues[12]). Populations of certain fish species in rivers in the United Kingdom have been shown to be affected by endocrine disruptors from sewage-treatment effluents.[13,14]

In each of these examples, a combination of laboratory and field methodologies, focusing on individual- and population-level endpoints, was needed to provide convincing evidence that contaminants were the cause of the population declines. Similar multidisciplinary and multilevel approaches are needed to investigate population declines of reptile species.

Reptiles are particularly poorly represented in ecotoxicologic studies.[2–4,8,15,16] Only a few examples document contaminant effects on reptiles at the population level. The best known of these are several studies of the effects of organochlorine and other compounds on freshwater turtles in the Great Lakes region[17–21] and studies of the effects of a mixture of contaminants on alligators in Florida (reviewed by Guillette[22]). There are no extensive studies of contaminant effects in any reptiles besides turtles and alligators. Clearly there is a need for more information on ecotoxicology across the entire spectrum of reptile species.

B. *Why Study Reptiles? Implications for Ecotoxicology*

Reviews of toxicology in vertebrates have proposed a number of reasons why reptiles should be included in more studies of contaminant effects.[1,16,23] Some of the reasons are related to aspects of reptilian metabolism. Reptiles are cold blooded (poikilothermic) and as such have lower rates of catabolism and depuration than do birds and mammals. Thus, they may maintain higher body burdens of contaminants. Also, many reptile species are known to store significant amounts of body fat, which may serve to bioaccumulate lipophilic contaminants. Furthermore, a number of reptiles are predators or scavengers that occupy relatively high positions in trophic food chains. This results in an increased probability of higher exposure to persistent contaminants as a result of biomagnification.

Others reasons for studying reptile toxicology relate to unique attributes of their life history. Many reptiles are late maturing and extremely long lived. This provides an opportunity to assess the effects of contaminants in species that have been exposed for long periods and to determine the responses of populations of long-lived species to toxins. Also, a number of reptile species are sedentary and exhibit strong site tenure relative to birds and mammals. This makes them ideal for studies involving effects of local contaminants. Moreover, a variety of reptiles have temperature-dependent sex determination, a process that may be particularly sensitive to perturbation by chemicals that mimic natural hormones (endocrine disruptors). One of the most pressing reasons for focusing on reptiles in studies of contaminants is the rapid decline of many reptile populations[16,24] (see also Chapter 2 of this text).

C. Properties of Populations

Biologists usually recognize three major levels of organization above the organism level: populations, communities, and ecosystems. A population comprises all individuals of the same species in a defined area, a community is the collection of all populations of different species in that area, and an ecosystem is the combination of the community with all the abiotic factors of the area. Each level of organization includes the properties of lower levels but in addition has its own emergent properties, which are not present at previous levels of organization. For example, a population has a density and a dispersion pattern, a community has a species composition and interspecific interactions, and an ecosystem has nutrient cycles and energy flow.

The term *population* is defined as the collection of all individuals of the same species in a given area at a given time. Note that both spatial and temporal boundaries must be defined when referring to a population (e.g., keeled earless lizards on South Padre Island in 1980). For natural populations, there is the presumption that members of the same population share a common gene pool (i.e., are interbreeding). The exact geographic limits of a natural population are difficult to discern because it is often unclear where interbreeding is occurring and where it is not. This is particularly true for a geographically widespread species. The concept of metapopulations has gained increasing attention. A metapopulation comprises a group of local populations that have gene flow between them as a result of immigration and emigration. It is important to note that most studies are not on entire natural populations; rather, they are on geographic areas chosen by the investigator. These local populations are subsets of the actual natural population, and the data collected may not necessarily reflect the dynamics of the natural population. Careful study design and sampling are needed to make sure that the results are representative of the natural population.

The study of the vital statistics of a population is termed *demography*. It is interesting that much of the early demographic theory and methodology came from the need to generate actuarial tables of human populations as tools for the insurance industry. As indicated previously, populations have a number of emergent properties not found at the individual level. These include a mean and variance for each of the various individual parameters, such as body size, clutch size, egg size, and hatchling size. Some of the more important population parameters are as follows: genetic variability, age structure, birth rate, death rate, generation time, sex ratio, density, and dispersion pattern. Note that some of these population parameters vary depending on the age of the individual. Thus, demographic analysis usually uses an age-specific approach. Particularly important are age-specific fertility and age-specific mortality. These two variables form the basis for a life table, which can be used to assess the overall ecologic health of a population and to determine management strategies for threatened or endangered populations.

D. *Impact of Contaminants on Populations*

Different environmental conditions favor different life history characteristics. The specific attributes of each population have been shaped by natural selection as the population adapts to local environmental conditions. The observed phenotypes for any population are the result of both long-term evolutionary (ultimate) factors and current environmental and physiologic (proximate) factors. Ultimate factors have acted over time to set the range for a given variable that is found in the population as a whole. In contrast, proximate factors act to determine the exact value observed for each individual at the present time. Contaminants are proximate factors and affect life history variables in the same way as other proximate factors, acting on individuals to shape the phenotypic expression of a given variable. The changes in the individual values of these variables as a result of exposure to contaminants will then affect the population dynamics accordingly.

There are several major ways that contaminants may affect the life history parameters of populations. The effects can be through direct interference with an individual's physiologic function, resulting in changes in growth, energy storage or reproduction, or even death. Direct effects on the physiology of the individual are easier to detect than indirect effects and also correspond with the data that have been traditionally collected in toxicologic research. Contaminants also may affect the population indirectly by altering the food supply, levels of predators or competitors, or the habitat that the organism occupies. Indirect effects, although equally important from a population growth perspective, are difficult to separate from the effects of other proximate variables, such as climate change or habitat destruction. Consequently, indirect effects may go unrecognized and the overall effects of contaminants on the population may be underestimated.

II. Life History Characteristics

A. Life History Strategies

Natural selection operates through differential reproductive success. A number of life history variables affect reproductive success, both directly and indirectly. However, because there is only a finite amount of energy to divide among these various factors, there are trade-offs to be made among the different life history variables.[25–27] For any given resource, it is a zero-sum game. Increases in allocation for one variable result in decreases for another. For example, egg number can be increased, but this may come at the expense of egg quality.

The particular suite of life history variable settings observed for a given population is referred to as its life history strategy. Given the number of life history variables, there is potentially a vast array of strategies that could be used by individual populations. However, the trade-offs involved in determining the values of life history variables place constraints on the actual number of strategies found in nature. It is convenient and useful to attempt to generalize by identifying certain patterns that are found in natural populations, especially because it is reasonable to assume that contaminants may have different effects on populations that express different life history patterns.

One of the more widely used generalizations regarding life history strategies is that of r- and K-selection.[25,26,28] This concept holds that species can be grouped into two major categories based on shared suites of life history variables. The r-strategists are opportunistic species that have adopted a suite of life history variables that are focused on maximizing reproduction (r is the intrinsic rate of natural increase in population growth equations and is based on reproductive potential). The r-strategists are characterized as occupying variable or unpredictable climates, having high early mortality rates, and having population sizes that are variable over time. Selection in these species favors rapid development, high reproductive rates, early reproduction, small body size, semelparity, and many small offspring.[25] They usually have short life spans. In contrast, the K-strategists are equilibrium species that have evolved a suite of characteristics focused on stabilizing population fluctuations at or near the environmental carrying capacity (carrying capacity is termed K in the logistic equation for population growth). They are characterized as occupying constant or predictable climates, having mortality that is constant for all ages or is concentrated late in the life span, and having population sizes that are relatively constant over time. Selection in the K-strategists favors slower development, strong competitive ability, delayed reproduction, larger body size, iteroparity, and fewer high-quality offspring.[25] They usually have long life spans. The suites of characteristics represented by r- and K-strategies are best viewed as ends of a linear continuum rather than a dichotomy. The terms are more useful in a comparative sense, relative to other organisms.[26]

Winemiller and Rose[29] proposed another model that places organisms into categories according to their suite of life history characteristics. Their model was based on characteristics found in North American fishes. The three-dimensional model has axes of juvenile survivorship, age at maturity, and fecundity. The result is a triangular continuum with three extreme strategies, termed *opportunistic*, *equilibrium*, and *periodic*. Equilibrium species have lower fecundity, older age at maturity, and higher juvenile survivorship. Thus, they are similar to K-selected species. Opportunistic species combine rapid growth with early maturity (similar to r-selected species) but have low fecundity and iteroparity. Periodic species have late maturity and high fecundity, combined with lower quality offspring that have high juvenile mortality.

It is generally agreed that both the r-K linear continuum[25] and the triangular continuum[29] models of life history strategies are useful for making generalizations about, and comparisons of, different populations. They are also valuable as frameworks for assessing the effects that contaminants can have on different types of populations.[9,30]

B. *Life History Strategies of Reptiles*

Regarding reptiles, either of the aforementioned models can be used to describe their life history strategies. Small lizards for the most part are early maturing and short lived, which would make them more r-selected under the r-K continuum. However, they usually produce multiple clutches that are relatively small, which fits well with the opportunistic strategy of the triangular continuum. Populations of these species are predicted to be strongly affected by changes in reproductive rate. In contrast, large squamates and most turtles and crocodilians combine late maturity with high longevity. This places them well on the K-selected end of the r-K continuum. However, they also tend to produce large clutches of eggs and have low juvenile survivorship. This suite of characteristics is best represented by the periodic strategy of the triangular continuum.[9] Populations of these species are predicted to be strongly affected by changes in adult survivorship.

The following is a list of several of the life history variables that might be particularly valuable in ecotoxicology studies. Included is information on the ecologic importance of these variables, their usual values in reptiles, and the ways that contaminants might affect them.

1. *Clutch Size or Offspring Number*

Clutch size or offspring number is the number of offspring produced per reproductive event. Offspring number frequently varies with age, resulting in age-specific fecundity. Thus, not all reproducing adults contribute equally to the future reproductive success of the population. Knowledge of age-specific fecundity is important in extrapolating individual adverse effects of contaminants to the level of population growth. Age-specific reproductive output is also a major concern from a wildlife-management perspective because it can help determine which age classes should be most strongly protected.

The primary trade-off with offspring number is offspring quality. Taking offspring quantity and quality together, the trade-off results in a continuum, from one extreme of many low-quality offspring with a presumably low chance of survival to the other extreme of one or a few high-quality offspring with a greater chance of individual survival.

For reptiles, offspring number shows considerable variation, from one egg in *Anolis* and some geckos, to dozens or even hundreds of eggs for some turtles and crocodilians. As with most vertebrates, there is a positive relationship between offspring number and body size, both within and among taxonomic groups.[26,31,32] Variation in offspring number among reptiles is greater than that found in mammals and birds, but it is substantially less than that found in fishes. The more limited clutch sizes found in mammals and birds are probably related to the extensive amount of parental care usually found in these taxa, which can place upper limits on offspring number.

The effect of contaminants on clutch size could conceivably come from direct interference with the regulatory pathways determining the number of eggs produced, as might occur with endocrine disruptors (as discussed in detail in Chapter 6). Alternatively, exposure to contaminants could result in lowering the energy stores that are available for reproduction. This could occur through increased overall stress on the individual, decreased feeding activity, or altered metabolic pathways. Lower energy stores will result either in lower clutch size, lower egg quality, or both. It may even result in delay of reproduction to a more suitable time.

The effect of reduced clutch size on the overall growth of a given population depends on the life history strategy of the organism in question. Species with an r-selected strategy (e.g., small lizards) are predicted to be more strongly affected by lowered reproductive rates than are species with K-selected strategies (e.g., large squamates, turtles, and crocodilians). Thus, a reduction in clutch size as a result of short-term contaminant exposure might have a severe negative effect on a population of small lizards in a given area but might have little or no effect on a population of turtles in the same area.

2. *Egg or Offspring Quality*

As indicated earlier, there is a trade-off between egg number and egg quality. It has been shown that increased offspring number within a species is correlated with decreased egg or offspring quality.[26,32] Furthermore, decreased egg quality is manifested in decreased hatchling quality[33] and ultimately decreased hatchling survivorship. It is predicted that for any given population, there is an optimal clutch size that results in maximal reproductive success; however, proximate factors result in deviations from the optimal number.

Contaminants are among the proximate factors that can influence offspring quality. As stated earlier, any contaminant that reduces available energy stores will reduce the overall contribution that an individual can make to reproduction. This reduction can be manifested in reductions in egg number, egg quality, or both. Contaminants can also affect development in such a manner as to cause embryonic malformations, which decrease offspring quality. For example, snapping turtle eggs from more contaminated sites (PCBs, dioxins, and furans) in the Great Lakes showed lower hatching rates and greater levels of deformities than those from lesser contaminated sites,[17] suggesting that certain contaminants have the potential to alter offspring quality and survivorship in reptiles.

3. *Frequency of Reproduction*

In the extreme, there are two categories of reproducers: (1) semelparous species that put all their effort into a single reproductive episode (e.g., salmon, many annual plants, and most insects), and (2) iteroparous species that reproduce multiple times (e.g., most perennial plants and most vertebrates). Note that iteroparity may range from two reproductive events for some species to many events for others. Iteroparous reproduction also may be in the form of continuous breeding or repeated seasonal breeding. Regarding trade-offs, semelparous species are able to put significant energy into their single reproductive event, often producing extremely large numbers of offspring.[26] However, if environmental conditions are not favorable at the time of offspring production, the entire reproductive effort of a semelparous individual may be futile. In other words, it is boom or bust. In contrast, iteroparous species have lower fecundity for any given event, but they are able to spread their reproductive effort over a broader period and therefore increase the chance that conditions will be favorable for survival of at least some of their offspring. Besides the trade-off of lower fecundity for a given reproductive episode, iteroparous species are also taking a chance that they will survive long enough to reproduce again.

Reptiles are mostly iteroparous.[34] Those with short life spans (e.g., small lizards) usually reproduce multiple times in a single season but for only one or a few seasons, whereas those with longer life spans (large lizards, snakes, turtles, and crocodilians) reproduce over a span of many years, although not necessarily every year. Some species of reptiles are extremely long lived and may reproduce repeatedly over several decades (e.g., marine turtles). It has been shown for freshwater turtles that clutch frequency varies considerably among years and is the most important determinant of reproductive output in these turtles.[35] Unknown proximate environmental factors (e.g., micro-environmental or nutritional) were suggested to be the determining factors for the observed variation in clutch frequency. Certainly, contamination could be among those proximate factors.

Timing of reproductive events is also an important variable in determining reproductive success. In seasonal climates, offspring from later clutches are often at a disadvantage relative to their earlier siblings. This is partly a result of lack of time to secure energy stores needed for overwintering and partly a result of poor competitive abilities compared with the larger offspring from earlier clutches. Thus, it is valuable to know how many clutches are produced per individual for a given population and also when those clutches are produced.

The frequency and timing of reproductive events for an individual may be affected by contaminant exposure, and this can have implications for overall population growth. Contaminants that act as endocrine disruptors could presumably interfere with the signals that initiate reproduction, delaying or preventing reproduction during the exposure period. Also, any contaminant that lowers energy stores could decrease the frequency of reproduction. Some species store energy over months or even years in preparation for reproduction and will not undergo the reproductive process if they have insufficient energy stores. Note that alterations in reproductive frequency could be manifested in fewer reproductive episodes within a given breeding season or over successive breeding seasons. Reduction in clutch frequency would have similar effects on population growth as a reduction in clutch size because both together are determinants of fecundity for an individual.

4. *Age and Size at First Reproduction*

An organism may reproduce in its first year of life or may delay reproduction for a period of years or even decades (e.g., sea turtles). The advantage of early reproduction is that there is an increased probability of surviving to reproduce. However, early reproducers have little time to grow. This means that their body size is relatively small and the number or size of eggs they can produce and carry is accordingly low. In contrast, late reproducers may not survive to reproduce and must put energy into growth and maintenance that could have gone to reproduction, but they achieve a larger body size and can produce more or larger eggs per reproductive event.

Reptiles vary from species that reproduce only months after hatching (small lizards) to species that wait more than a decade before producing offspring (large turtles and crocodilians). As is found in other groups, smaller reptiles tend to reproduce sooner than larger reptiles, supporting the concept that early reproduction comes at the expense of body size. Contaminants could affect the age at first reproduction either directly or indirectly. Direct effects could come from endocrine-disrupting chemicals altering the signals that initiate reproductive maturation (see Chapter 6). Indirect effects could result from increased stress or decreased energy stores interfering with the ability to reproduce. Stress in a well-known inhibitor of reproduction.

IV. Difficulties with Reptile Toxicology at the Population Level

There is a clear need for toxicologic studies to include more data on population effects; however, there are some difficulties encountered in determining the effects of contaminants on populations.

A. Separating Effects of Contaminants from Other Factors

One of the biggest problems is separation of the effects of contaminants from those of other proximate factors. This is especially true if the effect of the contaminant is indirect, acting on another factor such as food availability. Separation of contaminant effects from those of other proximate factors usually requires historical knowledge of the demographics for the population in question or requires comparison with another closely related population that is not exposed to contaminants. However, identification and use of a so-called pristine site can itself be problematic. Studies that compare contaminated and pristine sites often have exposure data only on the contaminant of interest and not on other factors that may have caused the observed differences between the two populations. For example, various aspects of the morphology and physiology of juvenile alligators in a highly contaminated lake (Apopka) in Florida have been compared with the same measures of alligators in other sites that are less contaminated with certain pesticides.[22] Differences were found in several of these parameters (e.g., sex: steroid ratios, phallus size) between Lake Apopka alligators and those from the less contaminated sites. Although there is compelling evidence that pesticides are the cause of the observed morphologic and physiologic differences, it is still possible that other unknown factors vary among the sites and may have caused the observed differences.

Studies comparing two populations of the same species must also take into account natural variation between populations. Members of the same species in different populations may have different phenotypes as a result of adaptation to local environmental conditions. The presence of this so-called ecophenotypic variation can confound attempts to compare contaminant effects between populations because it is difficult to separate variation resulting from contaminants from variation resulting from local natural environmental factors. For example, two populations of the keeled earless lizard (*Holbrookia propinqua*) that occur only a few kilometers apart from each other exhibit dramatically different phenotypes[36] including different average clutch sizes and body sizes. Similar differences between geographically close populations have been reported for many species, especially in situations where the two populations are separated by a geographic barrier or where there are altitudinal differences. Thus, in studies comparing two populations, it is particularly informative to have historical demographic data for both populations.

B. *Selecting Appropriate Endpoints*

Another difficulty in determining the effects of contaminants on populations is related to using reliable and measurable endpoints. Some demographic parameters are difficult to measure accurately, requiring substantial amounts of time and effort. This is especially true for long-lived species and for species that have patchy distributions across a broad range. Also, some parameters are more useful than others from the standpoint of risk assessment. It is important that researchers choose endpoints that are both reliable and useful when designing ecotoxicologic studies.

Abundance is probably the first parameter to come to mind when thinking about contaminant effects on populations. Certainly, population size is of the utmost concern in terms of protecting endangered species. Nevertheless, abundance is not always a reliable or useful endpoint in ecotoxicologic studies. Measuring abundance can be difficult, often requiring marking and recapturing of large numbers of individuals to achieve an accurate population estimate. The assumptions needed for valid estimates are sometimes difficult to meet, and even under the best of circumstances the variances of the estimate can be quite large. Also, many populations show considerable interannual variation in abundance, confounding attempts to determine effects of contaminants. Furthermore, abundance is not always a good indicator of the direction of population growth, especially in the early stages of perturbation.

Because of the problems associated with using abundance in determining the effects on populations, it has been suggested that life history parameters be the main focus of ecotoxicologic studies.[9,30] Information on growth rates, age structure, age-specific fecundity, and age-specific mortality may be more useful than abundance for assessing the status of a population. Furthermore, reliable measures of these parameters often can be more readily obtained with smaller samples and less effort.

Clutch size is one of the easiest variables to measure in wild populations, making it an ideal parameter to include in studies of ecotoxicology. Furthermore, clutch size is arguably one of the most relevant life history characteristics regarding toxicology because it is directly related to recruitment of new members of the population. For each species, the average size and the range of clutch sizes have been determined by evolutionary factors; however, the exact clutch size for each individual at each reproductive event is determined by proximate factors (excepting geckos and anoles), including the effects of contaminants.

In addition to clutch size, information on clutch frequency is needed to determine total reproductive output for individuals. However, unlike clutch size, data on clutch frequency are relatively hard to obtain in the wild, usually requiring intensive field studies that involve multiple recaptures of individuals. This makes it particularly difficult to determine the effect of contaminants on total reproductive effort and on frequency of reproduction.

C. *Unique Aspects of Reptile Populations*

Some aspects of reptile ecology make it difficult to obtain reliable data on population parameters. Species with a long generation time (e.g., many turtles and crocodilians) are particularly troublesome for gathering population data, with years or decades required to obtain good data. It is difficult for researchers to find funding for a study that may require 10 years or more of data collection to obtain reliable numbers. Also, individuals of many reptiles are widely dispersed, making it more difficult to collect population data than for species that congregate in large numbers. Finally, some species of reptiles are not well suited for captivity, precluding the laboratory studies that can be so important in determining cause and effect in ecotoxicology studies.

The paucity of information on reptiles also makes ecotoxicologic studies difficult. Prior ecologic data are simply not available for many populations of reptiles that now may be at risk from contaminants. For species that are endangered or threatened, it is especially difficult to gather population information without causing further impact.

D. Current Needs and Future Directions

Recent reviews of reptile toxicology[16,23,37] have provided a number of suggestions on research needs in this area. The following is a list of some of the more salient needs regarding reptile ecotoxicology.

1. Need for Reptile Contamination Studies to Include Population Parameters

Data on toxicology of reptiles is generally lacking, and much of what is available documents tissue levels of contaminants and not their effects. There is especially a paucity of information on effects of contaminants on reptile populations. Future studies should include as much population data as possible, particularly data on life history variables such as growth, age-specific reproductive parameters, and age-specific mortality.

2. Need for Nondestructive Sampling

Toxicologic and ecologic research sometimes requires collection and sacrifice of individuals. However, this should be minimized as much as possible for ethical and practical reasons, especially because many reptile populations appear to be declining. One way to minimize sacrifice is to use nondestructive sampling methods. New technologies and an increased interest in nonlethal methods have resulted in novel procedures to obtain valuable contaminant data.[38–41] These methods need to be put into routine practice in reptile ecotoxicology studies.

3. *Need to Identify Reptile Model Organisms to Serve as Sentinels, Surrogates, and Bioindicators*

Unlike birds, mammals, and fish, there are no standard toxicology models for reptiles. There is an urgent need to identify, validate, and utilize reptile models in field and laboratory ecotoxicology studies.[10,37] This includes identification of appropriate sentinel or bioindicator species for monitoring of ecosystems, surrogate species to provide data that may be useful for understanding risks to threatened or endangered species, and model species that can provide information on reptile toxicology in general and on certain types of reptiles in particular.

V. Candidate Reptile Species for Use as Sentinels

To begin to address the need for reptilian models, I have compiled a list of reptile species that may be useful for studies of ecotoxicology. In determining suggestions for model species, the following attributes were considered to be an asset: wide geographic distribution; commercial availability; readily adaptable to laboratory studies; used by humans for food or goods; and finally, large amounts of collateral data on the life history, ecology, physiology (particularly reproductive), or contaminant exposure. The suggestions for model species provided in the following paragraphs include only North American species. Some information provided on life history attributes and ranges of the various species was taken from Conant[42] or from Goin and colleagues.[34] Also included is some information on groups of interest to conservationists but that are not being recommended as model species because of their endangered or threatened status.

A. Chelonia (Turtles)

There is increasing concern about declining turtle populations worldwide.[43] Turtles suffer from a number of anthropogenic insults that collectively are taking a toll on their numbers. These include habitat destruction; loss of nesting sites; exploitation by humans for food, medicines, and pets; and environmental contaminants. Chapter 2 provides more detail on current global threats to turtle populations.

All turtle species are oviparous, with typically large clutch sizes. In general, marine turtles produce the largest clutches (some greater than 100 eggs per clutch) and terrestrial turtles produce the smallest (the extreme being a single egg per clutch). However, eggs from terrestrial species usually are larger. Turtle eggs can be soft or hard shelled, depending on the species. The nests of many turtles are relatively easy to locate, and the eggs are readily collected. Large clutches allow for sampling of contaminants from some eggs and incubation of other eggs from the same nest for assessing effects of contaminants on hatching rates and various hatchling parameters.[17,18]

Turtles usually have the periodic type of life history strategy that places a premium on adult survivorship. Thus, any factor that selectively removes adults from the populations (e.g., pet trade, use as food) can have serious consequences on turtle populations. Also, turtle populations may be slow in recovering after a population decline.[44,45] Turtles store considerable amounts of body fat that can serve as a reservoir for lipid-soluble contaminants. These lipids are used for reproduction and can be passed on to the eggs during the yolking process. Adult female turtles have been shown to have lower levels of accumulated contaminants than do adult males from the same site, providing indirect evidence that females undergo depuration during reproduction.[18] Some turtle species have temperature-dependent sex determination, whereas others have genotypic sex determination.

Many turtle species exhibit a high degree of site tenure, making them ideal for studies of local contamination. Also, some turtles appear to maintain high population densities in known contaminated sites, leading to the suggestion that they may be relatively insensitive to effects of toxins. Turtles occupy a variety of habitats, from marine and freshwater to terrestrial. Each of these lifestyles results in very different routes and levels of exposure to environmental contaminants; therefore, model species are suggested for each of these groups.

1. *Freshwater Aquatic Turtles*

Freshwater aquatic turtles are potentially exposed to aquatic contaminants at both the egg and adult stages. The eggs may be exposed to contaminants through maternal deposition and by contaminants in the nesting matrix. Adults can be exposed to contaminants through their food and from their aquatic medium because they are known to cycle water through their pharynx and cloaca. Freshwater aquatic turtles may be carnivorous, omnivorous, or herbivorous, depending on the species. It is likely that the more carnivorous species are at greatest risk of significant exposure to persistent contaminants that biomagnify.

The ecology of several species of freshwater turtles has been extensively studied,[31,35,44–46] resulting in a good demographic framework from which to investigate contaminants. Furthermore, compared with other reptiles, a substantial amount of data is available on contaminant levels present in freshwater turtles,[8,15,47] and comprehensive attempts have been made to determine the effects of contaminant exposure on population parameters of some species.[17–21]

Freshwater turtles as a group are declining worldwide, with exploitation by humans for food and the pet trade being among the major causes.[48] However, significant effects of contaminants on turtle populations cannot be ruled out.

1.1. Red-Eared Slider Turtle (Trachemys scripta). The red-eared slider turtle — common to lakes, ponds, and streams — is geographically widespread throughout the central and southeastern United States. Subspecies extend into Mexico and Central America. There is a substantial amount of data available on the life history and ecology of this turtle.[46] This species has temperature-dependent sex determination, and studies have investigated contaminant effects on this process.[49] This turtle is generally omnivorous, but individuals may shift their diet more toward herbivory as they move from juvenile to adult.

Red-eared slider turtles are common in the pet trade. Hatchlings and small juveniles (<10 cm) are illegal for sale in the United States as a result of concerns over *Salmonella*, but larger juveniles and adults are frequently seen in pet stores. Other attributes that make *T. scripta* useful or important as a model species include the following: it is established as a laboratory model, it is easy to care for in captivity, eggs and adults are available commercially, and these turtles are widely used as a food source (particularly in Asia). Also, red-eared slider turtles often occupy extremely contaminated sites, such as sewage-treatment ponds, where they appear to thrive. Some data are available on various aspects of *T. scripta* reproductive ecology and physiology.[50–53]

1.2. Painted Turtle (Chrysemys picta). The painted turtle species shares many of the same attributes as *T. scripta* and similarly may serve as a useful model. The painted turtle has a more northerly and easterly distribution in the United States than *T. scripta*. Taken together, the ranges of the two species cover a significant proportion of North America.

1.3. Snapping Turtle (Chelydra serpentina). The snapping turtle is a bottom-dwelling species found in lakes and ponds throughout the eastern two thirds of the United States from southern Canada to southern Texas. The demographics of this turtle are well known[45] and there are considerable amounts of data on exposure of this species to contaminants. Furthermore, there are several detailed studies of contaminant effects on snapping turtle populations.[17,18,20] This large turtle is widely used as a food source in the United States, which should place additional interest in both its preservation and in tissue levels of contaminants. The turtle is omnivorous, taking a substantial portion of its diet from scavenging and capturing of freshwater fish and invertebrates. This increases the probability of biomagnification of persistent contaminants. Their bottom-dwelling habits presumably expose snapping turtles to contaminants in sediment. They exhibit temperature-dependent sex determination, and the effects of certain contaminants on this process have been investigated.[19,54]

1.4. *Softshell Turtles* (Apalone *spp.*). Softshell turtles possess many of the same attributes as snapping turtles and should also make good models for studies of contamination effects on populations. They are omnivorous, with a significant portion of their diet consisting of fish and invertebrates. They are fully aquatic, usually occupying rivers. The two major species (spiny and smooth) together have a range that occupies much of the eastern and middle portion of the United States from Canada to Texas. The effects of contaminants have been investigated for populations of spiny softshells in Canada.[21] It is interesting that the spiny softshell turtle (*Apalone spinifera*) has genetic sex determination.[55]

1.5. *Diamondback Terrapin* (Malaclemys terrapin). The diamondback terrapin is an exclusively estuarine turtle that occupies Atlantic and Gulf of Mexico coastal salt marshes and tidal basins, from Massachusetts to Texas. It has historically been collected extensively for food and is renowned for its flavor. According to Burke and colleagues,[48] it was once one of the most economically important reptiles in the world, with thousands of kilograms of turtles collected each year for a period of decades. This species is sexually dimorphic, with females larger than males, resulting in females being preferentially collected. Although currently not as extensively collected for food as in the past, diamondback terrapins are also under pressure as a result of habitat degradation, inadvertent collection in crab traps, human recreation, and other factors.[48,56]

2. *Sea Turtles*

Given the endangered or vulnerable status of virtually all sea turtle species, I am not recommending any species as a model organism. The major threats to sea turtle survival are overexploitation, habitat destruction, and accidental drowning in fishing gear[57] (see Chapter 2 for further detail). Contaminant exposure is thought to be generally low for sea turtles because the open ocean has a dilution effect on contaminants and because most sea turtles consume organisms that are at relatively low trophic levels. Nevertheless, potential health effects have been associated with even low contaminant levels in loggerhead sea turtles[58,59] and may be a problem in other species. Any studies of effects of contaminants on sea turtle populations must necessarily use nonlethal methodologies.

3. *Terrestrial Turtles*

Aquatic turtles are presumably at more risk from environmental contaminants than are terrestrial turtles. Nevertheless, it is important to monitor populations occupying the terrestrial environment for potential effects of contaminants. Terrestrial turtles usually are late maturing and long lived. They produce smaller clutches than marine turtles and most freshwater aquatic turtles; however, they tend to produce large eggs. Feeding habits of terrestrial turtles range from omnivorous to herbivorous.

3.1. Box Turtles (Terrapene *spp.*). Box turtles are prized as pets in the United States. They do well in captivity and are easy to care for. Although they are terrestrial and resemble small tortoises, they are actually more closely related to freshwater turtles.[60] The two most abundant species, the common box turtle (*Terrapene carolina*) and the ornate box turtle (*Terrapene oranata*), together range throughout the eastern and midwestern portions of the United States. They are omnivorous, primarily eating vegetation and small invertebrates. Box turtles commonly live to 30 or 40 years and may live much longer, at least in captivity.

There is a reasonable amount of information of the ecology and physiology of box turtles.[61,62] Furthermore, there are a few studies of the effects of contaminants on these species.[63,64]

3.2. Desert Tortoises (Gopherus *spp.*). Desert tortoises (genus *Gopherus*) are under severe pressure from a variety of anthropogenic factors[60] and as a consequence are protected in most parts of their ranges. Thus, I am not recommending any of these species as model organisms. They usually occupy arid regions of the southwest (excepting the gopher tortoise in Florida) and are exclusively herbivorous. This combination of attributes means that they are less likely to be exposed to significant levels of environmental contaminants and also are unlikely to be subject to much bioaccumulation and biomagnification of persistent contaminants. However, these turtles are of extreme interest from a conservation standpoint. As with sea turtles, any studies should use nonlethal methodologies.

B. *Lacertes (Lizards)*

Lizards are among the most studied reptiles in terms of life history strategies.[65–67] Lizards in the United States are usually small, early maturing, relatively short-lived species having either medium or small clutch sizes. They are iteroparous, with most producing multiple clutches over one or several reproductive seasons. Lizards usually are habitat specialists with a high degree of site tenure. They are also usually food generalists, with their preferred prey being an assortment of small invertebrates, mostly insects. As terrestrial animals that consume prey from a relatively low position on the trophic ladder, lizards are presumably not exposed to high levels of contaminants relative to aquatic reptile species. However, a number of lizard species occupy habitats in or near agricultural areas that may be heavily treated with pesticides. Furthermore, their eggs are deposited in moist soil and may be at risk from exposure to soil contaminants. Thus, lizards may be useful in monitoring the impact of agricultural contaminants. Also, several species of lizard are commensal with humans, providing a unique opportunity to assess the impact of contaminants around homes and gardens.

1. *Fence and Prairie Lizards (Sceloporus undulatus and S. occidentalis)*

The genus *Sceloporus* consists of dozens of species that are often called swifts or spiny lizards. It is interesting that some species are oviparous and others are viviparous. The species *S. undulatus* is made up of four closely related subspecies that together occupy a range spanning the southern half of the United States from the Atlantic coast to the Rocky Mountains. The two eastern subspecies are more arboreal and are called fence lizards, whereas the two western subspecies are primarily terrestrial and are called prairie lizards. Western fence lizards, *Sceloporus occidentalis*, also known as blue-bellied lizards, are very similar to *S. undulatus* except that they are located west of the Rocky Mountains. Both species of small lizards are early maturing, with most of their growth in the first year, but with relatively short life spans of only a few years. These oviparous lizards produce multiple clutches (2 to 3 per season) of about 3 to 15 eggs. For both species there is a considerable amount of variation in life history attributes among populations.

Much information is available on the reproductive physiology and life history of *S. undulatus* and *S. occidentalis*.[68–76] Much of this work is recent because there is currently an active effort to establish *S. undulatus* and *S. occidentalis* as model species for reptile toxicology.[76]

2. *Mediterranean Gecko (Hemidactylus turcicus)*

Studies of species that are found in close association with humans may provide an opportunity for field experiments that can assess the effects of exposure to contaminants found in and around homes. The Mediterranean gecko, *Hemidactylus turcicus,* was introduced to the United States only in the past century, but they are now well established throughout the states bordering the Gulf of Mexico. These small geckos live primarily on and inside of buildings. They even deposit their eggs inside the walls and attics of houses as well as in garages and other accessible structures. There is some information on the ecology and reproduction of this species.[77–81] Mediterranean geckos are early maturing, reproducing in the year after hatching. However, unlike most small lizards, they may survive for 5 or more years. These geckos are iteroparous, producing as many as three clutches per year, but all clutches are exactly two eggs. They are insectivorous, taking virtually all their prey from off of or inside of the buildings they occupy. They have strong site tenure, rarely leaving the building they are hatched in. Mediterranean geckos are extremely adaptable to captivity. As a nocturnal species, they require none of the lighting and heating equipment essential for the health of most other reptiles. Furthermore, their calcareous-shelled eggs require no substrate for incubation, allowing for very high hatching success under laboratory conditions.

3. *Green Anole (Anolis carolinensis)*

The green anole is another example of a lizard that is closely associated with humans. Green anoles are found in the Atlantic and Gulf of Mexico coastal states from North Carolina to Texas. Anoles can be found on buildings but are more common on the vegetation immediately surrounding the structures. They produce multiple clutches per season of one egg per clutch. They are very common as pets (often mistakenly called chameleons) and, as such, have well-established husbandry. A substantial amount of information is available on their ecology and reproductive physiology.[82–90] Green anoles have been proposed as a suitable model for laboratory studies of reptile reproduction and behavior.[91] A closely related species, the brown anole (*Anolis sagrei*) has been introduced to Florida and is considered an exotic pest. It might be useful for ecotoxicologic studies requiring collection of significant numbers of individuals.

C. *Serpentes (Snakes)*

Snakes have also been suggested as suitable models for studies of contaminants for several reasons.[92,93] First, as obligate predators, they are high on the trophic food web. This presumably results in greater exposure to contaminants via biomagnification. Also, many snake species exhibit a high degree of site tenure, allowing for better correlation of observed effects with local levels of contamination. A number of snake species are fully or partly aquatic, making them excellent for studies of aquatic contaminants.[93] Furthermore, many snakes are easy to keep in captivity, facilitating laboratory studies. In general, snakes are long lived and late maturing and produce clutches repeatedly over a period of years or even decades. There are both oviparous and viviparous species of snakes. There are some data on levels of contaminants in snakes.[3]

1. *Water Snakes (Nerodia spp.)*

The genus *Nerodia* consists of the New World water snakes. Any of four species (*N. sipedon*, *N. fasciata*, *N. erythrogaster*, and *N. rhombifera*) would make for a good toxicology model because they are all primarily aquatic, widely distributed, and reasonably abundant within their ranges. Taken together, the ranges of these species extend east to west from the Atlantic coast to the Midwest and north to south from southern Canada to the states bordering the Gulf of Mexico. All are carnivores, feeding on both invertebrates and vertebrates in and around their aquatic habitats. Some information is available on the reproductive physiology[41,94,95] and on levels in and effects of contaminants on these species.[3,40,93,96–98]

2. *Garter Snakes (Thamnophis spp.)*

The genus *Thamnophis* includes garter snakes and ribbon snakes. Together, the various species range throughout the entire United States as well as into Canada and Mexico. The eastern garter snake (*T. sirtalis*) has the broadest range, but the plains garter snake (*T. radix*) and the wandering garter snake (*T. elegans*) also occupy substantial areas. The closely related ribbon snakes (*T. sauritus* and *T. proximus*) are also widely distributed. Garter snakes are familiar visitors to backyard gardens, and they are often captured and kept as pets. In arid regions, they usually are found near water. In more mesic areas, they are found in a variety of habitats, including those associated with humans. Ribbon snakes tend to be semiaquatic throughout their ranges. The prey of both garter and ribbon snakes includes a variety of amphibians (frogs and salamanders), fish, small birds and mammals, worms, slugs, and leeches.

The garter snakes are probably the most studied snakes from the standpoint of reproductive ecology and physiology, particularly regarding control of reproductive behavior.[99–109] There is also some information on contaminant levels in *Thamnophis*.[8]

D. Crocodylia (Alligators and Crocodiles)

Crocodiles usually are large-bodied, aquatic reptiles that live in tropical regions throughout the world. They have a life history strategy based on late maturity, extreme longevity, and large clutches produced over a period of years. They exhibit parental care and have temperature-dependent sex determination. They are fearsome predators, even occasionally preying on humans, and they are valued for their skin and meat. Many populations are threatened or endangered worldwide. There has been increased attention on the effects of contaminants on crocodilian populations.[22,110,111]

1. American Alligator (Alligator mississippiensis)

The range of the American alligator in the United States includes the states bordering on the Atlantic Ocean from North Carolina southward and all the states bordering the Gulf of Mexico. Their abundance varies widely within this range. The reproductive ecology and physiology of alligators has been well documented,[112–114] and contaminant effects on certain populations of this species have been extensively investigated.[22, 110, 115–117]

Literature Cited

1. Albers, P.H., Heinz, G.H., and Hall, R.J., Approaches for assessment of terrestrial vertebrate responses to contaminants: moving beyond individual organisms, in *Environmental Contaminants in Terrestrial Vertebrates: Effects on Populations, Communities, and Ecosystems*, Albers P.H., Heinz, G.H., and Ohlendorf, H.M., Eds., SETAC Press, Pensacola, FL, 2000, 109.
2. Campbell, K.R. and Campbell, T.S., Lizard contaminant data for ecological risk assessment, *Rev. Environ. Contam. Toxicol.*, 165, 39, 2000.
3. Campbell, K.R. and Campbell, T.S., The accumulation and effects of environmental contaminants on snakes: a review, *Environ. Monit. Assess.*, 70, 253, 2001.
4. Campbell, K.R. and Campbell, T.S., A logical starting point for developing priorities for lizard and snake toxicology: a review of available data, *Environ. Toxicol. Chem.*, 21, 894, 2002.
5. Fox, G.A., Perturbations in terrestrial vertebrate populations: contaminants as a cause, in *Environmental Contaminants in Terrestrial Vertebrates: Effects on Populations, Communities, and Ecosystems*, Albers, P.H., Heinz, G.H., and Ohlendorf, H.M., Eds., SETAC Press, Pensacola, FL, 2000, 19.
6. Linder, G. and Grillitsch, B., Ecotoxicology of metals, in, *Ecotoxicology of Amphibians and Reptiles*, Sparling, D.W., Linder, G., and Bishop, C.A., Eds., SETAC Press, Pensacola, FL, 2000, 793.

7. Pauli, B.D. and Money, S., Ecotoxicology of pesticides in reptiles, in *Ecotoxicology of Amphibians and Reptiles*, Sparling, D.W., Linder, G., and Bishop, C.A., Eds., SETAC Press, Pensacola, FL, 2000, 269.
8. Portelli, M.J. and Bishop, C.A., Ecotoxicology of organic contaminants in reptiles: a review of the concentrations and effects of organic contaminants in reptiles, in *Ecotoxicology of Amphibians and Reptiles*, Sparling, D.W., Linder, G., and Bishop, C.A., Eds., SETAC Press, Pensacola, FL, 2000, 495.
9. Rose, K.A., Brewer, L.W., Barnthouse, L.W., Fox, G.A., Gard, N.W., Mendonca, M., Munkittrick, K.R., and Vitt, L.J., Ecological responses of oviparous vertebrates to contaminant effects on reproduction and development, in *Reproductive and Developmental Effects of Contaminants in Oviparous Vertebrates*, Di Giulio, R.T. and Tillitt, D.E., Eds., SETAC Press, Pensacola, FL, 1999, 225.
10. Clark, J., Dickson, K., Geisy, J., Lackey, R., Mihaich, E., Stahl, R., and Zeeman, M., Using reproductive and developmental effects data in ecological risk assessments for oviparous vertebrates exposed to contaminants, in *Reproductive and Developmental Effects of Contaminants in Oviparous Vertebrates*, Di Giulio, R.T. and Tillitt, D.E., Eds., SETAC Press, Pensacola, FL, 1999, 363.
11. Sparling, D.W., Bishop, C.A., and Linder, G., The current status of amphibian and reptile ecotoxicological research, in *Ecotoxicology of Amphibians and Reptiles*, Sparling, D.W., Linder, G., and Bishop, C.A., Eds., SETAC Press, Pensacola, FL, 2000, 1.
12. Fairbrother, A., Ankley, G.T., Birnbaum, L.S., Bradbury, S.P., Francis, B., Grey, L.E., Hinton, D., Johnson, L.L., Peterson, R.E., and Van Der Kraak, G., Reproductive and developmental toxicology of contaminants in oviparous animals, in *Reproductive and Developmental Effects of Contaminants in Oviparous Vertebrates*, Di Giulio, R.T. and Tillitt, D.E., Eds., SETAC Press, Pensacola, FL, 1999, 283.
13. Jobling, S., Beresford, N., Nolan, M., Rodgers-Gray, T., Brighty, G.C., Sumpter, J.P., and Tyler, C.R., Altered sexual maturation and gamete production in wild roach (*Rutilus rutilus*) living in rivers that receive sewage effluents, *Biol. Reprod.*, 66, 272, 2002.
14. Jobling, S., Coey, S., Whitmore, J.G., Kime, D.E., Van Look, K.J.W., McAllister, B.G., Beresford, N., Henshaw, A.C., Brighty, G., Tyler, C.R., and Sumpter, J.P., Wild intersex roach have reduced fertility, *Biol. Reprod.*, 67, 515, 2002.
15. Hall, R.J., Effects of environmental contaminants on reptiles: a review, Washington DC, US Fish and Wildlife Service, *Spec. Sci. Rept. Wildl.*, No. 128, 1, 1980.
16. Hopkins, W.A., Reptile toxicology: challenges and opportunities on the last frontier in vertebrate ecotoxicology, *Environ. Toxicol. Chem.*, 19, 2391, 2000.
17. Bishop, C.A., Brooks, R.J., Carey, J.H., Ng, P., Norstrom, R.J., and Lean, D.R., The case for a cause-effect linkage between environmental contamination and development in eggs of the common snapping turtle (*Chelydra s. serpentina*) from Ontario, Canada, *J. Toxicol. Environ. Health*, 33, 521, 1991.
18. Bishop, C.A., Brown, G.P., Brooks, R.J., Lean, D.R.S., and Carey, J.H., Organochlorine contaminant concentrations in eggs and their relationship to body size, and clutch characteristics of the female common snapping turtle (*Chelydra serpentina serpentina*) in Lake Ontario, Canada, *Arch. Environ. Contam. Toxicol.*, 27, 82, 1994.

19. de Solla, S.R., Bishop, C.A., Van Der Kraak, G., and Brooks, R.J., Impact of organochlorine contamination on levels of sex hormones and external morphology of common snapping turtles (*Chelydra serpentina serpentina*) in Ontario, Canada, *Environ. Health Perspect.*, 106, 253, 1998.
20. de Solla, S.R., Bishop, C.A., and Brooks, R.J., Sexually dimorphic morphology of hatchling snapping turtles (*Chelydra serpentina*) from contaminated and reference sites in the Great Lakes and St. Lawrence River basin, North America, *Environ. Toxicol. Chem.*, 21, 922, 2002.
21. de Solla, S.R., Fletcher, M.L., and Bishop, C.A., Relative contributions of organochlorine contaminants, parasitism, and predation to reproductive success of eastern spiny softshell turtles (*Apalone spiniferus spiniferus*) from southern Ontario, Canada, *Ecotoxicology*, 12, 261, 2003.
22. Guillette, L.J., Contaminant-associated endocrine disruption in reptiles, in *Ecotoxicology of Amphibians and Reptiles*, Sparling, D.W., Linder, G., and Bishop, C.A., Eds., SETAC Press, Pensacola, FL, 2000, 595.
23. Sparling, D.W., Bishop, C.A., Pauli, B.D., and Money, S., Epilogue: Lessons to be learned, in *Ecotoxicology of Amphibians and Reptiles*, Sparling, D.W., Linder, G., Bishop, C.A., Eds., SETAC Press, Pensacola, FL, 2000, 811.
24. Gibbons, J.W., The global decline of reptiles, déjà vu amphibians, *BioScience*, 50, 653, 2000.
25. Pianka, E.R., On r- and K selection, *Am. Nat.*, 104, 592, 1970.
26. Pianka, E.R., *Evolutionary Ecology*, Harper and Row, New York, 1983.
27. Stearns, S.C., The evolution of life history traits: a critique of the theory and a review of the data, *Ann. Rev. Ecol. Syst.*, 8, 145, 1977.
28. McArthur, R.H. and Wilson, E.O., *The Theory of Island Biogeography*, Princeton University Press, Princeton, NJ, 1967.
29. Winemiller, K.O. and Rose, K.A., Patterns of life-history diversification in North American fishes: implications for population regulation, *Can. J. Fish Aquat. Sci.*, 49, 2196, 1992.
30. Sample, B.E., Rose, K.A., and Suter, G.W. II, Estimation of population-level effects on wildlife based on individual-level exposures: influence of life-history strategies, in *Environmental Contaminants in Terrestrial Vertebrates: Effects on Populations, Communities, and Ecosystems*, Albers, P.H., Heinz, G.H., and Ohlendorf, H.M., Eds., SETAC Press, Pensacola, FL, 2000, 225.
31. Gibbons, J.W., Greene, J.L., and Patterson, K.K., Variation in reproductive characteristics of aquatic turtles, *Copeia*, 1982, 766, 1982.
32. Lack, D., *The Natural Regulation of Animal Numbers*, Oxford University Press, New York, 1954.
33. Sinervo, B., The evolution of maternal investment in lizards: an experimental and comparative analysis of egg size and its effects on offspring performance, *Evolution*, 44, 279, 1990.
34. Goin, C.J., Goin, O.B., and Zug, G.R., *Introduction to Herpetology*, 3rd ed., W.H. Freeman and Company, San Francisco, 1978.
35. Gibbons, J.W., Reproductive patterns in freshwater turtles, *Herpetologica*, 38, 222, 1982.
36. Selcer, K.W. and Judd, F.W., Variation in the reproductive ecology of *Holbrookia propinqua* (Sauria: Iguanidae), *Tex. J. Sci.*, 34, 125, 1982.
37. Meyers-Schone, L., Ecological risk assessment of reptiles, in *Ecotoxicology of Amphibians and Reptiles*, Sparling, D.W., Linder, G., and Bishop, C.A., Eds., SETAC Press, Pensacola, FL, 2000, 793.

38. Cobb, G.P., Bargar, T.A., Pepper, C.B., Norman, D.M., Houlis, P.D., and Anderson, T.A., Using chorioallantoic membranes for non-lethal assessment of persistent organic pollutant exposure and effect in oviparous wildlife, *Ecotoxicology*, 12, 31, 2003.
39. Jackson, B.P., Hopkins, W.A., and Baionno, J., Laser ablation-ICP-MS analysis of dissected tissue: a conservation-minded approach to assessing contaminant exposure, *Environ. Sci. Technol.*, 37, 2511, 2003.
40. Hopkins, W.A., Roe, J.H., Snodgrass, J.W., Jackson, B.P., Kling, D.E., Rowe, C.L., and Congdon, J.D., Nondestructive indices of trace element exposure in squamate reptiles, *Environ. Pollut.*, 115, 1, 2001.
41. Secor, S.M. and Nagy, T.R., Non-invasive measure of body composition of snakes using dual-energy X-ray absorptiometry. *Comp. Biochem. Physiol. A*, 136, 379, 2003.
42. Conant, R., *A Field Guide to the Reptiles and Amphibians of Eastern and Central North America*, 2nd ed., Houghton Mifflin Company, Boston, 1975.
43. Klemens, M.W., *Turtle Conservation*, Smithsonian Institution Press, Washington, DC, 2000.
44. Congdon, J.D., Dunham, A.E., and von Loben Sels, R.C., Delayed sexual maturity and demographics of Blanding's turtles (*Emydoidea blandingii*): implications for conservation and management of long-lived organisms, *Conserv. Biol.*, 7, 826, 1993.
45. Congdon, J.D., Dunham, A.E., and von Loben Sels, R.C., Demographics of common snapping turtles (*Chelydra serpentina*): implications for conservation and management of long-lived organisms, *Am. Zool.*, 34, 397, 1994.
46. Gibbons, J.W., *Life History and Ecology of the Slider Turtle*, Smithsonian Institution Press, Washington, DC, 1990.
47. Meyers-Schone, L. and Watson, B.T., Turtles as monitors of chemical contaminants in the environment, *Rev. Environ. Contam. Toxicol.*, 135, 93, 1994.
48. Burke, V.J., Lovich, J.E., Gibbons, J.W., Conservation of freshwater turtles, in *Turtle Conservation*, Klemens, M.W., Ed., Smithsonian Institution Press, Washington, DC, 2000, 156.
49. Willingham, E. and Crews, D., The red-eared slider turtle: an animal model for the study of low doses and mixtures, *Am. Zool.*, 40, 421, 2000.
50. Rhen, T., Willingham, E., Sakata, J.T., and Crews, D., Incubation temperature influences sex-steroid levels in juvenile red-eared slider turtles, *Trachemys scripta*, a species with temperature-dependent sex determination, *Biol. Reprod.*, 61, 1275, 1999.
51. Selcer, K.W. and Leavitt, W.W., Progesterone downregulates progesterone receptor, but not estrogen receptor, in the estrogen-primed oviduct of a turtle (*Trachemys scripta*), *Gen. Comp. Endocrinol.*, 83, 316, 1991.
52. Selcer, K.W. and Palmer, B.D., Estrogen downregulation of albumin and a 170-kDa serum protein in the turtle, *Trachemys scripta*, *Gen. Comp. Endocrinol.*, 97, 340, 1995.
53. Willingham, E., Embryonic exposure to low-dose pesticides: effects on growth rate in the hatchling red-eared slider turtle, *J. Toxicol. Health A*, 12, 257, 2001.
54. Portelli, M.J., de Solla, S.R., Brooks, R.J., and Bishop, C.A., Effect of dichlorodiphenyltrichloroethane on sex determination of the common snapping turtle (*Chelydra serpentina serpentina*), *Ecotoxicol. Environ. Saf.*, 43, 284, 1999.
55. Greenbaum, E. and Carr J.L., Sexual differentiation in the spiny softshell turtle (*Apalone spinifera*), a species with genetic sex determination, *J. Exp. Zool.*, 290,190, 2001.

56. Burger, J. and Garber, S., Risk assessment, life history strategies and turtles: could declines be prevented or predicted?, *J. Toxicol. Environ. Health*, 46, 483, 1995.
57. Meylan, A.B. and Ehrenfeld, D., Conservation of marine turtles, in *Turtle Conservation*, Klemens, M.W., Ed., Smithsonian Institution Press, Washington, DC, 2000, 96.
58. Keller, J.M., Kucklick, J.R., Stamper, M.A., Harms, C.A., and McClellan-Green, P.D., Associations between organochlorine contaminant concentrations and clinical health parameters in loggerhead sea turtles from North Carolina, USA, *Environ. Health Perspect.*, 112, 1074, 2004.
59. Keller, J.M., Kucklick, J.R., and McClellan-Green, P.D., Organochlorine contaminants in loggerhead sea turtle blood: extraction techniques and distribution among plasma and red blood cells, *Arch. Environ. Contam. Toxicol.*, 46, 254, 2004.
60. McDougal, J., Conservation of tortoises and terrestrial turtles, in *Turtle Conservation*, Klemens, M.W., Ed., Smithsonian Institution Press, Washington, DC, 2000, 180.
61. Henry, P.F., The eastern box turtle at the Patuxent Wildlife Research 1940s to the present: another view, *Exp. Gerontol.*, 38, 773, 2003.
62. Penick, D.N., Congdon, J., Spotila, J.R., and Williams, J.B., Microclimates and energetics of free-living box turtles, *Terrapene carolina*, in South Carolina, *Physiol. Biochem. Zool.*, 75, 57, 2002.
63. Holcomb, C.M. and Parker, W.S., Mirex residues in eggs and livers of two long-lived reptiles (*Chrysemys scripta* and *Terrapene carolina*) in Mississippi, 1970-1977, *Bull. Environ. Contam. Toxicol.*, 23, 369, 1979.
64. Tangredi, B.P. and Evans, R.H., Organochlorine pesticides associated with ocular, nasal, or otic infection in the eastern box turtle (*Terrapene carolina carolina*), *J. Zool. Wildl. Med.*, 28, 97, 1997.
65. Dunham, A.E., Miles, D.B., Resnick, D.N., Life history patterns in squamate reptiles, in *Biology of the Reptilia*, Vol. 16, Gans, C., Ed., Alan R. Liss, New York, 1988, 441.
66. Tinkle, D.W., Wilbur, H.M., and Tilley, S.G., Evolutionary strategies in lizard reproduction, *Evolution*, 24, 55, 1970.
67. Vitt, L.J. and Congdon, J.D., Body shape, reproductive effort, and relative clutch mass in lizards: resolution of a paradox, *Am. Nat.*, 112, 595, 1978.
68. Angilletta, M.J., Sears, M.W., and Winters, S.R., Seasonal variation in reproductive effort and its effect on offspring size in the lizard *Sceloporus undulatus*, *Herpetologica*, 57, 365, 2001.
69. Brasfield, S.M., Weber, L.P., Talent, L.G., and Janz, D.M., Dose-response and time course relationships for vitellogenin induction in male western fence lizards (*Sceloporus occidentalis*) exposed to ethinylestradiol, *Environ. Toxicol. Chem.*, 21, 1410, 2002.
70. Brasfield, S.M., Bradham, K., Wells, J.B., Talent, L.G., Lannno, R.P., and Janz, D.M., Development of a terrestrial vertebrate model for assessing bioavailability of cadmium in the fence lizard (*Sceloporus undulatus*) and in ovo effects on hatchling size and thyroid function, *Chemosphere*, 54, 1643, 2004.
71. Burnham, D.K., Lackey, A., Manering, M., Jaensson, E., Pearson, J., Tyler, D.O., Melson, D., and Talent, L.G., Effects of 17alpha-ethinylestradiol on immune parameters in the lizard *Sceloporus occidentalis*, *Environ. Toxicol.*, 18, 211, 2003.

72. Carsia, R.V. and John-Alder, H., Seasonal alterations in adrenocortical cell function associated with stress-responsiveness and sex in the eastern fence lizard (*Sceloporus undulatus*), *Horm. Behav.*, 43, 408, 2003.
73. Dunlap, K.D., External and internal influences on indices of physiological stress: II. Seasonal and size-related variations in blood composition in free living lizards, *Sceloporus occidentalis*, *J. Exp. Zool.*, 272, 85, 1995.
74. Dunlap, K.D. and Wingfield, J.C. External and internal influences on indices of physiological stress. I. Seasonal and population variation in adrenocortical secretion of free-living lizards, *Sceloporus occidentalis*, *J. Exp. Zool.*, 271, 36, 1995.
75. Niewiarowski, P.H., Angilletta, M.J., Jr., and Leache, A.D., Phylogenetic comparative analysis of life-history variation among populations of the lizard *Sceloporus undulatus*: an example and prognosis, *Evol. Int. J. Org. Evolut.*, 58, 619, 2004.
76. Talent, L.G., Dumont, J.N., Bantle, J.A., Janz, D.M., and Talent, S.G., Evaluation of western fence lizards (*Sceloporus occidentalis*) and eastern fence lizards (*Sceloporus undulatus*) as laboratory reptile models for toxicological investigations, *Environ. Toxicol. Chem.*, 21, 899, 2002.
77. Girling, J.E., Cree, A., and Guillette, L.J., Jr., Oviducal structure in four species of gekkonid lizards differing in parity model and eggshell structure, *Reprod. Fertil. Dev.*, 10, 139, 1998.
78. Girling, J.E., Guillette, L.J., Jr., and Cree, A. Ultrastructure of the uterus in an ovariectomized gecko (*Hemidactylus turcicus*) after administration of exogenous estradiol, *J. Exp. Zool.*, 286, 76, 2000.
79. Rose, F.L. and Barbour, C.D., Ecology and reproductive cycles of the introduced gecko, *Hemidactylus turcicus*, in the southern United States, *Am. Midl. Nat.*, 79, 159, 1968.
80. Selcer, K.W., Life history of a successful colonizer: the Mediterranean gecko, *Hemidactylus turcicus*, in southern Texas, *Copeia*, 956, 1986.
81. Selcer, K.W., Seasonal variation in fatbody and liver mass of the introduced Mediterranean gecko, *Hemidactylus turcicus*, in Texas, *J. Herpetol.*, 21, 74, 1987.
82. Adkins, E. and Schlesinger, L., Androgens and the social behavior of male and female lizards (*Anolis carolinensis*), *Horm. Behav.*, 13, 139, 1979.
83. Andrews, R.M., Oviposition frequency of *Anolis carolinensis*, *Copeia*, 259, 1985.
84. Crews, D., Interrelationships among ecological, behavioral, and neuroendocrine processes in the reproductive cycle of *Anolis carolinensis* and other reptiles, *Adv. Study Behav.*, 11, 1, 1980.
85. Forbes, T.R., The development of the reproductive system of a lizard, *Anolis carolinensis*, *Am. J. Anat.*, 98, 139, 1956.
86. Jones, R.E., Guillette, L.J., Jr., Summers, C.H., Tokarz, R.R., and Crews, D., The relationship among ovarian condition, steroid hormones, and estrous behavior in *Anolis carolinensis*, *J. Exp. Zool.*, 227, 145, 1983.
87. Lovern, M.B., McNabb, F.M.A., and Jenssen, T.A., Developmental effects of testosterone on behavior in male and female green anoles (*Anolis carolinensis*), *Horm. Behav.*, 39, 139, 2001.
88. Lovern, M.B., Holmes, M.M., Fuller, C.O., and Wade, J., Effects of testosterone on the development of neuromuscular systems and their target tissues involved in courtship and copulation in green anoles (*Anolis carolinensis*), *Horm. Behav.*, 45, 295, 2004.

89. Orrell, K.S., Congdon, J.D., Jenssen, T.A., Michener, R.H., and Kunz, T.H., Intersexual differences in energy expenditure of *Anolis carolinensis* lizards during breeding and postbreeding seasons, *Physiol. Biochem. Zool.*, 77, 50, 2004.
90. Winkler, S.M. and Wade, J., Aromatase activity and regulation of sexual behaviors in the green anole lizard, *Physiol. Behav.*, 64, 723, 1998.
91. Lovern, M.B., Holmes, M.M., and Wand, J., The green anole (*Anolis carolinensis*): a reptilian model for laboratory studies of reproductive morphology and behavior, *I.L.A.R. J.*, 45, 54, 2004.
92. Bauerle, B., Spencer, D.L., and Wheeler, W.. The use of snakes as pollution indicator species, *Copeia*, 366, 1975.
93. Hopkins, W.A., Rowe, C.L., and Congdon, J.D., Elevated trace element concentrations and standard metabolic rates in banded water snakes (*Nerodia fasciata*) exposed to coal combustion wastes, *Environ. Toxicol. Chem.*, 18, 1258, 1999.
94. Riley, D. and Callard, I.P., An estrogen receptor in the liver of the viviparous watersnake, *Nerodia*: characterization and seasonal changes in binding capacity, *Endocrinology*, 123, 753, 1988.
95. Riley, D., Kleis-San Francisco, S.M., and Callard, I.P., A plasma steroid hormone binding protein in the viviparous water snake, *Nerodia*, *Gen. Comp. Endocrinol.*, 71, 419, 1988.
96. Bishop, C.A. and Rouse, J.D., Chlorinated hydrocarbon concentrations in plasma of the Lake Erie water snake (*Nerodia sipedon insularum*) and northern water snake (*Nerodia sipedon sipedon*) from the Great Lakes basin in 1998, *Arch. Environ. Contam. Toxicol.*, 39, 500, 2000.
97. Fontenot, L.W., Noblet, G.P., Akins, J.M., Stephens, M.D., and Cobb, G.P., Bioaccumulation of polychlorinated biphenyls in ranid frogs and northern water snakes from a hazardous waste site and a contaminated watershed, *Chemosphere*, 40, 803, 2000.
98. Hopkins, W.A., Roe, J.H., Snodgrass, J.W., Staub, B.P., Jackson, B.P., and Congdon, J.D., Effects of chronic dietary exposure to trace elements on banded water snakes (*Nerodia fasciata*), *Environ. Toxicol. Chem.*, 21, 906, 2002.
99. Garstka, W.R., Tokarz, R.R., Diamond, M., Halpert, A., and Crews, D., Behavioral and physiological control of yolk synthesis and deposition in the female red-sided garter snake (*Thamnophis sirtalis parietalis*), *Horm. Behav.*, 19, 137, 1985.
100. Hoffman, L.H., Placentation in the garter snake, *Thamnophis sirtalis.*, *J. Morphol.*, 131, 57, 1970.
101. Krohmer, R.W., The male red-sided garter snake (*Thamnophis sirtalis parietalis*): reproductive pattern and behavior, *I.L.A.R. J.*, 45, 54, 2004.
102. Krohmer, R.W., Grassman, M., and Crews, D., Annual reproductive cycle in the male red-sided garter snake, *Thamnophis sirtalis parietalis*: field and laboratory studies, *Gen. Comp. Endocrinol.*, 68, 64, 1987.
103. Lerner, D.T. and Mason, R.T., The influence of sex steroids on the sexual dimorphism in the red-spotted garter snake, *Thamnophis sirtalis concinnus*, *Gen. Comp. Endocrinol.*, 124, 218, 2001.
104. Mendonca, M.T. and Crews, D., Control of attractivity and receptivity in female red-sided garter snakes, *Horm. Behav.*, 40, 43, 2001.
105. Moore, I.T. and Mason, R.T., Behavioral and hormonal responses to corticosterone in the male red-sided garter snake, *Thamnophis sirtalis parietalis*, *Physiol. Behav.*, 72, 669, 2001.

106. Moore, I.T., Lerner, J.P., Lerner, D.T., and Mason, R.T., Relationships between annual cycles of testosterone, corticosterone, and body condition in male red-spotted garter snakes, *Thamnophis sirtalis concinnus*, *Physiol. Biochem. Zool.*, 73, 307, 2000.
107. Moore, I.T., Green, M.J., and Mason, R.T., Environmental and seasonal adaptations of the adrenocortical and gonadal responses to capture stress in two populations of the male garter snake, *Thamnophis sirtalis*, *J. Exp. Zool.*, 289, 99, 2001.
108. Shine, R., LeMaster, M.P., Moore, I.T., Olsson, M.M., and Mason, R.T., Bumpus in the snake den: effects of sex, size, and body condition on mortality of red-sided garter snakes, *Evol. Int. J. Org. Evolut.*, 55, 598, 2001.
109. Whittier, J.M., Mason, R.T., and Crews, D., Plasma steroid hormone levels of female red-sided garter snakes, *Thamnophis sirtalis parietalis*: relationship to mating and gestation, *Gen. Comp. Endocrinol.*, 67, 33, 1987.
110. Gunderson, M.P., Bermudez, D.S., Bryan, T.A., Degala, S., Edwards, T.M., Kools, S.A., Milnes, M.R., Woodward, A.R., and Guillette, L.J., Jr., Variation in sex steroids and phallus size in juvenile American alligators (*Alligator mississippiensis*) collected from 3 sites within the Kissimmee–Everglades drainage in Florida (USA), *Chemosphere*, 56, 335, 2004.
111. Wu, T.H., Rainwater, T.R., Platt, S.G., McMurry, S.T., and Anderson, T.A., Organochlorine contaminants in Morelet's crocodile (*Crocodylus moreletti*) eggs from Belize, *Chemosphere*, 40, 671, 2000.
112. Guillette, L.J., Jr., Woodward, A.R., Crain, D.A., Masson, G.R., Palmer, B.D., Cox, M.C., Xiang, Q.Y., and Orlando, E.F., The reproductive cycle of the female American alligator (*Alligator mississippiensis*), *Gen. Comp. Endocrinol.*, 108, 87, 1997.
113. Katsu, Y., Bermudez, D.S., Braun, E.L., Helbing, C., Miyagawa, S., Gunderson, M., Kohno, S., Bryan, T.A., Guillette, L.J., Jr., and Iguchi, T, Molecular cloning of the estrogen and progesterone receptors of the American alligator, *Gen. Comp. Endocrinol.*, 136, 122, 2004.
114. Lance, V.A., Alligator physiology and life history: the importance of temperature, *Exp. Gerentol.*, 38, 801, 2003.
115. Guillette, L.J., Jr., Gross, T.S., Masson, G.R., Matter, J.M., Percival, H.F., and Woodward, A.R., Developmental abnormalities of the gonad and abnormal sex hormone concentrations in juvenile alligators from contaminated and control lakes in Florida, *Environ. Health Perspect.*, 102, 680, 1994.
116. Guillette, L.J., Jr., Pickford, D.B., Crain, D.A., Rooney, A.A., and Percival, H.F., Reduction in penis size and plasma testosterone concentrations in juvenile alligators living in a contaminated environment, *Gen. Comp. Endocrinol.*, 101, 32, 1996.
117. Matter, J.M., McMurry, C.S., Anthony, A.B., and Dickerson, R.L., Development and implementation of endocrine biomarkers of exposure and effects in American alligators (*Alligator mississippiensis*), *Chemosphere*, 37, 1905, 1998.

Index

A

B

C

D

E

F

G

T

U

V

W

X

Y

Z